ir Quality Handbook

Air Quality Handbook

Edited by **Bernie Goldman**

R CALLISTO
REFERENCE

New York

Published by Callisto Reference,
106 Park Avenue, Suite 200,
New York, NY 10016, USA
www.callistoreference.com

Air Quality Handbook
Edited by Bernie Goldman

International Standard Book Number: 978-1-63239-064-6 (Hardback)

Printed in the United States of America.

Contents

Preface

I am honored to present to you this unique book which encompasses the most up-to-date data in the field. I was extremely pleased to get this opportunity of editing the work of experts from across the globe. I have also written papers in this field and researched the various aspects revolving around the progress of the discipline. I have tried to unify my knowledge along with that of stalwarts from every corner of the world, to produce a text which not only benefits the readers but also facilitates the growth of the field.

This book provides an introduction to various aspects of air quality. It contains several chapters dealing with air pollution; a topic of importance for experts in universities and manufacturing plants dealing with environmental problems. The book consists of an analysis of a variety of geographic regions and an evaluation of diverse actions related to these areas. A detailed analysis gives us the reasons for air pollution and its consequences on society. The main sources of air pollutants and their effects on the environment are also discussed. This book provides techniques and tools for measurement of pollution according to each place. Other significant aspects included in the book are the actions of governmental authorities and academic sectors for solving environmental problems resulting from air pollution.

Finally, I would like to thank all the contributing authors for their valuable time and contributions. This book would not have been possible without their efforts. I would also like to thank my friends and family for their constant support.

Editor

A Technology Assessment Tool for Evaluation of VOC Abatement Technologies from Solvent Based Industrial Coating Operations

Dhananjai S. Borwankar, William A. Anderson and Michael Fowler

Additional information is available at the end of the chapter

1. Introduction

Ground level ozone detrimentally effects local air quality. In mammals, ground level ozone causes damage to the upper respiratory tract (Farley, 1992), while in plants, it can inhibit photosynthetic activity through complex interactions effecting plant structure (Cape, 2008). Ground level ozone is primarily formed through the photochemical reaction of volatile organic compounds (VOCs) and nitrogen oxides (NOx) (Farley, 1992). Therefore, in an effort to improve local air quality, countries and various regions around the world have begun implementing regulations and guidance documents to curb VOC emission rates.

Canada, the United States, the European Union, and the United Kingdom are among those countries or regions which have implemented guidance documents, mandatory emission limits, or mandatory emission reporting. These tools allow governments to increase the business risk associated with releasing VOC emissions. The premise is that by increasing the business risk associated with emitting these compounds will compel industries to handle their emissions with greater responsibility. VOCs are typically emitted from industrial operations that utilize products such as solvents, thinners, degreasers, cleaners, lubricants, coatings and liquid fuels (Doble & Kumar, 2005). Therefore industries targeted by these types of regulatory measures include industrial printing, coating and painting operations, metal finishing facilities, and petrochemical operations.

The ideal (and most effective) pollution treatment solutions normally involve removing the pollution source from the process. This would involve substitution of current VOC based coatings with low or no VOC based coatings. Secondary pollution abatement strategies involve alteration of the process itself to reduce the overall contaminant processed and released. An example of this would involve altering (updating) of spray equipment to increase the transfer efficiency of the coating being applied onto the part. Unfortunately, due to very long and costly approval procedures for new processes and materials, these types of solutions tend to be prohibitive to implement, and therefore industries tend to use the remaining category of pollution treatment solutions, "end of pipe" or "add on" solutions.

There are many "end of pipe" technologies currently available that can effectively reduce VOC emissions, however, in many cases implementation of these technologies will substantially increase facility costs, downtime, and/or maintenance. Subsequent to implementation, facility managers are found having to invest significant time and resources to handle and operate a system that by itself is not "a value added" process, and hence does not assist the overall productivity of the facility. The result is the implementation of a technology that functions as designed, but does not function in an optimal way with respect to efficiency or productivity.

It is imperative that facility managers are provided with appropriate tools to determine the overall effect implementation of VOC abatement strategies will have on facility performance and their resources prior to implementation. The primary purpose of this work is to outline a methodology (Technology Assessment Tool) to rate various VOC abatement technologies according to the constraint of meeting current and foreseeable future legislative requirements; and the criteria of lifecycle costs and operational flexibility.

In this chapter, a brief review of current mainstream VOC abatement technologies is presented first, with emphasis upon design considerations, operational patterns and criteria, and economics. The discussion then outlines the Technology Assessment Tool methodology that is proposed for the purposes of rating current VOC abatement systems. Finally, a case study using the Technology Assessment Tool methodology will be presented to outline its use in a typical application.

2. Background

2.1. Summary of technologies

VOC abatement systems are divided into two main categories; destruction, and recovery (Doble & Kumar, 2005). Destruction technologies involve oxidation of the VOC substances to their most oxidized form; namely, carbon dioxide, and water (for hydrocarbons containing chlorine or sulphur, the exhaust will also include HCl and SO_2) (Baukal, 2004). Recovery technologies simply remove the contaminant from the exhaust stream for recovery or further treatment. Figure 1 outlines this division, and includes subcategories for each type of system.

2.1.1. Destructive technologies

Destructive technologies use oxidative processes to break down complex VOC compounds. These technologies can be further sub-divided into thermal oxidation and biological oxidation.

A Technology Assessment Tool for Evaluation of VOC Abatement Technologies from Solvent Based Industrial
Coating Operations

3

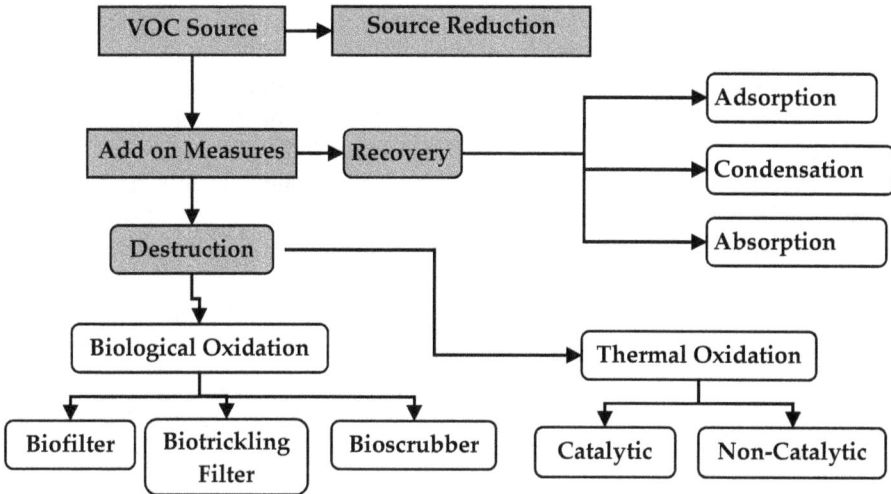

Figure 1. Categorization of VOC abatement technology systems.

Thermal Oxidation. Thermal oxidation (incineration) is the process of raising and maintaining the temperature of a combustible substance above its auto-ignition temperature in the presence of oxygen to complete its conversion to carbon dioxide and water (Baukal, 2004; Moretti, 2002). This process is quite effective, and virtually any gaseous organic stream may be safely incinerated given the proper design, engineering, and maintenance conditions (EPA, 2002).

Design parameters are a function of the feed stream composition, and consist of residence time, combustion chamber temperature, and turbulence (Moretti, 2002; Lewandowski, 2000). Knowledge of these parameters will provide enough information to develop a system lifecycle cost. To minimize capital and operational costs, it is always recommended that the designer attempt to lower the total volume of air that is being sent to the thermal oxidizer. This could mean either concentrating the emission before it enters the thermal oxidizer, or incorporating some sort of air re-circulation system in the process to minimize clean air sent to the system (Shelley et al., 1999). There are practical upper limits to the level of concentration that can be achieved, as most municipal/regional fire prevention bodies and insurance companies limit the concentration of emission going through thermal oxidizers to below 25% of the lower explosive limit (LEL) (Moretti, 2002; Lewandowski, 2000; Gibson, 1999). This is a preventative measure used to minimize the risk of fires or explosions within the system, and can be considered part of the design constraint.

Another aspect of the design is the recovery of heat. Since these systems are operating between temperatures of 650°C to 1,100°C it is standard practice to design the system to retain as much heat as possible (Moretti, 2002). One method involves using heat exchangers to transfer heat from the exhaust side of the oxidizer to the feed stream. When this type of heat recovery is used the thermal oxidizer is called a Recuperative Thermal

Oxidizer (Lewandowski, 2000). The second method of recovering heat involves the use of multiple beds packed with a ceramic type insulating material. Prior to entering the combustion chamber the feed stream is preheated as it is sent through one or more of these heated ceramic beds (Lewandowski, 2000). When the reaction is complete, the hot exhaust stream coming from the combustion chamber is sent through one of the cold ceramic beds exchanging heat with this bed to prepare it for the next volume of feed (Moretti, 2002). The bed packing normally has a very high rate of heat recovery, and can last between 5 to 10 years (Baukal, 2004). These systems are called Regenerative Thermal Oxidizers (RTOs). Both the recuperative and regenerative thermal oxidizers can be referred to as RTOs. Figures 2 and 3 illustrate both styles of RTO. For a typical hydrocarbon emission stream, a characteristic residence time would range between 0.5 sec to 2 sec, with temperatures between 650 to 1,100°C (1,200 - 2,000°F) (Baukal, 2004; Lewandowski, 2000).

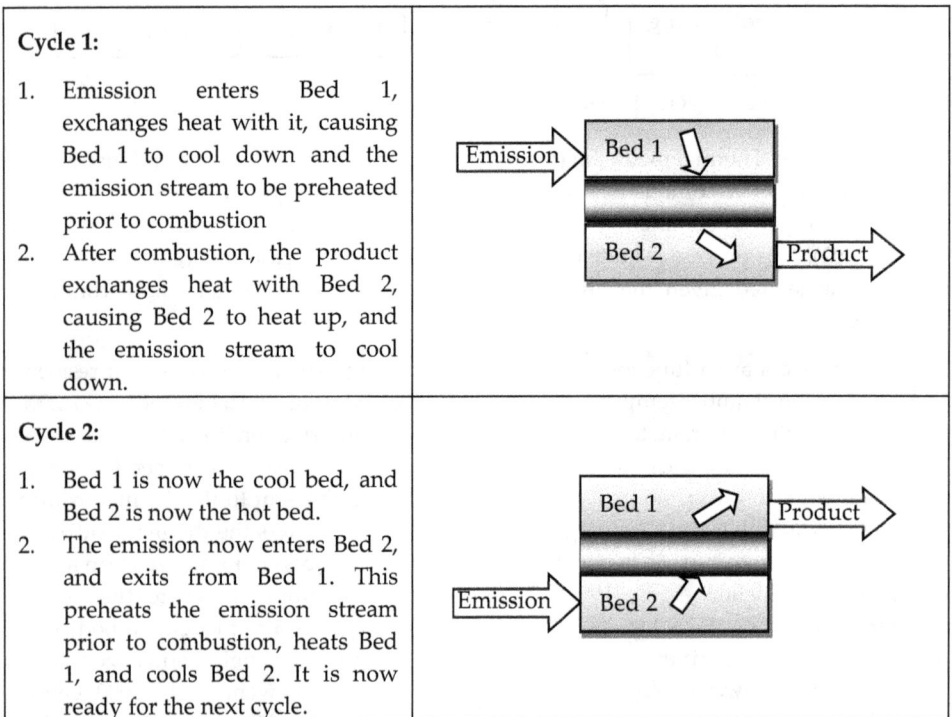

Cycle 1:	
1. Emission enters Bed 1, exchanges heat with it, causing Bed 1 to cool down and the emission stream to be preheated prior to combustion 2. After combustion, the product exchanges heat with Bed 2, causing Bed 2 to heat up, and the emission stream to cool down.	
Cycle 2:	
1. Bed 1 is now the cool bed, and Bed 2 is now the hot bed. 2. The emission now enters Bed 2, and exits from Bed 1. This preheats the emission stream prior to combustion, heats Bed 1, and cools Bed 2. It is now ready for the next cycle.	

Figure 2. Illustration of the Regenerative type thermal oxidizer operating/heat exchange cycle

Catalytic thermal oxidation operates on the same principle as thermal oxidation, except catalysts are used to increase the reaction kinetics such that combustion can occur at a lower temperature and shorter residence time. Limitations exist on which types of exhaust streams this process can be used, as certain combustion by-products from contaminants may poison the catalyst (for example, corrosive by-products such as HCl) (Lewandowski, 2000).

A Technology Assessment Tool for Evaluation of VOC Abatement Technologies from Solvent Based Industrial
Coating Operations

5

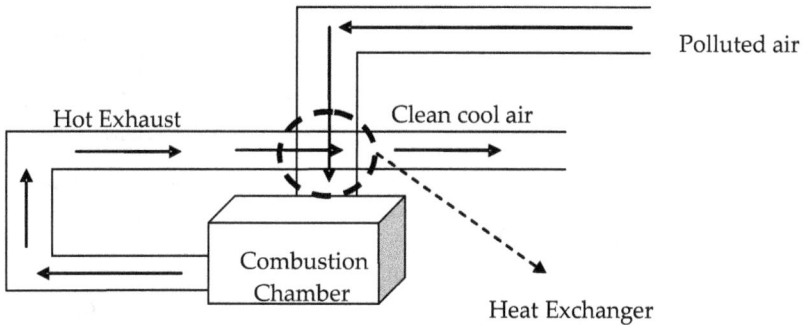

Figure 3. Recuperative type thermal oxidizer heat exchange cycle.

Overall, thermal oxidation is an attractive VOC abatement option because it can be used for complex mixtures of compounds, and it can provide very high levels of control. On the downside, once these systems are designed, they are set for a specific residence time, flow rate range and contaminant type, all of which result in a somewhat inflexible technology (Baukal, 2004; Moretti, 2002). Despite having efficient heat recovery systems, auxiliary fuel costs and electrical consumption arising from air circulation add significantly to the operational costs.

Biological Oxidation. Biological oxidation processes use microbial populations that are able to utilize volatile organic compounds as the primary source for both their catabolic (respiration) and anabolic (growth) requirements (Delhomémine & Heitz, 2005). The basic concept is to immobilize microorganisms (bacteria and fungi) in a packed porous bed or media through which nutrients and pollutants may flow (Doble & Kumar, 2005; Devinny et al., 1999). Nutrients, primarily nitrogen and phosphorus, are supplied to the culture through a mixed solution, while the pollutants are allowed to flow through the media. The immobilized microbial population will utilize the pollutants as their primary carbon source for growth and metabolism, oxidizing the VOC components to form carbon dioxide, water, salts and biomass (Devinny et al., 1999). The outcome is a relatively safe process that results in degradation of VOC compounds to carbon dioxide, water, nitrogen oxides, and salts. The main criteria to ensure successful operation lies in controlling the microbial population health and growth rates. This can be done by ensuring the following are met:

1. Carbon, oxygen, water, and nutrient sources are provided to cells to meet their catabolic and anabolic requirements,
2. Carbon, oxygen and nutrient sources provided to the cells are able to reach the cells; and,
3. Biomass and wastes produced as a result of oxidation; do not accumulate in or around the microbial population (Delhoménie & Heitz, 2005).

Figure 4 provides an illustrative example of a typical biofilter operation.

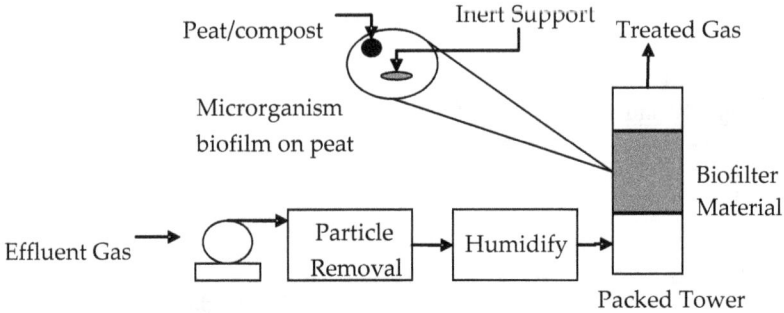

Figure 4. Illustrative depiction of a typical biofilter operation.

These three requirements can be met by understanding the contaminant type (and concentration), the substrate used to support cell growth, the physiochemical parameters of the system, and nutrient and moisture requirements of the bed (Delhoménie & Heitz, 2005).

The biodegradability of a pollutant type is dependent on the VOC transfer rate to the biofilm, and the VOC biodegradation rate of the microbial population (Delhoménie & Heitz, 2005). The VOC transfer rate depends upon three basic processes: transport of the VOC and oxygen from the gas phase to the liquid phase, transport of the VOC, oxygen and nutrients from the liquid phase to the surface of the biofilm, and simultaneous diffusion and biotransformation of VOC, oxygen and nutrients within the biofilm (Malhautier et al., 2005). In comparison to the reaction within the cell, diffusion and convection can be on the order of 1000 times slower then the cellular reaction rate (Delhoménie & Heitz, 2005). This indicates that mass transport (diffusion & convection) activities in and around the microbial population could be the limiting factor for the overall biodegradation. For VOCs in particular, this occurs because industrial paint exhausts are composed of solvents with poor water solubility. This solubility relates to poor liquid-gas phase interactions, leading to poor pollutant-biofilm absorption rates, thereby potentially limiting overall degradation rates (Delhoménie & Heitz, 2005).

The support medium is one of the most integral parts of these systems because it provides a safe and habitable environment for microbial populations to grow. The ideal media would have the following characteristics:

1. Resistance to compaction (high tensile strength),
2. Good moisture and nutrient holding capacity,
3. High surface area for bacterial attachment and improved VOC mass transfer,
4. Suitable surface for bacterial attachment (rough, porous, large surface area and hydrophilic); and,
5. pH buffering capacity (Devinny et al., 1999).

Media such as compost, peat, and soil have excellent water retention capacities, but may be prone to compaction events (Delhoménie & Heitz, 2005). Compaction has a tendency to

create fixed pathways throughout the media, limiting gas and nutrient distribution resulting
in inefficient conversions, or worse, cell death. Inorganic medias have high tensile strengths
which can help minimize compaction, however, microbial populations have trouble
adhering to these surfaces (particularly metals and glasses), and can sometimes fall and
block gas diffusion pathways (Doble & Kumar, 2005; Malhautier et al., 2005). Furthermore,
inorganic media do not store/hold moisture near as well as organic media do, and therefore
will require microbial inoculation and robust systems in place to ensure moisture and
nutrient contents in the media remain suitable for the microbial population. Table 1
summarizes the most important properties for various biofilter media (Devinny et al., 1999).

Biological oxidation can be very cost effective provided the system has been designed
properly. The main consideration for these systems is in understanding the properties of the
emission stream and matching a biological system that can handle the requirements.
Unfortunately, because of the complex processes that occur within the biofilter, the design
process for these systems can be much more tedious then for other systems (Devinny et al.,
1999). High residence times correlate to large bed sizes, or in economic terms, large facility
space requirements. Lastly, because of the nature of these systems, increased operator
training and effort is required to ensure smooth operation.

Characteristic	Compost	Peat	Soil	Inert Materials (e.g. perlite)	Synthetic Material
Indigenous microbial population	High	Medium-low	High	None	None
Surface Area	Medium	High	Low-medium	High	High
Air permeability	Medium	High	Low	Medium-High	Very High
Assimilable nutrient content	High	Medium-High	High	None	None
Sorption Capacity	Medium	Medium	Medium	Low-High	None-High
Lifetime	2-4 years	2-4 years	> 30 years	> 5 years	> 15 years
Cost	Low	Low	Very Low	Medium-High	Very High
General Applicability	Easy, Cost effective	Medium, water control problems	Easy, low-activity biofilters	Needs nutrients, may be expensive	Prototype

Table 1. Summary of Important Properties of Common Biofilter Materials.

2.1.2. Recovery technologies

Adsorption, absorption, and condensation are among the leading recovery technologies
used to separate VOC emissions from process exhaust streams. An added consideration for
these technologies is what to do with the captured contaminant once the process has been
completed. In some cases, it is feasible to recover the pollutant with sufficient purity as a

commodity for resale. In other cases, a final disposal step is required to complete the abatement process.

Adsorption. Adsorption of VOCs involves passing the contaminant-air (adsorbate) mixture through a solid porous bed (adsorbent) to be selectively held by physical attractive forces (EPA, 2002). When the bed becomes saturated, the contaminant-air stream is switched to another bed for adsorption, while the saturated bed is desorbed (regenerated) by passing hot inert gas or steam through the bed, or by reducing the pressure sufficiently to create a vacuum (U.S. Army Corps of Engineers [USACE], 2005; Knaebel, 2007). The removed mixture of inert gas and contaminant, or steam and contaminant is then sent to another process (RTO, decanter, or distillation column) for final disposal or recovery. The design of these systems depends on the chemical characteristics of the VOC (e.g., polarity, molecular weight, size, and other chemical reactivity characteristics), the physical properties of the inlet stream (temperature, pressure, and volumetric flow rate), and the physical properties of the adsorbent (Moretti, 2002; USACE, 2005; Knaebel, 2007).

Together these factors outline how well a particular adsorbate may adsorb onto an adsorbent. In practice there have been many isotherm models postulated that can accurately describe the adsorption-desorption process within specific ranges of temperatures and pressures. Therefore, as long as the pressure and concentrations of contaminants are within the appropriate levels, efficient systems may be designed.

Absorption. Absorption involves the selective transfer of a gas to a specific solvent depending on the solubility of the gas in the liquid, and the mass transfer of the gas to the gas-liquid interface (Cooper & Alley, 2002). The diffusional component is made up of both molecular and turbulent diffusion, for which turbulent diffusion is orders of magnitude higher. For this reason, absorption systems are designed to maximize turbulence during absorption by flowing both the liquid solvent and gas contaminant through a randomly packed solid column (Cooper & Alley, 2002). The basic structure of the column and packing material is designed to increase the contact surface area increasing the overall mass transfer of the absorbate to the solvent. A typical system operation would be to percolate the absorbent liquid through the top of the column down, and allow the contaminant gas to pass from the bottom of the column upwards.

Condensation. Separating VOCs by condensation can be accomplished by one of two processes: holding the temperature constant and increasing the pressure (compression condensation), or holding the pressure constant and lowering the temperature (refrigeration condensation), but most condensation systems are of the refrigeration type (Moretti, 2002). For efficient operation these systems are limited to VOCs with boiling points above 38°C and relatively high concentrations above 5,000 ppm (Khan & Ghoshal, 2000; Kohl & Nielsen, 1997).

3. Technology assessment: Methodology

The primary purpose of the assessment is to provide facility managers guidance in choosing the best available abatement technology while also outlining the effect the

A Technology Assessment Tool for Evaluation of VOC Abatement Technologies from Solvent Based Industrial
Coating Operations

9

chosen system will have on their facility. In order to accomplish this, each technology must be assessed against key operational parameters that will gauge overall technology effectiveness. These operational parameters will be a function of the design criteria and constraints, which in turn will rely heavily on process characteristics and the emission composition and type. The overall comparison will be made by rating different abatement technologies against criteria including: meeting current and foreseeable future legislative requirements, lifecycle costs and operational flexibility. The basic assessment flow is seen in Figure 5.

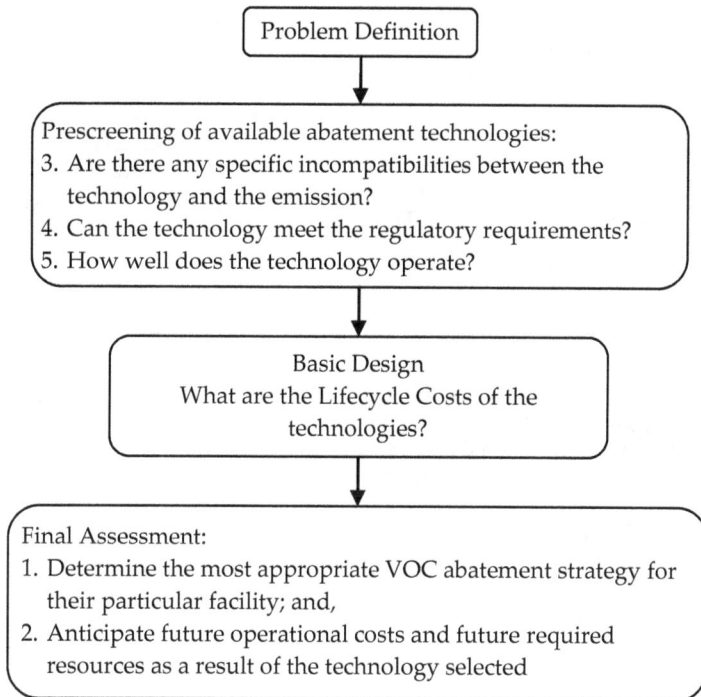

Problem Definition

Prescreening of available abatement technologies:
3. Are there any specific incompatibilities between the
   technology and the emission?
4. Can the technology meet the regulatory requirements?
5. How well does the technology operate?

Basic Design
What are the Lifecycle Costs of the
technologies?

Final Assessment:
1. Determine the most appropriate VOC abatement strategy for
   their particular facility; and,
2. Anticipate future operational costs and future required
   resources as a result of the technology selected

Figure 5. VOC abatement technology selection flow sheet.

3.1. Problem definition

The problem definition is the step in which all preliminary data required to assess the issue is found. This means defining four key elements: the emission source, the emission itself, legal requirements that must be met, and any budgetary constraints. In almost all cases, this data is readily available at the facility, and can be compiled quickly.

3.2. Prescreening

As outlined in Figure 6, the prescreening step answers three questions:

1.   Are there any specific incompatibilities between the technology and the emission?
2.   Will the technology meet the current regulatory requirements?
3.   How well does the technology operate (technology robustness, and technology flexibility)?

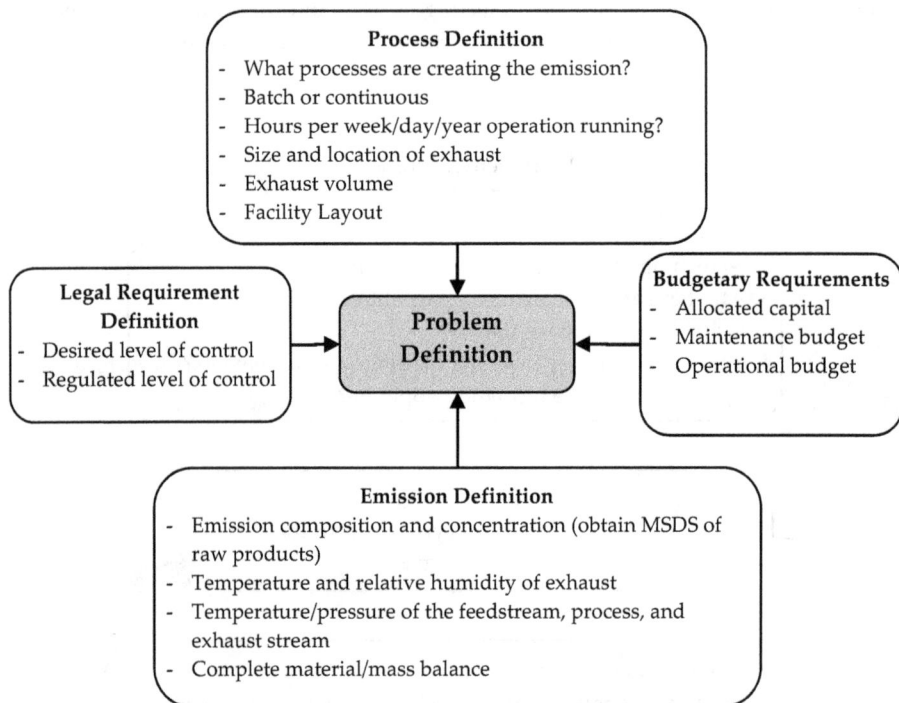

Figure 6. Information recommended when defining industrial emission problems.

The idea behind the prescreening step is to compare the information compiled from the problem definition with all the limitations found in literature on the abatement technologies considered. This will allow facility management to discard technologies from consideration that don't meet their prescribed criteria, and thus shorten the number of technologies required to undergo the basic design step.

For instance, if it was found that the emission contained significant amounts of acidic gases (such as chlorine or sulfur compounds), certain regenerative catalytic thermal oxidation techniques may not be feasible (unless a scrubber step was implemented prior to the combustion chamber) as the catalysts would be poisoned (Baukal, 2004).

Can technology meet current regulatory requirements?

The concern is if the system can lower the emission rate below the regulatory threshold, and if future regulatory requirements are identified, can the system maintain the facility's emission output below this future threshold. In Ontario Canada, it is considered good industry practice to design systems to reduce emissions below 50% of the current threshold ensuring the technology will be able to be used for the entire lifespan of the system. In this work, to obtain a quantitative measurement of how well each technology handles regulatory requirements, the matrix outlined in Table 2 will be used.

Score	Description
3	Meets current and foreseeable future regulations and requirements
2	Meets current regulations and requirements
1	Does not meet current regulations and requirements

Table 2. Matrix outlining regulatory scoring requirements.

How well does the technology operate?

The ability of an abatement technology to smoothly incorporate process or facility changes in its operation is a function of how the basic technology works, and not the size of the facility or the emission composition used for the design phase. This is the Operational Flexibility of the technology, and can be assessed by examining the following factors:

1. Start-up times
2. Continuous vs Intermittent Operation
3. Load Tolerance
   a. Ability to handle fluctuations in emission concentration or emission types; and,
   b. Ability to handle fluctuations in total flow

The scoring matrices listed in Tables 3 through 5 outline how each type of abatement technology will be quantifiably assessed for operational flexibility.

Score	Description
3	Start-up quicker then ½ production shift
2	Start-up quicker then 1 production shift
1	Start-up takes longer than 1 production shift

Table 3. Scoring system used to assess start-up timing for each type of abatement technology.

Score	Description
3	Intermittent Operation or cycling ok
2	Moderate Toleration to Intermittent operation
1	Does not tolerate intermittent operation

Table 4. Scoring system used to assess operational flexibility in terms of continuous vs intermittent operation for the abatement technologies.

Score	Flow or Concentration
3	Tolerates fluctuations
2	Tolerates minor fluctuations
1	Does not tolerate flucuations

Table 5. Scoring system used to assess operational flexibility in terms of load tolerance for the abatement technologies.

3.3. Basic design

Once the prescreening step is complete, the list of feasible technologies for further investigation will be relatively small. A basic design for each of the technologies under consideration can now be completed. The purpose of the basic design is to determine the size and main operational requirements (utilities) of the system. The size will be used to obtain a good estimate of the capital costs, and the operational requirements will be used to determine the annual operating costs. Together, with the implementation of proper engineering economics, a lifecycle cost for the abatement system can be determined.

When determining lifecycle costs there are many things that need to be considered to ensure that a full assessment can be made. For this particular assessment the lifecycle cost calculation will include the following:

1. Capital costs
2. Resource costs (natural gas, electricity, water)
3. Labour and maintenance costs
4. Anticipated major maintenance costs
5. Land use costs; and,
6. Anticipated lifespan

It is important to note that land use costs are very complex in nature and are highly dependent upon the geographic location, whether new land was required to be purchased, or existing land could be used, and what type of construction would be required. For this reason, the land use costs used in the following case study are based on data found for the City of Toronto for industrial land ($110/m^2). Other locations may have significantly different land costs.

3.4. Final assessment

Up to this point, a prescreening assessment has been completed to eliminate technologies not able to meet the following minimum constraints. Furthermore, qualitative operational criteria and the ability of an abatement technology to meet legislative requirements have been quantified into numerical values for objective comparisons, and a basic lifecycle cost estimate has been created. At this stage the information is compiled into a meaningful way so facility management will be able to make a clear and informed decision on the abatement strategy they wish to pursue. In this analysis, there is only one constraint and two criteria to

A Technology Assessment Tool for Evaluation of VOC Abatement Technologies from Solvent Based Industrial
Coating Operations

13

base the decisions on. These scores can all be multiplied together to get an overall score assessing non-economic performance, and the lifecycle costs can be used to consider economic considerations. To illustrate the methodology, a case study using this Technology Assessment Tool for a typical facility is presented in the following section.

4. Technology assessment: Case study

4.1. Problem definition

Overall Prescreening Summary. The information in Table 6 illustrates the types of contaminants in the emission stream, the approximate concentration, the total flow rate, and the type of operation. Temperature for the coating preparation and spray application is typically between 20⁰C and 25⁰C. Humidity can vary depending on the operation, but it is assumed the relative humidity (RH) of the system is approximately 75%. It is also assumed that the target emission load reduction is 50% of current levels.

Industrial coating operations using solvent based coatings are typically composed of high volume low concentration exhaust streams of components with relatively low boiling points. Condensation systems are limited to operations above 5,000 ppm in concentration and to solvents with boiling points above 5,000 ppm (Khan & Ghoshal, 2000; Kohl & Nielsen, 1997). The concentration of the emissions listed in Table 6 are all well below this 5,000 ppm threshold. Under these conditions condensation systems do not operate efficiently because the energy required to condense the component emissions would be substantial.

Definition	Facility Information
Emission Definition	15% Xylene 35% Methyl Ethyl Ketone 40% Toluene 10% N-Butyl Acetate
Process Definition	1.) Process: Automotive trim parts painting 2.) Continuous process 3.) Annual Hours of Operation: 240 days @ 16 hrs/day annually Flow: 60,000 cfm Emission: 100 tonnes VOCs
Budget Definition	Lifecycle costs will be compared using net present value costs for each technology
*Legal Requirement.	Federal Reporting Program National Pollutant Release Inventory (NPRI), Ontario Regulations 419 and 127

*Legal requirements are based upon the province/state/country the facility resides in. In this case, we are considering the geographical location to be Southern Ontario, Canada.

Table 6. Problem definition for the three sample facilities considered.

Furthermore, industrial coating operations emit highly variable and complex emission streams in which a suitable absorbing liquid is difficult to find. The result is that in this case, absorption systems will not operate efficiently either, and therefore are not considered further.

In summary, under the conditions listed, only the following technologies will be reviewed for the facility listed in Table 6: RTO and RCO (Regenerative and Recuperative types), Biological Oxidation, and Carbon Adsorption.

4.2. Basic design: Thermal oxidation assessment

Regulatory Requirements. Thermal oxidation systems are rated with destruction efficiencies that range between 95% - 99%. When properly designed and operated, a thermal oxidation system will always be able to remove the contaminants below the prescribed emission limit. Even if regulatory threshold values were to decrease substantially, these rated destruction efficiencies ensure that the system would be able to maintain a consistent reduction. The only foreseeable problem is if regulatory thresholds were established for carbon dioxide, carbon monoxide, and/or nitrogen oxides. In this case, because the thermal reaction involves the transformation of VOCs to these specific contaminants, the facility would have to ensure that they are monitoring these emissions as well. Based upon the matrix outlined in Table 2, the numerical value associated with this criterion is 3.

Lifecycle Costs. In order to determine the lifecycle costs for thermal oxidation abatement systems, the capital and operational costs must be established. For the sample facility, the operational and capital costs were calculated using the EPA methodologies (EPA, 2002). Finally, to update 1994 capital costs to more current figures (2009 dollar values), the appropriate Marshall & Swift equipment cost indices were used (Cooper and Alley, 2011).

The capital and operational costs to purchase and operate the various thermal abatement systems outlined here are described in Table 7.

As expected, the recuperative type RTO, which is the least technologically advanced has a substantially lower capital investment, but requires significantly more money to operate due to its poorer energy recovery capabilities. Similarly, the RCO (catalytic system), has the highest capital cost, but a significantly lower operating cost, since it operates at lower temperatures.

Therefore, as a business case, the choice between which thermal system to use now becomes a choice between the Regenerative type RTO or the Recuperative RTO. A lifecycle analysis was completed to differentiate between each system and obtain a true cost comparison. The lifecycle analysis was determined using a discount rate of 3% which is outlined in the NIST Handbook 135 Life-cycle costing Manual for the Federal Energy Management Program (Fuller & Petersen, 1996), and the results are shown in the bottom row of Table 7, as a net present value (or cost in this case), over a ten year period.

Therefore, although the Recuperative technology is a substantially lower capital investment, and the catalytic oxidation system is overall the most economically feasible system for these particular emission scenarios.

Purchased and installed equipment costs:	Factor	Recuperative	Regenerative	Catalytic
Incinerator (1994)		197,400	810,000	662,474
Incinerator (2009)	1.496	295,310	1,211,760	991,061
Instrumentation, taxes, freight	0.28	82,687	339,293	277,497
Direct installation costs included in purchase				
Subtotal Direct Installation Costs		377,977	1,551,053	1,268,558
Indirect Costs	0.31	91,546	375,646	307,229
Total Installation Costs		469,544	1,926,698	1,575,787
Direct Annual Costs				
Utilities Fuel		1,755,700	658,500	348,000
Electricity		267,200	264,700	238,500
Labour Supervisor 15% of labour		1,248	1,248	1,248
Labour		8,320	8,320	8,320
Maintenance		23,477	96,335	78,789
Subtotal Direct Costs		2,052,945	1,029,103	674,857
Overhead on labour	0.6	5,741	5,741	5,741
Taxes, insurance, depreciation	0.12	56,345	216,905	189,094
Subtotal Indirect Costs		62,086	222,646	194,835
Total Annual Costs		2,118,031	1,251,749	869,693
Lifecycle Cost		18,523,103	12,548,252	8,948,543

Notes: a.) The operational costs calculated here do not include the cost to keep the systems hot during non-production working times (8 hours per day).

Table 7. Thermal treatment systems cost analysis.

Operational Flexibility. The operational flexibility of thermal oxidation systems is moderate. Start-up times are dependent upon the time it takes to get the combustion chamber to reach the combustion temperature. From a dormant state, thermal systems typically take 4 to 6 hours to reach combustion temperatures.

In terms of load tolerance, thermal oxidative systems are generally good because they can destroy virtually all VOC compounds provided the combustion temperature is above the auto-ignition temperature of all species in the mixture. The problem comes when varying the emission concentration or total flow rate. In practice, some VOC mixtures have enough energy to sustain combustion, provided heat recovery of the system is at a sufficient level. Kohl and Nielsen provide a figure outlining this relationship, and from it, one can interpolate that at a heat recovery percentage of 95%, the VOC concentration would need to be approximately 3% of the LEL to sustain combustion (Kohl & Nielsen, 1997). This means, in theory no supplementary natural gas is required, thereby lowering operational costs. For example, if the VOC mixture was initially at a concentration of 3% of the LEL, and was lowered due to production changes, large amounts of supplementary natural gas will be required, and operational costs could become prohibitive (this is the case for the three sample facilities described here).

On the other hand, if emission concentrations are raised significantly due to production variations (above 25% LEL), dilution air may be required to be added to the system. This would increase the overall flow rate, changing design requirements, which will result in an in-efficient destruction of the VOC compounds.

Therefore, in terms of concentration changes, the system has a fairly wide range of tolerance. Conversely, as previously mentioned, thermal oxidation systems are not suited to highly variable flow rates. The design of the system incorporates the flow rate to establish appropriate residence times and mixing of the exhaust stream. As a result, changing the flow rate may result in the incomplete combustion of the emission. Simply put, thermal oxidation systems are intolerant to flow rate changes and tolerant to changes in concentration.

Lastly, thermal oxidation systems are limited to processes that continually operate. If continuous operation is not possible, systems can be run intermittently if the system is kept hot during off hours. This is accomplished by continuously combusting natural gas in the oxidation chamber. This prevents the bed cycling between hot and cold temperatures, which will prevent cracking of the heat exchanger beds. Overall, operational flexibility scores for thermal technologies are outlined in Table 8:

Abatement System	Start-up times	Continuous vs intermittent operation	Flexibility: Flow tolerance	Flexibility: Conc. tolerance
RTO (Recup.)	2	2	1	3
RTO (Regen.)	2	2	1	3
RCO	2	2	1	3

Table 8. Overall operational flexibility scores for thermal oxidation abatement technologies.

4.3. Basic design: Biological oxidation assessment

Regulatory Requirements. For the three sample facilities in this analysis the removal efficiency (termed destruction efficiency for RTO's) for biological oxidation processes is substantially lower than that of thermal oxidation and adsorption abatement techniques. Despite this, biological oxidation systems can meet regulatory thresholds. Meeting regulatory thresholds only refers to reducing the given facility's emission output for a specific contaminant below the prescribed limit. If the facility in question was emitting pollutants barely above the facility limits, then the level of reduction required would be quite small. For this reason, a removal efficiency of 50% or lower may be satisfactory. The only issue this may present is if regulatory bodies substantially lower emission limits in the future.

Lifecycle Costs. Operational costs are normally quite low for these systems because the temperature and pressure will remain close to ambient conditions. The largest energy input is the power required to circulate the emission through the bed. Therefore, provided the bed is maintained appropriately, the operational cost should not be high. The problem comes when assessing the overall space requirement for these systems. Mass transfer limitations require that contact areas (bed surface areas) are large. After associating the required space with an appropriate monetary value (lease or purchase cost), the lifecycle cost of this system can become substantial. Tables 9 and 10 summarize the capital and operational costs. The

Item	Value or Cost
Site Costs	
Height x Length x Width (m)	1.1 x 40 x 40
Area (m²)	1600
Gravel Requirements height (m):	0.30
Gravel Requirements (m³)	480
Total Volume Required (m³)	2179
Total Site Cost(equipment, labour, overhead)	$15,500
Media Costs	
Assume $37/m³ for compost	80,623
Assume $33/m³ for gravel	71,907
Assume $2.5/m³ installation costs	5,448
Total Media Costs	$158,000
Equipment Costs	
Blower and tower humidifier:	13,500
Piping, design, electrical, installation (18% of capital costs):	48,574
Liner Cost	193,965
Total Equipment Costs	256,045
Subtotal Equipment + Site + Media Costs	429,545
Land Cost ($110 per m²)	1,188,336
Total Capital Costs	**$1,617,800**

Table 9. Open Bed Biofilter Capital Cost Summary

Electricity	
Superficial Velocity (m/s)	1.20
Pressure drop (Pa/m)	7598
Blower horsepower (HP)	288
Blower (kWh)	1245209
Annual Pump Electrical (assume 2 HP) (kWh)	10800
Total Cost (kWh x $0.08 per kWh)	$100,481
Water for humidification (m³)	3744
Cost of water $0.7/m³	$2,621
Labour (assume 2 hours per day @ $20/hr)	$14,600
Overhead (60% of Labour)	$10,074
Insurance, property taxes	$11,335
Depreciation (equipment only, not land)	$25,604
Media Replacement (3 year cycle annualized)	$52,659
Total Miscellaneous Cost:	$137,938
Total Annual Operating Costs:	**$241,040**
Lifecycle Cost	$2,644,000

Table 10. Annual Biofilter (Open Bed) Operational Costs

costing process used to calculate the biofilter capital and operational costs were taken from Devinny et al. (1999). Furthermore, the lifecycle analysis was determined using a discount rate of 3% (Fuller & Petersen, 1996) as previously, resulting in a net present value (cost) of $2,644,000 based on the 10 year period.

Operational Flexibility. The key to understanding biological oxidative technologies is in performing a thorough literature search to determine experimental parameters such as the removal efficiencies for specific contaminant and media configurations. These experimental parameters are then used to create a pilot design for testing. Once the pilot system is tested a better understanding of the overall performance can be established. In this analysis, the cost of performing trials is not explicitly outlined, however, it is considered as part of the engineering design fees. Operational considerations, such as system robustness, start-up times, and load tolerance can be tested and evaluated before the implementation of a full scale system is undertaken.

Biological systems are generally not as robust as other systems. Times required to achieve steady state performance may vary between several days to several months (Devinny et al., 1999). This is specifically due to the kinetics associated with biological processes and start-up conditions. Biofilters are not very tolerant of load variations either. In biofilters, the main source of energy for the microbial population's anabolic and metabolic processes comes

A Technology Assessment Tool for Evaluation of VOC Abatement Technologies from Solvent Based Industrial
Coating Operations

19

from the VOC that is fed to the biofilter (Delhoménie & Heitz, 2005). Therefore if there is a large decrease in VOC loading, the microbial population may begin to starve. If the microbial population does die, the entire bed would require replacement or re-inoculation. This in effect, would cause a significant disruption in production and could be a very expensive endeavor.

If the VOC load is increased one of two problems may occur:

1. The microbial population cannot destroy enough of the VOC emission to maintain the regulatory limit because they have reached their maximum elimination capacity; and/or,
2. The microbial population's growth rate rapidly increases, substantially increasing microbial wastes in the media, potentially creating a toxic environment for the microbial population, or increased pressure drop.

Both situations would result in significant performance issues that would require considerable resources to remedy.

Biofilters can be somewhat resilient to intermittent operation, provided some contaminant has been adsorbed into the media. This adsorbed contaminant can be utilized by the microbial population during downtime, or off shift periods. The actual level of resilience cannot be quantified unless pilot studies are completed. Table 11 outlines the Biofilter's operational flexibility scores.

Abatement System	Start-up times	Continuous vs intermittent operation	Flexibility: Flow tolerance	Flexibility: Conc. tolerance
Biofilter	1	3	1	1

Table 11. Operational flexibility scores for the biofilter.

4.4. Basic design: Adsorption system assessment

Regulatory Requirements. Similar to the thermal oxidation systems, removal efficiencies can be quite high (over 95%) and provided the system is sized appropriately, it should always be able to reduce the contaminant loading well below the regulatory threshold.

Lifecycle Costs. The principal operation of this system is solvent recovery. For the purposes of the sample facilities described, solvent recovery will not be considered possible because the quality of the product will likely be poor, however, it should still be examined, as the recovered solvent may have a second market providing an extra revenue stream for the facility.

As with the thermal abatement systems, the lifecycle costs can only be established after capital and operating costs are determined. In this case, capital costs were established using EPA methodologies (EPA, 2002), and the operational costs were established using recommended procedures (Cooper & Alley, 2002). Lifecycle costs were then established using a discount rate of 3% (Fuller & Petersen, 1996) as described previously. Table 12 describes the calculated capital and operational costs, concluding with the lifecycle cost as indicated for the previous technologies.

Purchased and installed equipment costs:			Factor	Cost ($)
Vessel Cost				55,100
Carbon Cost				21,700
Equipment Cost (1994 Costs)				473,500
Equipment Cost (2009 Costs)			1.496	708,356
Taxes, freight			0.18	127,504
Foundations, erection, electrical, piping, painting			0.44	311,677
Subtotal Direct Installation Costs				**$1,147,537**
Indirect Installation Costs			0.31	219,590
Total Installation Costs				**$1,367,127**
Direct Annual Costs				
Utilities	Steam (Gas Cost)			41,200
	Water Cost			66,200
	Electricity (system fan)			12,900
Labour	Labour	8 hr/week	$20/hr	7,680
	Supervision	15% of labour		1,152
Maintenance				68,356
Solvent Disposal		$300/drum		166,300
Subtotal Direct Costs				**$364,524**
Overhead (labour)			0.6	5,741
Taxes, insurance, depreciation			0.12	164,055
Subtotal Indirect Costs				**$169,796**
Total Annual Costs				**$534,320**
Lifecycle cost				**$5,885,169**

Table 12. Adsorption system cost analysis

Operational Flexibility. The adsorption process takes place at near ambient pressure and temperature. Therefore as long as enough carbon beds are available for operation, start-up times will be fast.

A Technology Assessment Tool for Evaluation of VOC Abatement Technologies from Solvent Based Industrial
Coating Operations

21

In terms of design, the carbon requirement dictates the overall load that can be treated, and once the system is sized it is very difficult to make alterations for different loading levels or different flow rates. Therefore if the load is increased substantially, the system will not be able to handle the increase in load unless the overall configuration of the system is changed. Cycling and regeneration times would have to be re-calculated, and in extreme cases these changes may not be feasible. On the other hand, if the load was reduced substantially, the system would simply operate inefficiently incurring higher than expected operational costs.

Moreover, changes in load composition pose other unique problems. The original design of these systems relies on the adsorption characteristics for the most difficult species to adsorb. If a new compound was added into the emission mixture with poorer adsorption characteristics, a breakthrough situation may occur during operation. The system would have to be redesigned according to the new compound's adsorption characteristics.

Conversely, this system can operate both continuously and intermittently. This offers the advantage of not using extra energy to operate the system when production is down, or to start-up the systems when production times fluctuate substantially.

Overall, adsorption systems are moderately tolerant to flow rate variations, fairly intolerant to changes in emission composition, and very tolerant to intermittent vs continuous operational strategies. Table 13 outlines the operational flexibility scores for adsorption based VOC abatement technologies.

Abatement System	Start-up times	Continuous vs intermittent operation	Flexibility: Flow tolerance	Flexibility: Conc. tolerance
Adsorption	3	3	2	1

Table 13. Operational flexibility scores for adsorption abatement systems.

4.5. Final analysis

The Technology Assessment presented here is a resource to aid facility managers in determining which VOC abatement strategy would be most applicable for their particular facility. To do this, criteria scores and lifecycle costs are compared. For the final comparison it is up to facility management to determine which criteria holds the most weight. A typical order of importance would be: Regulatory Compliance, Lifecycle Costs, and finally, Operational Flexibility.

4.5.1. Regulatory compliance

Thermal oxidation systems are rated with destruction efficiencies that range between 95% - 99%. Properly designed, a thermal oxidation system will always be able to remove the contaminants below the prescribed emission limit. Even if regulatory threshold values were to decrease substantially, these rated destruction efficiencies ensure that the system would maintain a consistent reduction. The only foreseeable issue is if regulatory thresholds were established for carbon dioxide, carbon monoxide, and/or nitrogen oxides.

For biological oxidation processes the elimination capacity of VOC's is substantially lower than that of thermal oxidation and adsorption abatement techniques. Despite this, biological oxidation systems can meet regulatory thresholds. Meeting regulatory thresholds means: reducing the given facility's emission output for a specific contaminant below the prescribed limit. In the example case, literature reports that elimination capacities range between 60% and 80%. Therefore if the emissions exhausted from the facility were up to 60% above the legislated limit, the Biofilter would be able to meet the legislated requirements. In this particular case, it is assumed that the emissions exhausted are within this range, and therefore the Biofilter can meet the current legislated requirements.

Similar to RTO's, removal efficiencies for Adsorption systems can be quite high. A typical adsorption system can maintain removal efficiencies up to 95% provided there are no VOCs in the mixture with a particularly poor adsorption isotherm for the adsorbent being used. This means that even if emission limits were to become more stringent, Adsorption systems should be able to meet. In summary, in terms of Regulatory Compliance, these systems are compared numerically in Table 14, based on the rating system outlined in Table 2.

Abatement System	Regulatory Compliance
Regenerative Thermal Oxidation	3
Biofilter	2
Adsorption	3

Table 14. Regulatory Compliance Score Summary for Abatement Technology Analysis of three Sample Facilities.

4.5.2. Lifecycle costs

The Lifecycle cost based on a 10 year life expectancy for each abatement system and each sample facility is summarized in Table 15.

Abatement System	Net Present Value (cost)
Regenerative Thermal Oxidation	$12,548,252
Biofilter	$2,644,000
Adsorption	$5,885,169

Table 15. Net Present Value (Costs) for Abatement Technology Analysis of three Sample Facilities (at 3%)

A Technology Assessment Tool for Evaluation of VOC Abatement Technologies from Solvent Based Industrial
Coating Operations

23

The Biofilter system is the most economical to operate, over the life span of the technology, while the thermal oxidation is the least. However, it must be emphasized that this may not be a general trend. These costs could vary significantly depending on the starting parameters such as exhaust flow rates and emission concentrations, the geographic area considered, or even purity and value of recovered solvent.

4.5.3. Operational flexibility

The numerical rating for each Operational Flexibility characteristic is listed in Table 16, where the technology with the highest total is the most favourable according to this rating scheme.

Abatement System	Start-up times	Continuous vs intermittent operation	Flexibility: Flow tolerance	Flexibility: Conc. Tolerance	Total
RTO (Regen.)	2	2	1	3	12
Biofilter	1	3	1	1	3
Adsorption	3	3	2	1	18

Table 16. Operational Flexibility Scores summarized for the Abatement Technology Analysis of three sample facilities.

Overall, the Adsorption systems appear to have the best operational flexibility rating. This is because they are easy and quick to start up, and they have the ability to treat both continuous and intermittent processes. Conversely, minor design and cycling considerations may be required if concentrations in the emission are changed, whereas major design changes may need to be considered should the emission composition change. This is because although adsorption systems can be designed to operate efficiently for a given set of parameters, it is typically unknown how well the system would perform with mixed species until laboratory and pilot studies are performed, or a well known isotherm has been established for that particular contaminant with the adsorbent being used.

RTO systems are relatively easy to operate, and are fairly tolerant to changes in emission concentration and composition. These systems do not handle fluctuations in flow rate well, and although they work efficiently in continuous operation, during intermittent operations system performance may be compromised as a result of how major temperature fluctuations affects the ceramic heat exchanger. This problem can be mitigated by keeping the system hot during off hours with a steady flow of natural gas. This of course will increase facility operational costs, and thus, the flexibility in this category was downgraded.

Operationally, the Biofilter system is potentially the least flexible. Start-up times could be quite long, and changes in emission flows or concentrations would have unpredictable effects on the microbial population. Essentially this means in comparison to the other technologies, the Biofilter system may be difficult to maintain in sustained operation.

5. Conclusion

The Technology Assessment created for the sample facility profile listed in Table 6 indicates the following:

1. The Adsorption system has a moderate lifecycle cost, and is the most operationally flexible. Major design changes would be required should changes in emission compositions or concentrations occur.
2. The RTO system has the highest lifecycle cost, and is moderately operationally flexible. Major design changes would be required should process flow rates be changed.
3. The Biofilter has the lowest lifecycle cost, but has the least operational flexibility, and may require a full design change if regulatory limits are significantly decreased.

For these sample facilities, it is likely facility management will choose between the Adsorption system and the RTO. The decision between the two systems will depend on the relative importance facility management places on lifecycle costs vs operational flexibility. Which is more important will depend on future business forecasts. For instance, if customer coating requirements are expected to change, such as coating compositions or concentrations, then the RTO may be a more suitable abatement system because it can essentially destroy any VOC with the same destruction efficiency and would be able to function as is despite these changes in the emission stream.

It is important to note that this case study Technology Assessment is only valid for these sample facilities, and this specific geographic area. At any given facility, emissions exhausted may have differing compositions, concentrations, and flow rates. Legislative requirements prescribed in various regions differ, and each geographic area will have different land values associated with them. Three examples of this are provided as follows:

1. If the VOC concentration was high enough to sustain combustion for the thermal oxidation systems, the operational costs would have been substantially lowered.
2. If the geographic area of the facility was associated with lower land values, the Biofilter would have substantially lower lifecycle costs.
3. If the purity of the solvent recovered by the Adsorption system was suitable for resale, the cost associated with disposal of the solvent would be negligible, and a new revenue stream could be developed that reduces the lifecycle costs.

These different scenarios outline the need for the Abatement Technology Assessment technique developed here. Using this systematic approach facility management will able to predict the functionality and cost associated with implementing abatement technologies on existing facilities. Overall this ensures the decision will be based upon the most pertinent factors associated with the facility.

A Technology Assessment Tool for Evaluation of VOC Abatement Technologies from Solvent Based Industrial
Coating Operations

25

Author details

Dhananjai S. Borwankar, William A. Anderson, and Michael Fowler
Department of Chemical Engineering, University of Waterloo, Canada

6. References

Baukal, C. E. (2004). *Industrial combustion pollution and control;* Environmental science and pollution control series: 27; Marcel Dekker, New York.

Cape, J. N. (2008). Interactions of forests with secondary air pollutants. *Environmental Pollution*, 155, pp. 391-397.

Cooper, C. D., Alley, F.C., (2002) *Air pollution control: a design approach;* Waveland Press: Prospect Heights, Ill.

Cooper, C. D., Alley, F.C., (2011). *Air pollution control: a design approach, 4th edition;* Waveland Press: Prospect Heights, Ill.

Delhoménie, M.; Heitz, M. (2005). Biofiltration of Air. *Critical Reviews in Biotechnology*, 25, pp. 53-72.

Devinny, J. S.; Deshusses, M. A.; Webster, T. S. (1999). *Biofiltration for air pollution control;* Lewis Publishers: Boca Raton, Fl.

Doble, M.; Kumar, A. (2005). *Biotreatment of Industrial Effluents.* Elsevier Butterworth-Heinemann, Oxford, U.K.

EPA. (2002). *EPA Air Pollution Control Cost Manual* (6th edition). U.S. Evironmental Protection Agency, Research Triangle Park, North Carolina.

Farley, J. M. (1992). Inhaled toxicants and airway hyperresponsiveness. *Annu. Rev. Pharmacol. Toxicol., 32*, pp. 67-88.

Fuller, S. K.; Petersen, S. R. (1996). *NIST Handbook 135: Life-Cycle Costing Manual for the Federal Energy Management Program;* National Institute of Standards and Technology, Washington, DC.

Gibson, D., (1999). A burning issue : Putting the heat on air pollutants. *Chem. Eng., 106*, pp. 45-53.

Khan, F. I.; Ghoshal, A. K. (2000). Removal of Volatile Organic Compounds from polluted air. *J. Loss Prevention in the Process Industries*, 13, pp. 527-545.

Knaebel, K., S. (2007). *A "How to" Guide for Adsorption Design;* Adsorbent Research Inc. Dublin, Ohio.

Kohl, A. L.; Nielsen, R. (1997). *Gas Purification* (5th edition). Gulf Professional Publishing, Houston, TX.

Lewandowski, D. A. (2000). *Design of thermal oxidation systems for volatile organic compounds;* Lewis Publishers: Boca Raton, Fla.; London.

Malhautier, L.; Khammar, N.; Bayle, S.; Fanlo, (2005). Biofiltration of volatile organic compounds. *J. Appl. Microbiol. Biotechnol.*, 68, pp. 16-22.

Moretti, E. C. (2002). Reduce VOC and HAP emissions. *Chem. Eng. Prog.*, 98, pp. 30-40.

Shelley, S.; Pennington, R., L.; Lisewski, M. (1999)., Get More From Your Regenerative Thermal Oxidizer. *Chemical Engineering*, 106, pp. 137.

U.S. Army Corps of Engineers. (2005). *Adsorption Design Guide*; DG 1110-1-2; University Press of the Pacific.

Chapter 2

Air Quality in Portal Areas:
An Index for VOCs Pollution Assessment

Davide Astiaso Garcia, Fabrizio Cumo and Franco Gugliermetti

Additional information is available at the end of the chapter

1. Introduction

Air quality in portal areas is usually compromised by local emissions related to many different typologies of commercial, industrial and touristic activities.

Therefore, in order to reduce the consequences on the environment and human health due to exhausted gas released near the water surface and the ground, the air quality of portal areas should be monitored in all the different subareas of a harbour, by dividing it according with the different intended use areas.

Thus, it will be possible to pinpoint the most critical zones of each considered port and consequentially to plan specific actions for improving air quality in those areas.

In particular, this chapter deals with the emissions of VOCs (Volatile Organic Compounds) which play a key role in the short term chemical composition of the troposphere, as well as in climate changes (Murrells & Derwent, 2007).

Are classified as VOCs, in fact, both hydrocarbons containing carbon and hydrogen as the only elements (alkenes and aromatic compounds) and compounds containing also oxygen, chlorine or other elements, such as aldehydes, ethers, alcohols, esters, chlorofluorocarbons (CFCs) and hydrochlorofluorocarbons (HCFCs). According with the Italian regulation (article 268 of the 152/2006 Legislative Decree) VOCs are those organic substances which have at 293.15 K (20°C) a vapour pressure greater then or equal to 0.01 kPa.

The contents of this chapter have been developed considering the Italian and European regulations, which applications will improve the environmental quality in portal areas.

Among these, the Marine Environmental Protection Committee (MEPC) of the International Maritime Organization (IMO) has developed a protocol of an International Convention for the Prevention of Pollution from Shipping (IMO/MARPOL 73/78) establishing a monitoring program for reducing emissions (IMO, 2008).

In the light of these considerations, the main goals of our chapter are:

- To elaborate an air quality index weighed on the atmospheric concentrations of all VOCs, in order to obtain a single number that expresses the overall VOCs pollution of an area.
- To validate this index through its application in some case studies areas
- To pinpoint the most critical areas of the analysed harbours in order to select BAT (Best Available Technologies) and best practices for mitigating VOCs concentrations and improving local air quality.

2. AQI$_{voc}$ index elaboration

In order to elaborate an Air Quality Index for VOCs, called AQI$_{voc}$, weighed on the atmospheric concentrations of all VOCs as well as on dangerousness and impact of each substance in atmosphere, we assigned to each VOC an environmental impact coefficient (α) interrelated with its emission limit value according with the Italian regulations

$$\alpha_i = \frac{V_{max}}{V_i} \qquad (1)$$

Where:
α_i = environmental impact coefficient for the i-th VOC
V_{max} = highest emission limit value among all VOCs
V_i = emission limit value for the i-th VOC

The values of the environmental impact coefficient were assigned in proportion to the emission limit value specified for each ith class, giving a coefficient of greater environmental impact where the regulatory limit value of emission in atmosphere is lower.

In particular, considering the Italian regulation, the Annex III of the fifth part of the 152/2006 Legislative Decree shows a classification of volatile organic compounds divided into five classes according to their impact on the environment; consequentially the same Decree assigns to each of the five classes a maximum value of emission in atmosphere (Table 1).

Pollutants classification (152/2006 Legislative Decree)	Emission limit in atmosphere (152/2006 Legislative Decree)	environmental impact coefficient α
Class I	5 mg/Nm³	120
Class II	20 mg/Nm³	30
Class III	150 mg/Nm³	4
Class IV	300 mg/Nm³	2
Class V	600 mg/Nm³	1

Table 1. Environmental impact coefficient for VOCs related to their normative classification and limit values for emissions

Briefly, the equation for the AQIvoc indices evaluation is the following:

$$AQi_{voc} = 100 \cdot \frac{1/\sum_{i=1}^{i=18} \alpha_i \cdot v_i}{\left(1/\sum_{i=1}^{i=18} \alpha_i \cdot v_i\right)_{max}} \qquad (2)$$

Where:

AQIvoc = Air quality index related to VOCs concentrations

α_i = environmental impact coefficient for the i-th VOC

v_i =atmospheric concentration of the i-th VOC detected in the analysed intended use area

$\left(1/\sum_{i=1}^{i=18} \alpha_i \cdot v_i\right)_{max}$ = highest value of $1/\sum_{i=1}^{i=18} \alpha_i \cdot v_i$ (related to the intended use portal area

with lower concentrations of VOCs pollution in the atmosphere)

Consequentially, these values, standardized in a range from zero to hundred, have a comparative nature, with the value 100 assigned to the lowest VOCs concentration detected (in a certain port, in a certain detected area and in a certain season).

3. Case studies: The harbours of Anzio, Formia, Terracina and Ventotene

In order to validate the above mentioned methodology four ports of the Lazio region have been selected as case studies (Fig. 1)

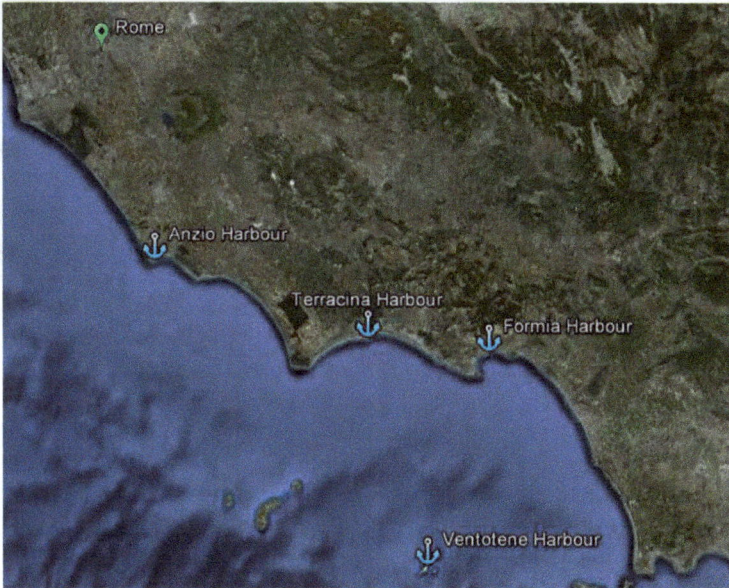

Figure 1. Geographic location of the harbours of Anzio, Formia, Terracina and Ventotene

In each port, divided into subareas according with its different intended use zones, was carried out an annual field data gathering, detecting VOC concentration in each season.

Lastly, an air quality matrix with VOC concentrations and AQIvoc values (for each intended use area in each season) has been elaborated for each analysed port.

4. Data gathering methods

In each of these four ports were monitored the concentrations in atmosphere of the following 18 VOCs: Dichloromethane; 2-Methylpentane; Hexane; Methylcyclopentane; Chloroform; 2-Methylhexane; Cyclohexane; Benzene; Heptane; Trichloroethylene; Methylcyclohexane; Toluene; Tetrachloroethylene; Ethylbenzene; m- p- xylene; o- xylene; 1,2,4-Trimethylbenzene; 1,2-Dichlorobenzene.

The concentrations of these substances were sampled seasonally in each intended use area of the four ports, leaving many radial diffusive samplers called "Radiello ®" (Bruno et al., 2008) for a period ranging between 7 and 10 days per season (Fig. 2)

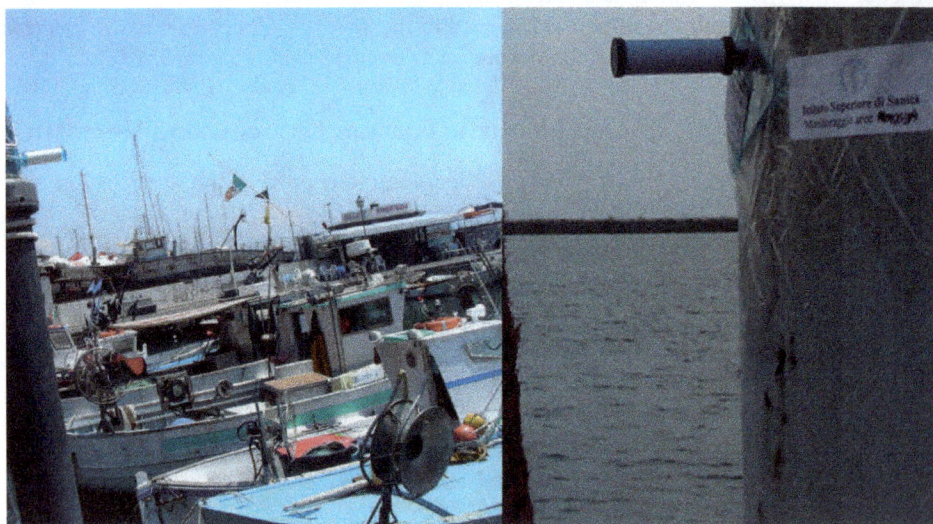

Figure 2. Two examples of VOCs monitoring in portal areas using radial diffusive samplers

5. Results

All the obtained VOCs concentrations have been compared spatially and seasonally in order to pinpoint portal areas and seasons where VOC pollution was higher.

5.1. VOCs concentrations results in the four case study harbours

The following graphics summarize the results obtained in each port, considering the seasons and the intended uses zones.

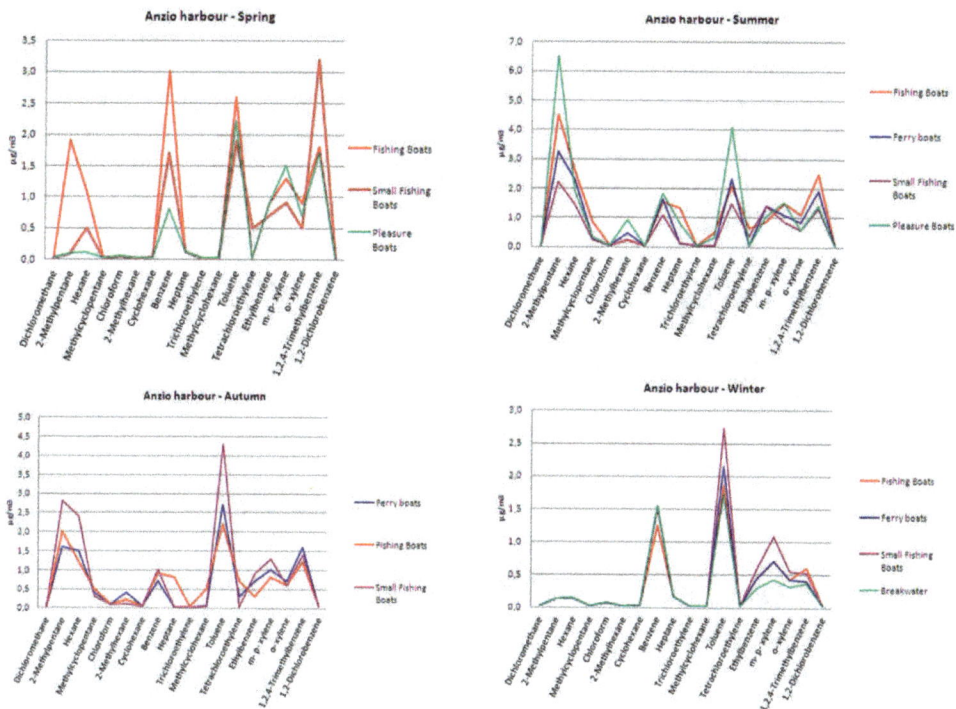

Figure 3. VOCs concentrations in atmosphere in the different seasons and intended use areas of the harbour of Anzio in 2010

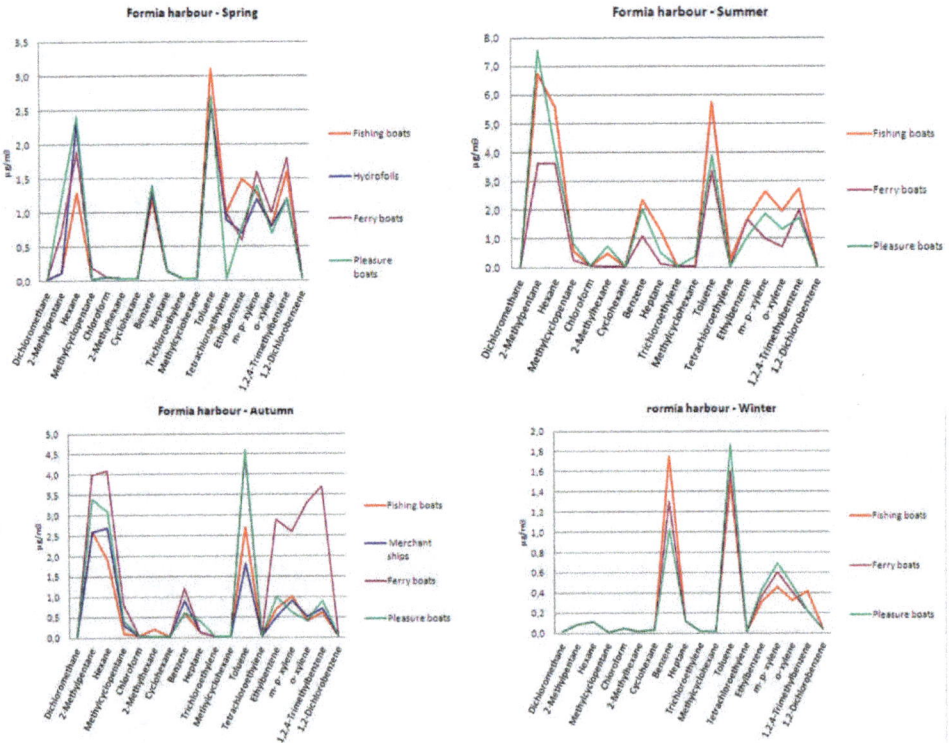

Figure 4. VOCs concentrations in atmosphere in the different seasons and intended use areas of the harbour of Formia in 2010

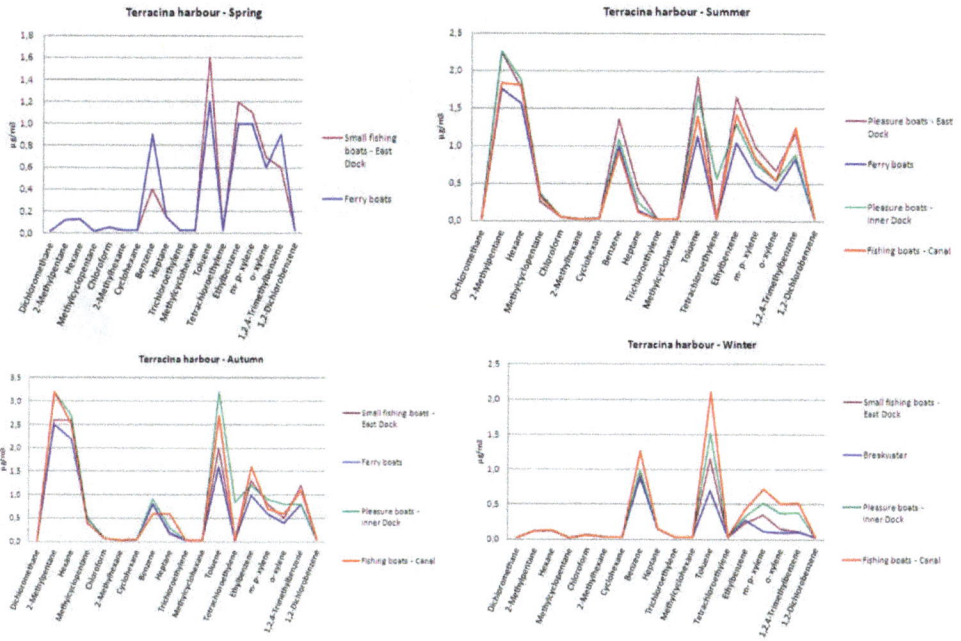

Figure 5. VOCs concentrations in atmosphere in the different seasons and intended use areas of the harbour of Terracina in 2010

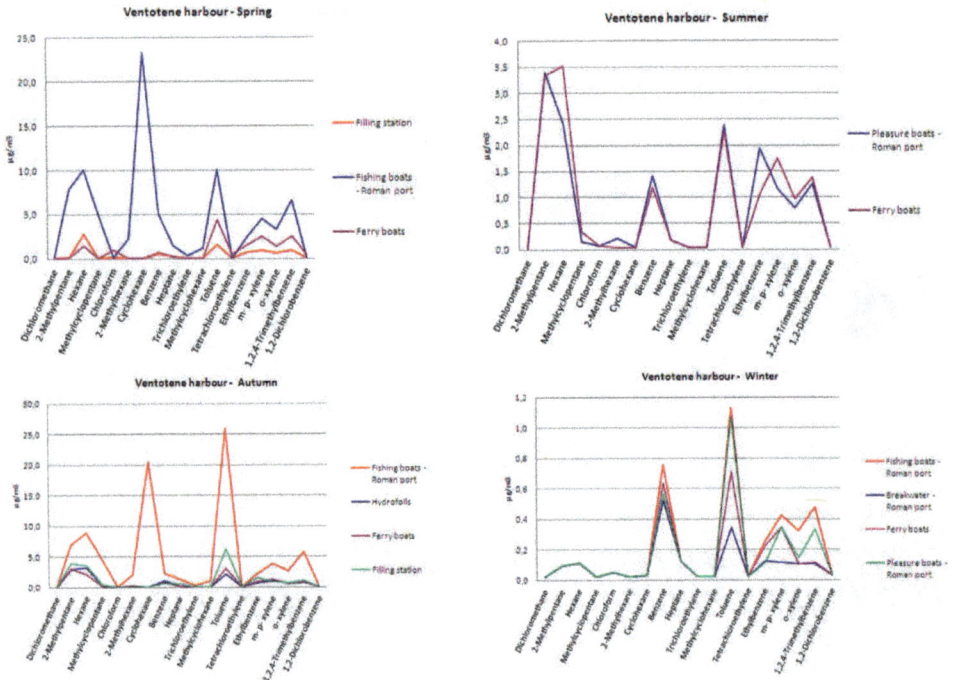

Figure 6. VOCs concentrations in atmosphere in the different seasons and intended use areas of the harbour of Ventotene in 2010

The sums of all the 18 VOCs concentrations recorded during each data gathering were compared, in order to rank in each port the intended use areas and the seasons according with their VOCs pollution (Tables 2 and 3). In order to have a reference value to define the "minimum" VOCs pollution of each port, a Radiello was placed in the most area distant from every sources of pollutant emissions (end of the breakwater) during the season with lower portal activity (winter). Unfortunately, it was no possible to obtain this value in the harbour of Formia because the Radiello in the breakwater has not been found at the end of the ten sampling days.

Anzio Harbour	
Season and intended use area	Total VOCs concentratio ns μ g/m3
Summer - Pleasure Boats	21,18
Summer - Fishing Boats	20
Summer - Ferry boats	15,6
Autumn - Small Fishing Boats	15,41
Spring - Fishing Boats	13,85
Autumn - Fishing Boats	12,12
Autumn - Ferry boats	11,88
Summer - Small Fishing Boats	11,34
Spring - Small Fishing Boats	10,32
Spring - Pleasure Boats	8,37
Winter - Small Fishing Boats	7,74
Winter - Ferry boats	6,44
Winter - Fishing Boats	6,04
Winter - Breakwater	5,41

Formia Harbour	
Season and intended use area	Total VOCs concentratio ns μ g/m3
Summer - Fishing boats	32,21
Autumn - Ferry boats	27,45
Summer - Pleasure boats	26,12
Summer - Ferry boats	17,67
Autumn - Pleasure boats	15,62
Spring - Ferry boats	13,18
Spring - Fishing boats	12,32
Spring - Pleasure boats	12,23
Spring - Hydrofoils	11,52
Autumn - Merchant ships	11,25
Autumn - Fishing boats	11,13
Winter - Fishing boats	5,3
Winter - Pleasure boats	5,23
Winter - Ferry boats	5,04

Table 2. Total VOCs concentrations in atmosphere in the different seasons and intended use areas of the harbours of Anzio and Formia

Terracina Harbour	
Season and intended use area	Total VOCs concentratio ns μ g/m³
Autumn – Pleasure boats – Inner Dock	15,6
Autumn – Fishing boats – Canal	14,29
Autumn – Small fishing boats – East Dock	12,76
Summer – Pleasure boats – East Dock	12,64
Summer – Pleasure boats – Inner Dock	11,72
Autumn – Ferry boats	10,89
Summer – Fishing boats – Canal	10,65
Summer – Ferry boats	9,02
Spring – Small fishing boats – East Dock	6,28
Spring – Ferry boats	6,28
Winter – Fishing boats – Canal	6,23
Winter – Pleasure boats – Inner Dock	4,78
Winter – Small fishing boats – East Dock	3,63
Winter – Breakwater	2,88

Ventotene Harbour	
Season and intended use area	Total VOCs Concentrati on μ g/m³
Autumn – Fishing boats – Roman port	86,93
Spring – Fishing boats – Roman port	83,06
Autumn – Filling station	19,02
Spring – Ferry boats	16,65
Summer – Ferry boats	16,24
Summer – Pleasure boats – Roman port	15,49
Autumn – Hydrofoils	12,65
Autumn – Ferry boats	12,63
Spring – Filling station	8,52
Winter – Fishing boats – Roman port	3,89
Winter – Pleasure boats – Roman port	3,1
Winter – Ferry boats	2,66
Winter – Breakwater – Roman port	1,84

Table 3. Total VOCs concentrations in atmosphere in the different seasons and intended use areas of the harbours of Terracina and Ventotene

5.2. AQIvoc indices in the four case study harbours

The AQIvoc index equation has been used for the elaboration of four air quality matrices that provide an overview of the air quality level within each one of the four portal areas. This approach allows to highlight those portal activities that have a major impact on air quality, and will be preparatory for the choose of which BAT or best practices is better to use for the mitigation of air pollution in each particular harbour.

In order to facilitate the reading of the comparison of the results, the AQIvoc values have been subdivided into three categories: low VOCs pollution, values under the twenty-fifth percentile (green boxes); average VOCs pollution, values between the twenty-fifth and the

ANZIO HARBOUR

Compound columns (left to right): Dichloromethane, 2-Methylpentane, Hexane, Methylcyclopentane, Chloroform, 2-Methylhexane, Cyclohexane, Benzene, Heptane, Trichloroethylene, Methylcyclohexane, Toluene, Tetrachloroethylene, Ethylbenzene, m-p-xylene, o-xylene, 1,2,4-Trimethylbenzene, 1,2-Dichlorobenzene, Σaxv, 1/Σaxv, AQIvoc

Season	intended use zones		Σaxv	1/Σaxv	AQIvoc
		α			
Spring	Fishing boats	v / αxv	43,71	0,0229	48,03
Spring	Small fishing boats	αxv / v	47,60	0,0210	44,11
Spring	Pleasure boats	αxv	37,85	0,0264	55,47
Summer	Fishing boats	αxv / v	64,98	0,0154	32,31
Summer	Ferry boats	αxv / v	59,39	0,0168	35,35
Summer	Small fishing boats	v / αxv	64,14	0,0156	32,73
Summer	Pleasure boats	αxv	57,16	0,0175	36,73
Autumn	Ferry boats	v / αxv	44,45	0,0225	47,23
Autumn	Fishing boats	αxv / v	44,19	0,0226	47,51
Autumn	Small fishing boats	αxv	47,55	0,0210	44,15
Winter	Fishing boats	v / αxv	34,81	0,0287	60,31
Winter	Ferry boats	v / αxv	37,52	0,0267	55,96
Winter	Small fishing boats	v / αxv	47,44	0,0211	44,26
Winter	Breakwater	αxv	28,69	0,0349	73,17

Figure 7. Air quality matrix for the evaluation of the AQIvoc values in Anzio harbour

FORMIA HARBOUR

Compound columns (left to right): Dichloromethane, 2-Methylpentane, Hexane, Methylcyclopentane, Chloroform, 2-Methylhexane, Cyclohexane, Benzene, Heptane, Trichloroethylene, Methylcyclohexane, Toluene, Tetrachloroethylene, Ethylbenzene, m-p-xylene, o-xylene, 1,2,4-Trimethylbenzene, 1,2-Dichlorobenzene, Σaxv, 1/Σaxv, AQIvoc

Season	intended use zones		Σaxv	1/Σaxv	AQIvoc
		α			
Spring	Hydrofoils	v / αxv	61,39	0,0163	34,20
Spring	Ferry boats	v / αxv	63,05	0,0159	33,30
Spring	Pleasure boats	v / αxv	39,87	0,0251	52,66
Spring	Fishing boats	v / αxv	88,79	0,0113	23,64
Summer	Pleasure boats	v / αxv	62,21	0,0161	33,75
Summer	Ferry boats	αxv / v	70,89	0,0141	29,61
Summer	Fishing boats	v / αxv	94,52	0,0106	22,21
Autumn	Fishing boats	v / αxv	35,58	0,0281	59,01
Autumn	Ferry boats	v / αxv	117,50	0,0085	17,87
Autumn	Pleasure boats	αxv / v	50,67	0,0197	41,43
Autumn	Merchant ships	v / αxv	29,00	0,0345	72,39
Winter	Pleasure boats	αxv / v	50,28	0,0199	41,75
Winter	Ferry boats	αxv / v	44,58	0,0224	47,09
Winter	Fishing boats	v / αxv	41,48	0,0241	50,61

Figure 8. Air quality matrix for the evaluation of the AQIvoc values in Formia harbour

Figure 9. Air quality matrix for the evaluation of the AQIvoc values in Terracina harbour

TERRACINA HARBOUR

Season	Intended use zones		Dichloromethane	2-Methylpentane	Hexane	Methylcyclopentane	Chloroform	2-Methylhexane	Cyclohexane	Benzene	Heptane	Trichloroethylene	Methylcyclohexane	Toluene	Tetrachloroethylene	Ethylbenzene	m- p- xylene	o- xylene	1,2,4-Trimethylbenzene	1,2-Dichlorobenzene	Σαxv	1/Σαxv	AQIvoc
	intended use zones	α	1	1	1	1	1	1	1	1	1	30	1	2	30	30	1	1	1	1			
Spring	Small fishing boats - East Dock	p	0.02	0.12	0.13	0.02	0.06	0.03	0.03	0.40	0.15	0.03	0.03	1.60	0.03	1.20	1.10	0.70	0.60	0.03			
		αxp	0.02	0.12	0.13	0.02	0.06	0.03	0.03	0.40	0.15	0.90	0.03	3.20	0.90	36.00	1.10	0.70	0.60	0.03	44.42	0.0225	47.26
	Ferry boats	p	0.02	0.12	0.13	0.02	0.06	0.03	0.03	0.90	0.15	0.03	0.03	1.20	0.03	1.00	1.00	0.60	0.90	0.03			
		αxp	0.02	0.12	0.13	0.02	0.06	0.03	0.03	0.90	0.15	0.90	0.03	2.40	0.90	30.00	1.00	0.60	0.90	0.03	38.22	0.0262	54.93
Summer	Pleasure boats - Inner Dock	p	0.02	2.36	1.87	0.36	0.05	0.02	0.03	1.09	0.24	0.03	0.02	1.67	0.56	1.29	0.77	0.54	0.88	0.03			
		αxp	0.02	2.36	1.87	0.36	0.05	0.02	0.03	1.09	0.24	0.60	0.02	3.34	16.80	38.70	0.77	0.54	0.88	0.03	67.62	0.0148	31.05
	Ferry boats	p	0.02	1.76	1.56	0.34	0.05	0.02	0.03	0.99	0.14	0.03	0.02	1.13	0.02	1.04	0.60	0.42	0.83	0.03			
		αxp	0.02	1.76	1.56	0.34	0.05	0.02	0.03	0.99	0.14	0.60	0.02	2.26	0.60	31.20	0.60	0.42	0.83	0.03	41.47	0.0241	50.62
	Fishing boats - Canal	p	0.02	1.83	1.81	0.32	0.05	0.02	0.03	0.94	0.12	0.02	0.02	1.39	0.02	1.41	0.83	0.54	1.25	0.03			
		αxp	0.03	1.83	1.81	0.32	0.05	0.02	0.03	0.94	0.12	0.60	0.02	2.78	0.60	42.30	0.83	0.54	1.25	0.03	54.09	0.0185	38.81
	Pleasure boats - East Dock	p	0.02	2.24	1.77	0.26	0.05	0.02	0.03	1.35	0.43	0.02	0.02	1.91	0.02	1.65	0.95	0.67	1.17	0.03			
		αxp	0.02	2.24	1.77	0.26	0.05	0.02	0.03	1.35	0.43	0.60	0.02	3.82	0.60	49.50	0.98	0.67	1.17	0.03	63.56	0.0157	33.03
Autumn	Pleasure boats - Inner Dock	p	0.02	3.20	2.70	0.50	0.07	0.03	0.04	0.90	0.30	0.03	0.03	3.20	0.84	1.20	0.90	0.80	0.80	0.04			
		αxp	0.02	3.20	2.70	0.50	0.07	0.03	0.04	0.90	0.30	0.90	0.03	6.40	25.20	36.00	0.90	0.80	0.80	0.04	78.85	0.0127	26.63
	Ferry boats	p	0.02	2.50	2.20	0.50	0.07	0.03	0.04	0.80	0.20	0.03	0.03	1.60	0.03	1.00	0.60	0.40	0.80	0.04			
		αxp	0.02	2.50	2.20	0.50	0.07	0.03	0.04	0.80	0.20	0.90	0.03	3.20	0.90	30.00	0.60	0.40	0.80	0.04	43.23	0.0231	48.56
	Small fishing boats - East Dock	p	0.02	2.60	2.60	0.50	0.07	0.03	0.04	0.80	0.17	0.03	0.03	2.00	0.03	1.30	0.80	0.50	1.20	0.04			
		αxp	0.02	2.60	2.60	0.50	0.07	0.03	0.04	0.80	0.17	0.90	0.03	4.00	0.90	39.00	0.80	0.50	1.20	0.04	54.20	0.0185	38.73
	Fishing boats - Canal	p	0.02	3.20	2.50	0.40	0.07	0.03	0.04	0.60	0.60	0.03	0.03	2.70	0.03	1.60	0.70	0.60	1.10	0.04			
		αxp	0.02	3.20	2.50	0.40	0.07	0.03	0.04	0.60	0.60	0.90	0.03	5.40	0.90	48.00	0.70	0.60	1.10	0.04	65.13	0.0154	32.93
Winter	Breakwater	p	0	0.1	0.13	0	0.1	0	0.03	0.9	0.2	0	0	0.7	0.03	0.28	0.1	0.1	0.1	0			
		αxp	0.00	0.21	0.20	0.01	0.00	0.00	0.00	0.88	0.02	0.02	0.00	1.58	0.02	8.74	0.07	0.04	0.09	0.00	11.89	0.0841	100.00
	Small fishing boats - East Dock	p	0	0.1	0.13	0	0.1	0	0.03	1	0.2	0	0	1.15	0.03	0.23	0.4	0.2	0.1	0			
		αxp	0.00	0.21	0.20	0.01	0.00	0.00	0.00	0.94	0.02	0.02	0.00	2.60	0.02	7.18	0.21	0.06	0.10	0.00	11.57	0.0864	100.00
	Pleasure boats - Inner Dock	p	0	0.1	0.13	0	0.1	0	0.03	1	0.2	0	0	1.52	0.03	0.33	0.5	0.4	0.4	0			
		αxp	0.00	0.21	0.20	0.01	0.00	0.00	0.00	0.97	0.02	0.02	0.00	3.44	0.02	10.30	0.31	0.16	0.32	0.00	15.97	0.0626	100.00
	Fishing boats - Canal	p	0	0.1	0.13	0	0.1	0	0.03	1.3	0.2	0	0	2.11	0.03	0.43	0.7	0.5	0.5	0			
		αxp	0.00	0.21	0.20	0.01	0.00	0.00	0.00	1.25	0.02	0.02	0.00	4.77	0.02	13.42	0.43	0.21	0.43	0.00	20.99	0.0476	100.00

Figure 9. Air quality matrix for the evaluation of the AQIvoc values in Terracina harbour

VENTOTENE HARBOUR

Season	Intended use zones		Dichloromethane	2-Methylpentane	Hexane	Methylcyclopentane	Chloroform	2-Methylhexane	Cyclohexane	Benzene	Heptane	Trichloroethylene	Methylcyclohexane	Toluene	Tetrachloroethylene	Ethylbenzene	m- p- xylene	o- xylene	1,2,4-Trimethylbenzene	1,2-Dichlorobenzene	Σαxv	1/Σαxv	AQIvoc
	intended use zones	α	1	1	1	1	1	1	1	1	1	30	1	2	30	30	1	1	1	1			
Spring	Ferry boats	u	0.02	0.13	1.40	0.03	0.97	0.03	0.04	0.70	0.17	0.03	0.03	4.40	0.50	1.50	2.50	1.30	2.50	0.40			
		αxu	0.02	0.13	1.40	0.03	0.97	0.03	0.04	0.70	0.17	0.90	0.03	8.80	15.00	45.00	2.50	1.30	2.50	0.40	79.92	0.0125	26.27
	Filling station	u	0.02	0.13	2.80	0.03	0.97	0.03	0.04	0.50	0.17	0.03	0.03	1.50	0.03	0.70	0.90	0.60	0.90	0.04			
		αxu	0.02	0.13	2.80	0.03	0.97	0.03	0.04	0.50	0.17	0.90	0.03	3.00	0.90	21.00	0.90	0.60	0.90	0.04	32.06	0.0312	65.48
	Fishing boats - Roman port	u	0.02	7.90	10.00	5.00	0.07	2.20	23.30	5.00	1.40	0.30	1.10	10.00	0.03	2.50	4.50	3.20	6.50	0.04			
		αxu	0.02	7.90	10.00	5.00	0.07	2.20	23.30	5.00	1.40	9.00	1.10	20.00	0.90	75.00	4.50	3.20	6.50	0.04	175.13	0.0057	11.99
Summer	Pleasure boats - Roman port	u	0	3.39	2.41	0.14	0.07	0.21	0.04	1.41	0.17	0.03	0.03	2.39	0.03	1.93	1.15	0.78	1.25	0.04			
		αxu	0.02	3.39	2.41	0.14	0.07	0.21	0.04	1.41	0.17	0.90	0.03	4.78	0.03	57.90	1.15	0.78	1.25	0.04	75.59	0.0132	27.77
	Ferry boats	u	0	3.33	3.52	0.34	0.07	0.03	0.04	1.19	0.17	0.03	0.03	2.28	0.03	1.05	1.74	0.96	1.37	0.04			
		αxu	0.02	3.33	3.52	0.34	0.07	0.03	0.04	1.19	0.17	0.90	0.03	4.56	0.90	31.50	1.74	0.96	1.37	0.04	50.71	0.0197	41.40
Autumn	Ferry boats	u	0	3.00	2.10	0.10	0.05	0.20	0.03	0.70	0.13	0.02	0.02	3.00	0.03	0.80	1.20	0.50	0.70	0.03			
		αxu	0.02	3.00	2.10	0.10	0.05	0.20	0.03	0.70	0.13	0.60	0.02	6.00	0.90	24.00	1.20	0.50	0.70	0.03	40.28	0.0248	52.12
	Hydrofoils	u	0	2.90	3.10	0.30	0.05	0.02	0.03	1.00	0.13	0.02	0.02	2.00	0.03	0.60	1.00	0.60	0.80	0.03			
		αxu	0.02	2.90	3.10	0.30	0.05	0.02	0.03	1.00	0.13	0.90	0.02	4.00	0.90	18.00	1.00	0.60	0.80	0.03	33.50	0.0299	62.67
	Filling station	u	0	3.80	3.50	0.40	0.05	0.02	0.03	0.60	0.50	0.02	0.02	6.10	0.03	1.40	0.90	0.60	1.00	0.03			
		αxu	0.02	3.80	3.50	0.40	0.05	0.02	0.03	0.60	0.50	0.90	0.02	12.20	0.90	42.00	0.90	0.60	1.00	0.03	67.17	0.0149	31.26
	Fishing boats - Roman port	u	0	6.90	8.80	4.40	0.05	1.90	20.35	2.20	1.20	0.25	0.95	25.95	0.03	2.00	3.70	2.50	5.70	0.03			
		αxu	0.02	6.90	8.80	4.40	0.05	1.90	20.35	2.20	1.20	7.50	0.95	51.90	0.90	60.00	3.70	2.50	5.70	0.03	179.00	0.0056	11.73
Winter	Fishing boats - Roman port	u	0	0.1	0.11	0	0.1	0	0.03	0.8	0.1	0	0	1.13	0.02	0.25	0.4	0.3	0.5	0			
		αxu	0.00	0.31	0.27	0.00	0.00	0.00	0.00	1.06	0.02	0.02	0.00	5.40	0.02	14.48	0.48	0.25	0.59	0.00	22.89	0.0437	91.70
	Breakwater - Roman port	u	0	0.1	0.11	0	0.1	0	0.03	0.5	0.1	0	0	0.34	0.02	0.12	0.1	0.1	0.1	0			
		αxu	0.00	0.31	0.27	0.00	0.00	0.00	0.00	0.73	0.02	0.02	0.00	1.63	0.02	6.95	0.13	0.08	0.13	0.00	10.28	0.0973	100.00
	Ferry boats	u	0	0.1	0.11	0	0.1	0	0.03	0.6	0.1	0	0	0.71	0.02	0.22	0.3	0.1	0.1	0			
		αxu	0.00	0.31	0.27	0.00	0.00	0.00	0.00	0.89	0.02	0.02	0.00	3.39	0.02	12.74	0.39	0.08	0.14	0.00	18.27	0.0547	100.00
	Pleasure boats - Roman port	u	0	0.1	0.11	0	0.1	0	0.03	0.6	0.1	0	0	1.07	0.02	0.11	0.3	0.1	0.3	0			
		αxu	0.00	0.31	0.27	0.00	0.00	0.00	0.00	0.79	0.02	0.02	0.00	5.11	0.02	6.37	0.39	0.11	0.41	0.00	13.83	0.0723	100.00

Figure 10. Air quality matrix for the evaluation of the AQIvoc values in Ventotene harbour

seventy-fifthpercentile (yellow boxes) and high VOCs pollution, values over the seventy-fifth percentile (red boxes).

Before that, the statistical distribution of the data has been considered, highlighting unusual observations (outliers and extreme values) by means of boxplot analysis.

In particular, the box-plot method analyzes the distribution of data considering the median, the interval between interquartiles, the outliers values and the extreme values of individual variables. The length of the box was considered as the range of values between interquartiles, or rather between the twenty-fifth and seventy-fifth percentile. Consequently, the outliers values are those that are at a distance between 1.5 and 3 boxes from the top or the bottom edge of the box, between the twenty-fifth and seventy-fifth percentile; at the same way, the extreme values are distant more than 3 boxes from the top or bottom edge of the box. The purpose of the method is therefore to identify these values in order to get a distribution composed by values statistically attributable to the same population.

The box-plot application proceeds step by step in order to be able to select all the extreme and outliers values up to a statistically homogeneous distribution of the population.

In this way, the 4 unusual values have been pinpointed and removed assigning to them the maximum IQAcov value of one hundred (Figure 11).

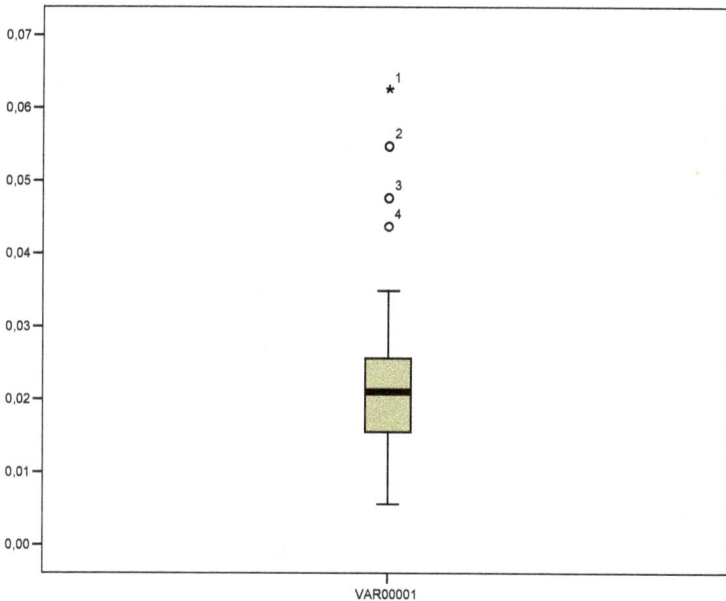

Figure 11. Boxplot statistical methods: 1 outliers (star) and 3 extreme values detection trough SPSS software

It was therefore possible to elaborate the 100 standardization of the IQAcov value, considering a statistically homogeneous distribution (Table 4).

	Valid		Missing		Total	
	N	Percent	N	Percent	N	Percent
VAR00001	51	92,7%	4	7,3%	55	100,0%

Table 4. Statistical data analysis of the 55 IQAvoc indices of the four case study ports

Moreover, the results have been registered in a GIS (Geographic Information System) database that contains a comparative spatial analysis of IQAvoc values in order to produce thematic maps able to pinpoint areas where the VOCs pollution was higher (some examples of these maps will be reported in the following maps).

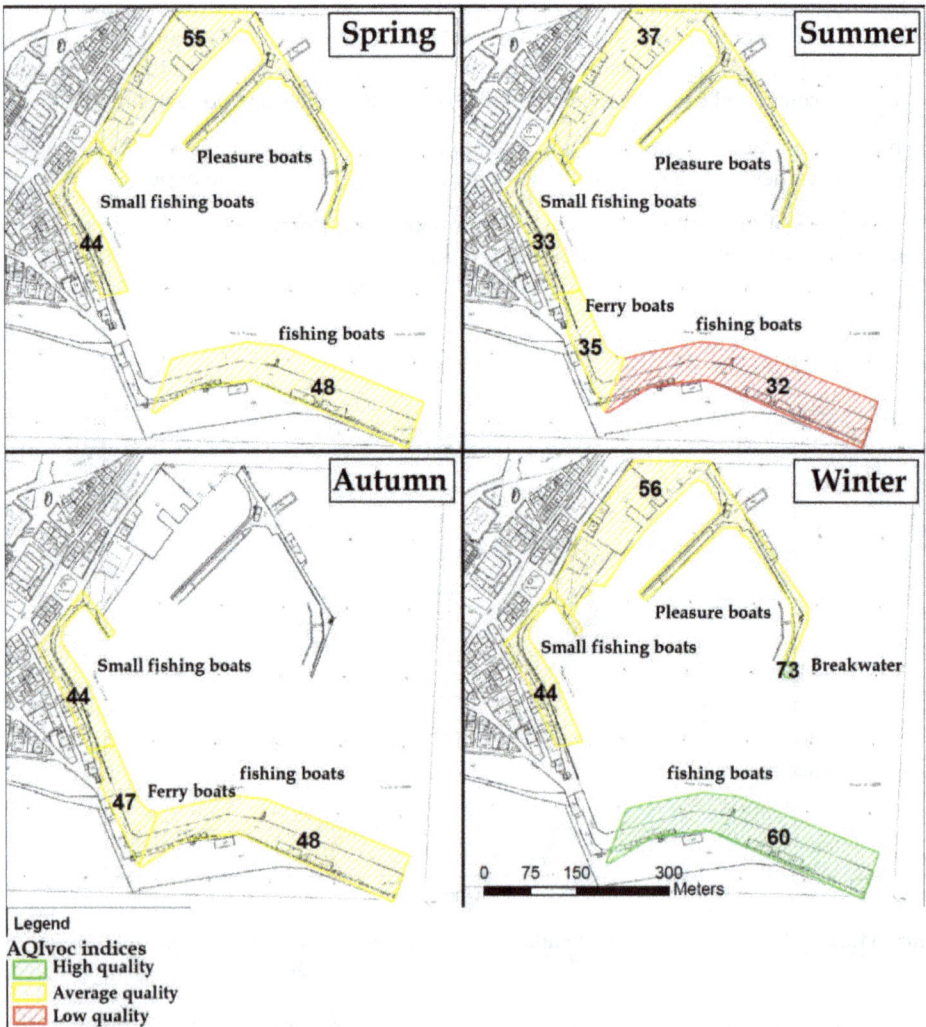

Figure 12. IQAvoc values of the harbour of Anzio in each season and intended use area

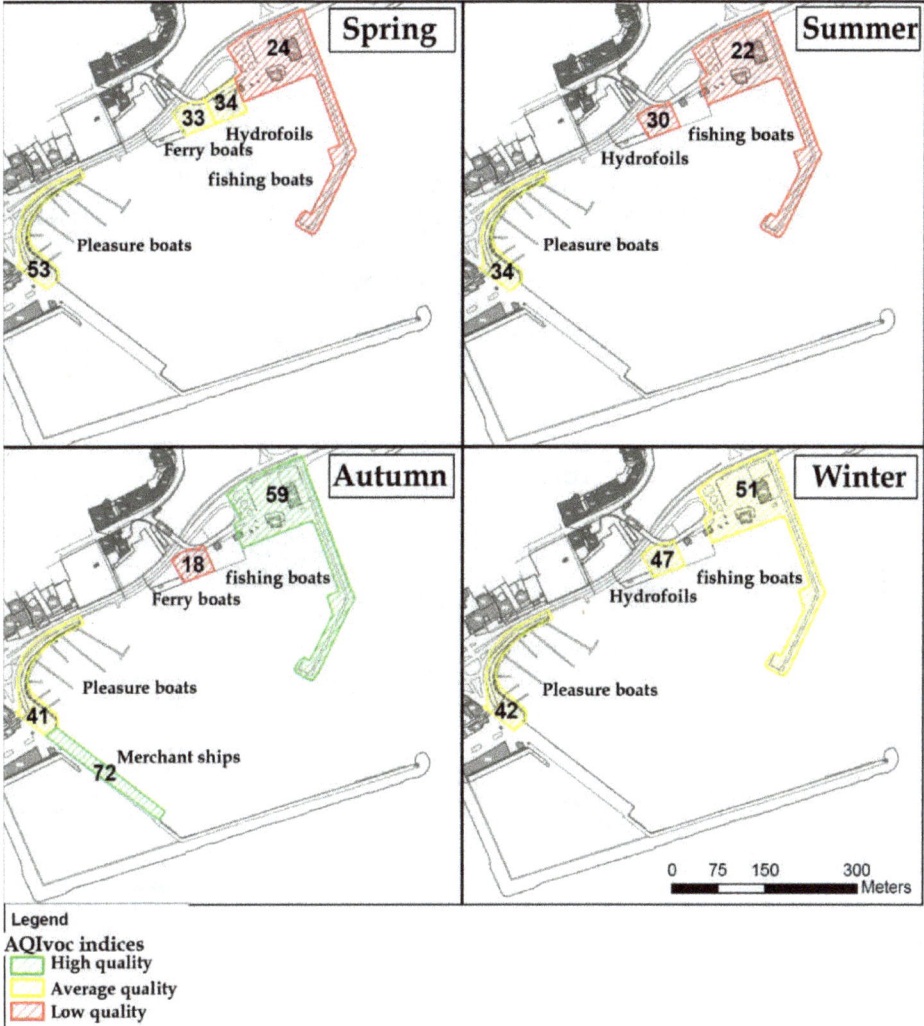

Figure 13. IQAvoc values of the harbour of Formia in each season and intended use area

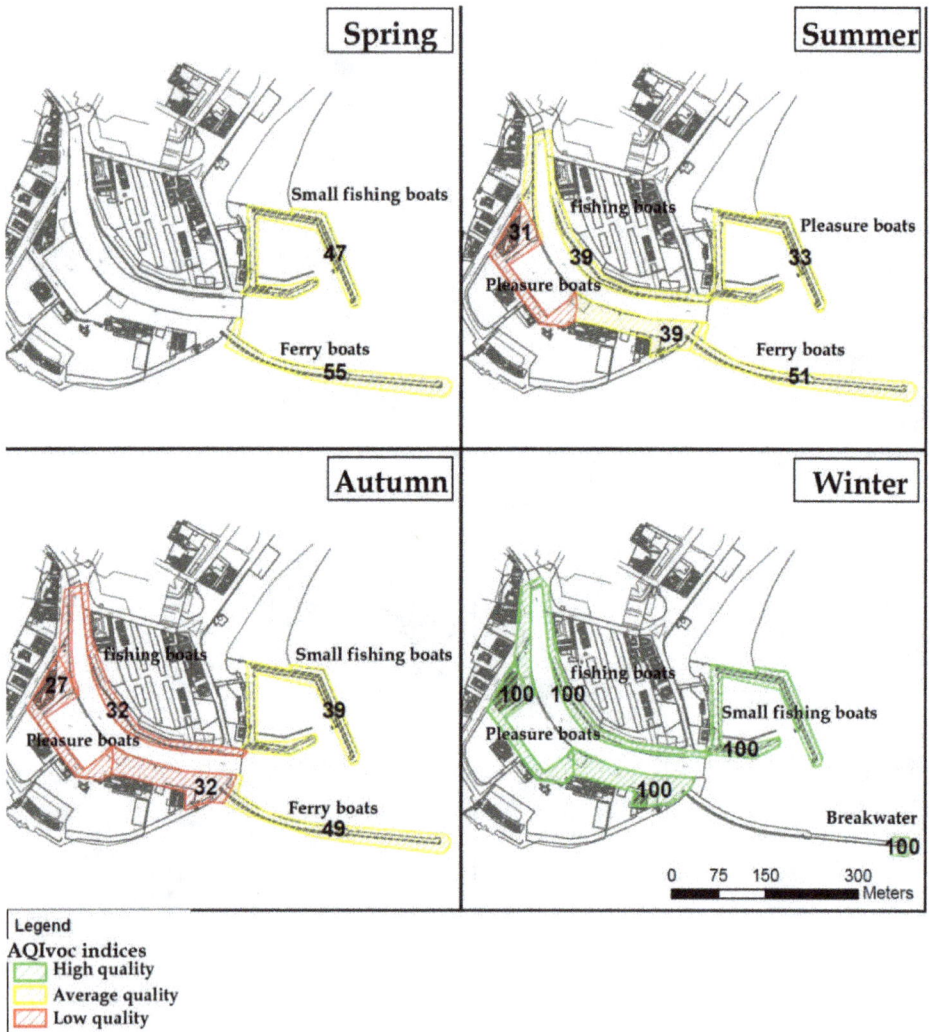

Figure 14. IQAvoc values of the harbour of Terracina in each season and intended use area

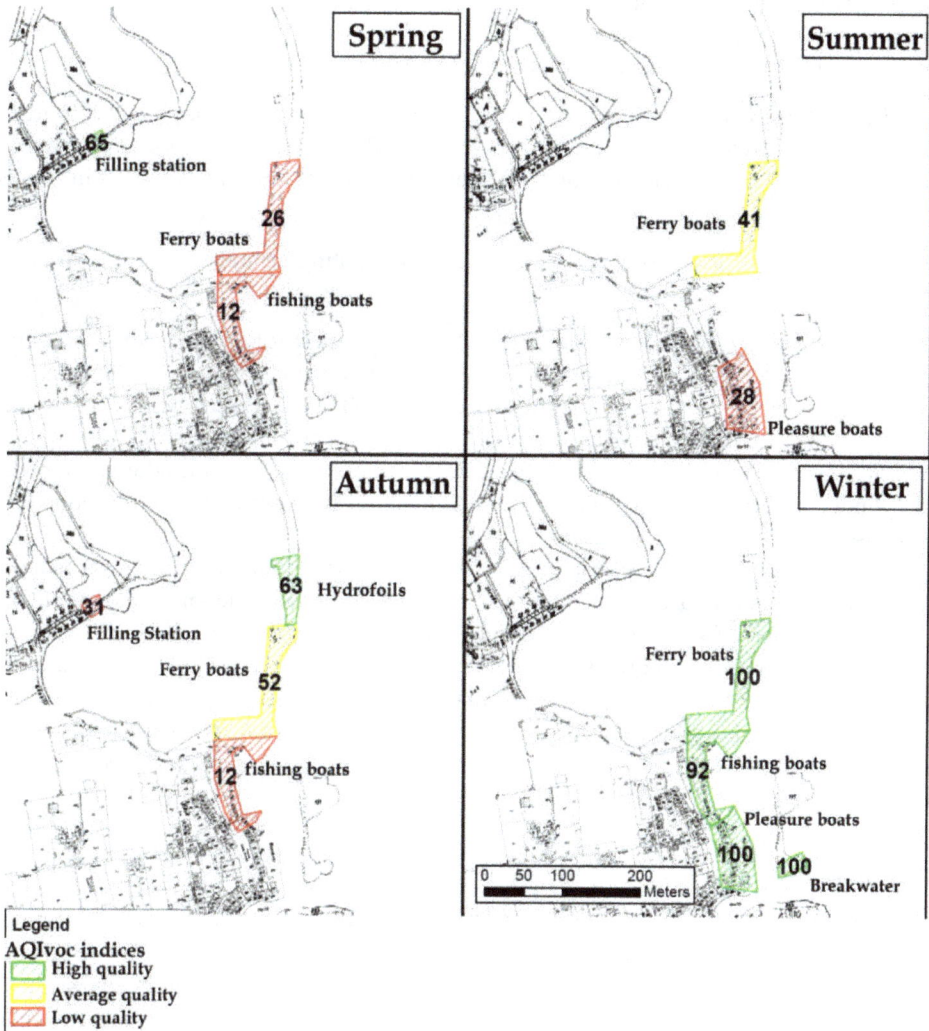

Figure 15. IQAvoc values of the harbour of Ventotene in each season and intended use area

6. Best practices for air quality improvement in portal areas

Aim of this paragraph is to illustrate some BAT (Best Available Technologies) and best practices for mitigating VOCs concentrations and improving local air quality in those portal areas characterized by high concentrations of pollutants.

The obtained results show that the main sources of pollutant emissions in the four analyzed harbours are the Internal Combustion Engines (ICE) of Ro/Pax ferries and hydrofoils, pleasure boats, fishing boats as well as cars and trucks circulating in the ports.

Excluding fishing boats, these sources of emission are all highly dependent on tourism activities which involve an increase of vehicular traffic in the port areas, an enhanced number of daily trips of ferries and hydrofoils and, last but not least, a heavy pleasure boats traffic.

Nowadays, the best practices and technologies for mitigating air pollution in portal areas are:

- SSE (Shore-Side Electricity) enables ships at port to use electricity from a local power grid through a substation at the port to power loading and unloading activities, electronic systems, fuel systems, discontinuing the use of their auxiliary engines. The emission reduction efficiency of this solution is about 94% for VOCs (De Jonge et al., 2005).
- DWI (Direct Water Injection) is a technology which consists in introducing into the cylinder a mixture of water and pressurized fuel which allows lower consumption and emissions (Wahlström at al., 2006).
- Use of low emission fuels: in particolar seaweeds hold a huge potential as a biofuel. Briefly, biofuels are used for fighting climate changes because the same amount of CO_2 that is released from combusting biofuels has previously been taken up from the atmosphere as the plant grows, thus not leading to any net increase in the concentration of CO_2 in the atmosphere (Opdal & Johannes, 2007).
- Optimization of combustion processes in ship engines by means of particolar devices able to optimize combustion disaggregating hydrocarbons (gas emission reduction up to 75-80%).

7. Conclusion

The use of the IQAcov index and the implementation of the best practices and technologies described in the last paragraph could be considered useful tools for monitoring and improving air quality in portal areas for stakeholders and decision makers such as port/maritime authorities, licensed port company operators and local and governmental authorities involved in port jurisdiction.

Indeed, as recommended by the European Sea Port Organisation (ESPO, 2003), among the main environmental objectives which the EU port sector should aim to achieve there is the increase of environmental awareness, the implementation of environmental monitoring and the use of best practices and technologies on environmental issues.

These targets, together with the promotion of environmental monitoring in ports, are fully included among the ESPO top ten environmental objectives that the European port sector should pursue.

Author details

Davide Astiaso Garcia and Franco Gugliermetti
DIAEE (Dipartimento di Ingegneria. Astronautica, Elettrica ed Energetica – Department of Astronautical, Electrical and Energy Engineering) of the Sapienza University of Rome, Italy

Fabrizio Cumo
DATA (Dipartimento Design, Tecnologia dell'Architettura, Territorio e Ambiente – Design, Architectural Technolog, Territory and Environment Department) of the Sapienza University of Rome, Italy

Acknowledgement

The authors wishes to thank all the members of this project for their active contribution to this research, in particular: Dr. Federica Barbanera, Dr. Daniele Bruschi, and Dr. Teresita Gravina.

We also wish to thank Dr. Sergio Fuselli, Dr. Marco De Felice and Dr. Roberta Morlino of the Istituto Superiore di Sanità (ISS) (Superior Health Institute) for their major contribution providing tools and analysis for VOCs concentration monitoring.

Moreover, a special thanks to the whole staff of Anzio, Formia and Terracina Coast Guards for supporting out team during the data gathering works in their portal areas.

Finally the whole research group wishes to thank the Regione Lazio, Dipartimento Istituzionale e Territorio, Direzione Regionale Trasporti – Area Porti - for the financial contribution given to this project.

8. References

Bruno, P.; Caselli, M.; de Gennaro, G.; Scolletta, L.; Trizio, L. & Tutino, M. (2008). *Assessment of the Impact Produced by the Traffic Source on VOC Level in the Urban Area of Canosa di Puglia (Italy)*. Water, Air, and Soil Pollution, DOI 10.1007/s11270-008-9666-3, 2008

De Jonge, E.; Hugi, C.; Copper, D. (2005). *Service Contract on Ship Emissions: Assignment, Abatement and Market-based Instruments. Task 2a – Shore-Side Electricity*. Final Report. European Commission Directorate General Environment. Entec UK Limited. August 2005.

European Sea Port Organisation (ESPO), (2003). *Environmental code of practice*. September 2003

IMO (International Maritime Organisation) (2008). Amendment MARPOL Annex IV; *Reduction emissions from ships*. Marine Environment Protection Committee (MEPC), 57° Session.

Murrells, T.; Derwent, R. G. (2007). Climate Change Consequences of VOC Emission Controls. *Report to The Department for Environment, Food and Rural Affairs, Welsh Assembly Government, the Scottish Executive and the Department of the Environment for Northern Ireland*. ED48749102. AEAT/ENV/R/2475 - Issue 3. September 2007.

Opdal, O. A.; & Johannes F. H. (2007). *Biofuels in ships*. ZERO Emission Resource Organisation-REPORT - December 2007

Wahlström, J.; Karvosenoja N. & Porvari, P. (2006). Ship emissions and technical emission reduction potential in the Northern Baltic Sea. Reports of Finnish Environment Institute 8/2006. Helsinki, 2006. ISBN 952-11-2277-3

The Influence of the Concealed Pollution Sources Upon the Indoor Air Quality in Detached Houses

Motoya Hayashi, Yishinori Honma and Haruki Osawa

Additional information is available at the end of the chapter

1. Introduction

There are many infiltration routes in Japanese traditional wooden houses. The equivalent leakage areas of recent houses have become smaller but the infiltration routes are left in the concealed spaces like beam spaces, crawl spaces and inside-wall spaces. The previous studies in test houses showed that these routes lead chemical compounds into the indoor spaces from the concealed spaces as in [1]. Therefore, the infiltration from concealed spaces was taken into consideration in the amendment of Japanese building standard law in 2003 as in [2], [3].

Common Japanese houses are composed of post-and-beam structures and the structures have many air leaks, but recently the airtightness and insulation of recent houses have been improved using insulation materials and films as shown in Figure1. However, many infiltration routes lurk in the spaces concealed inside the walls, the ceilings and the floors. In

Figure 1. Common wooden structure of Japanese house under construction

the concealed spaces, the pollutant sources for example, poly wood and glass wool are used. The crawl spaces are open to outside and insecticides are used to prevent termites and corrosion fungus in the spaces in most cases.

In order to prevent indoor air pollution, it is necessary to control emission rates of volatile organic compounds (VOC) from interior materials and to design effective ventilation routes. The prevention of infiltration of VOC from the concealed pollution sources is also needed as shown in Figure2. The authors made it clear that the concealed pollution sources influence indoor air quality and that the indoor concentrations increase rapidly when the exhaust ventilation system is used. To estimate the influence of the concealed pollution sources upon the indoor air quality, the ratios of the infiltration rates to the emission rates in the concealed spaces were used as an indicator. In a previous study reported in 2008, the ratios were measured in twelve detached houses in Japan and the status of the ratio was investigated using the measurement results and the results on the three test houses which had already been reported in 2005.

In this study, in order to find an effective method to reduce infiltration from the concealed spaces, the characteristics of the movement of chemical compounds in the concealed spaces and indoor spaces were investigated using building cut models and a simulation program named Fresh2006.

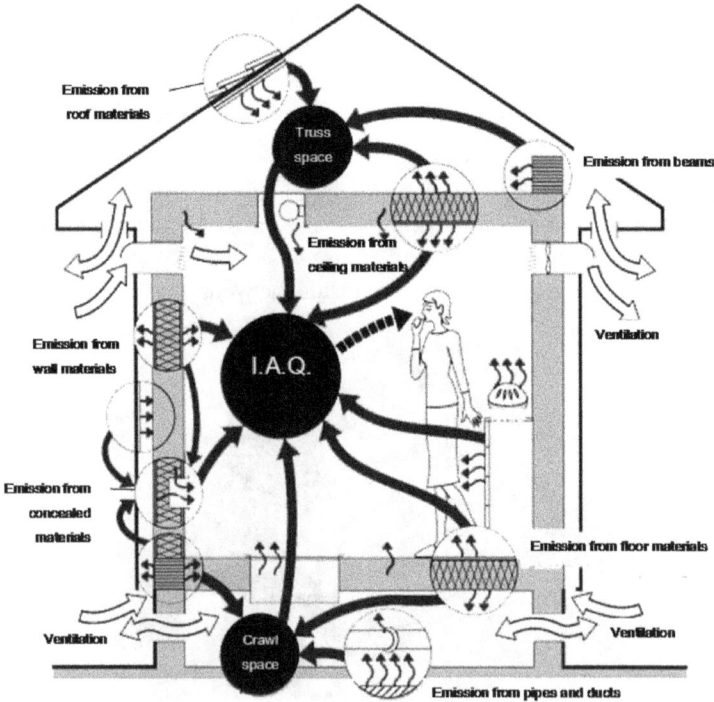

Figure 2. Infiltration of pollutants from the concealed spaces

Figure 3. Cut models of improved wooden post-and-beam structures and wooden stud structures

2. Measurement methods of infiltration ratio from concealed spaces

A ratio of the infiltration rate to the emission rate in the concealed space was used as an indicator. The ratios were calculated from the measurement results using two tracer gases. The influence of the concealed pollutant sources is explained basically using the following equations. The concentration in the indoor space C is given using equation 1 considering the infiltration from the concealed space to indoor space.

$$C = Et \, / \, Q + Co \qquad (1)$$

Where Et is the total emission rate of pollutants considering the infiltration from the concealed spaces, Co: the ambient concentration and Q ventilation rate.

Et is given as the next equation using κ: the infiltration ratio of pollutants from concealed spaces to indoor spaces.

$$Et = \kappa \, Ej + Ei \qquad (2)$$

Where Ej is the emission rate of pollutants in concealed spaces and Ei is the emission rate of pollutants in indoor spaces.

$$\kappa = \{(C-Co)\,Q - Ei\} / Ej \tag{3}$$

Two kinds of tracer gas were generated: one in the concealed space and the other in the indoor space. The concentrations were measured in the indoor space and the infiltration ratio was calculated using equation 3. At first, the ventilation rate Q was calculated using the generation rate and the concentration of the gas generated in indoor space. The infiltration ratio was calculated using the value of the ventilation rate Q and the data from another gas.

The ventilation rates, the infiltration ratios, the equivalent leakage areas per its floor areas, the air flow rate at the inlet or the outlet and the indoor and outdoor temperature and humidity were measured. The ventilation rates and the infiltration ratios were measured using two gases: sulphur hexafluoride (SF6) and freon22 (R22). A tracer gas was injected in the indoor space continuously. And the other tracer gas was injected to each concealed space as shown in Figure 5.

Figure 6 shows the change of tracer gas concentrations. When the concentrations became steady, the ventilation rates and the infiltration ratios were calculated using the stable concentrations and the injection rates were measured using mass flow controllers. These measurements were carried out without heating rooms to decrease the influence of temperature difference between indoor and outdoor upon the ventilation. The temperature differences were lower than ten degrees.

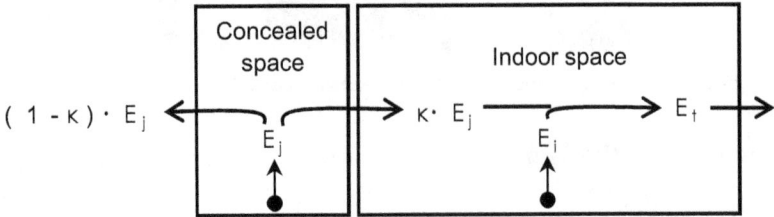

Figure 4. Influence of the concealed pollutant source

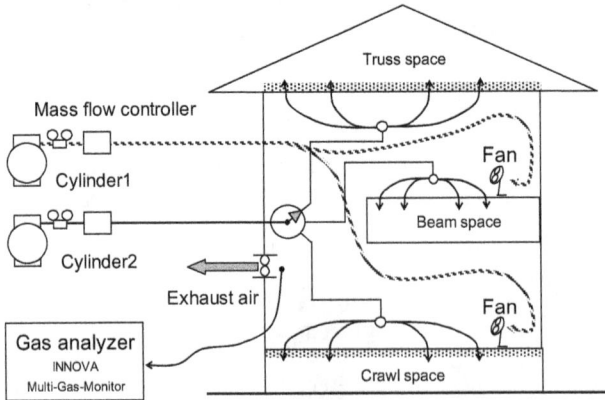

Figure 5. System to measure the ventilation rates and the infiltration ratios

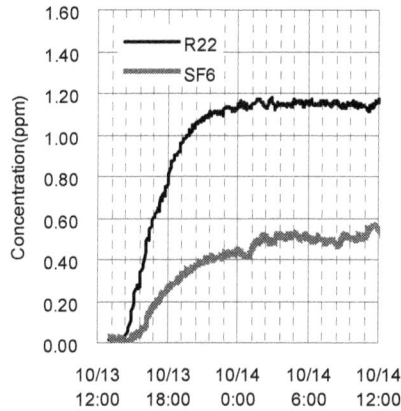

Figure 6. Change of concentrations of tracer gases

3. Simulation methods of the influence of infiltration upon I.A.Q.

The equivalent leakage areas in the concealed spaces were measured using cut models of wooden structures: a common post-and-beam wooden structure, an improved post-and-beam wooden structure built according to the latest building insulation code and a wooden (2 inch x 4 inch) stud structure built according to the latest building insulation code, which was established in 1999.

It was difficult to measure the equivalent leakage areas in the concealed spaces of real houses, so the cut models of these structures were constructed in a laboratory in Miyagigakuin Women's University. Figure 3 shows a cut model of improved post-and-beam wooden structure. The sizes of the elements of the cut model are the same as those of the elements of real houses but the height of the cut model was lower than a real house. The same size of cut model of improved post-and-beam wooden structure as a real house was built according to the latest insulation code was as the same size. These cut models include several leakages in the concealed spaces in these cut models. Cut models of a wooden (2 inch x 4 inch) stud structures were made in another method. Figure 3 shows divided cut models of wooden (2 inch x 4 inch) stud structures. Some partial cut models were made in order to save the space for measurement. These cut models were built by carpenters in the laboratory.

Leakage areas in the concealed spaces and the leakage areas between the concealed spaces and indoor spaces were measured using mass-flow controllers and pressure analyzers. Figure 7 shows the measurement system. When the leakage areas between cell1 and cell2 are measured, cell1 is controlled to be open to the outside and the air pressures of cell3 and cell4 are kept to be the same pressure as cell2. On this condition the air of cell2 goes only to cell1. The airflow rate from cell2 to cell1 accords the air flow rate through the mass-flow controller between the air tank and cell2, so the air flow rate can be known. Figure 8, Figure 9 and Figure 10 show the equivalent leakage areas per 1m of each structure.

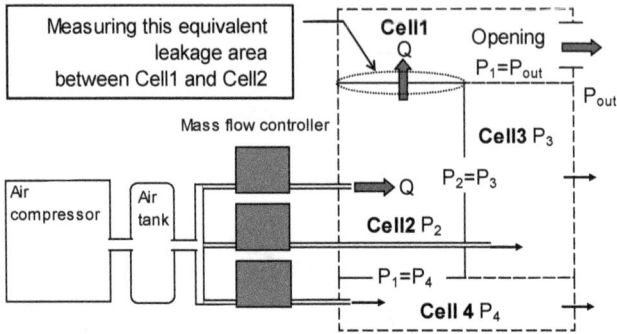

Figure 7. Measurement systems to measure leakage areas between cells

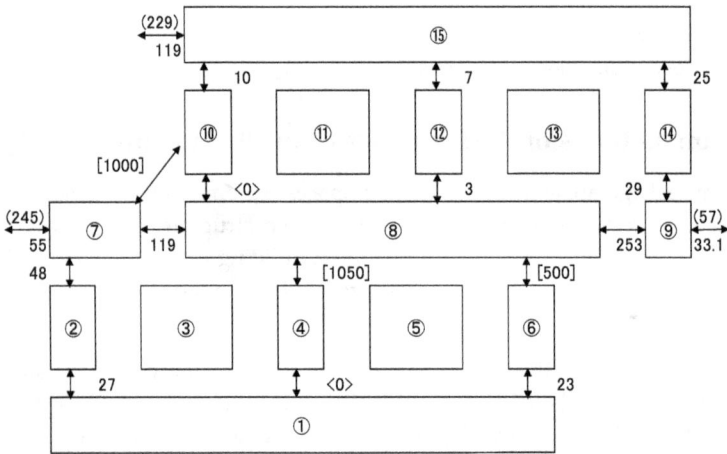

Figure 8. Leakage network of a common post-and-beam structure (cm²/m)

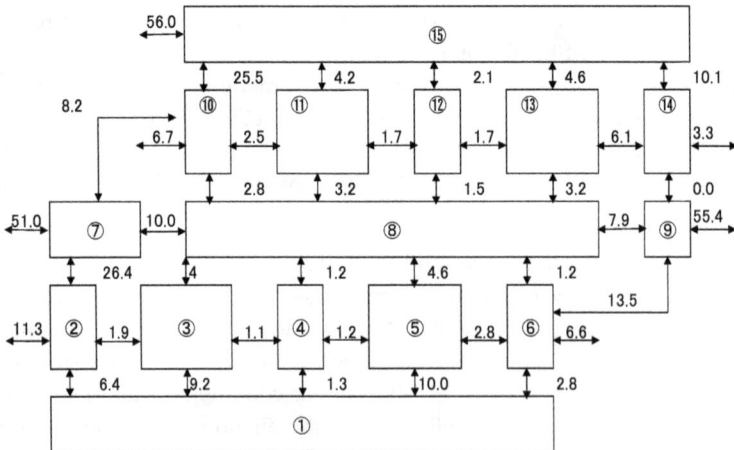

Figure 9. Leakage network of an improved post-and-beam structure (cm²/m)

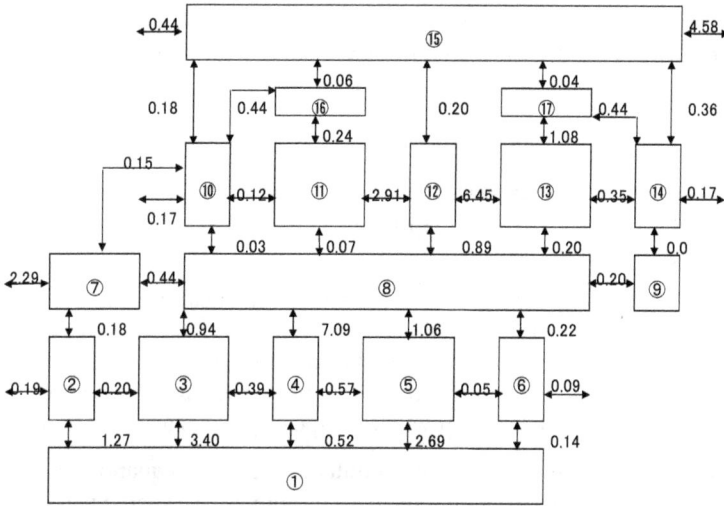

Figure 10. Leakage network of a wooden (2 inch x 4 inch) stud structure (cm²/m)

The movements of chemical compounds were calculated using a simulation program as in [4]. The program simulates the temperatures, the air flow rates, the concentrations and the generation rates of pollutants like formaldehyde, and carbon dioxide using the Japanese standard living schedule model by NHK as in [5] and the HASP weather data on Tokyo.

The simulation program was written in 1996, and was named 'Fresh96'. It was composed of the following three calculation methods.

1. Dynamic thermal calculation of temperature, heating and cooling loads: the calculation method was devised by Prof. Aratani, Hokkaido University in 1974. The initial responses of the thermal-flow rates are calculated and the functions of the responses are described as the following equation in order to increase the speed of the calculation.

$$h(t) = B_0 + \Sigma B_m e^{-\beta_m \cdot t} + q \cdot \delta(t) \tag{4}$$

Where, $h(t)$ the initial response of thermal-flow rate, B_0 the steady value of thermal flow rate, $q = \Sigma B_m / \beta_m$ and $\delta(t)$ Delta function.

The temperatures and the heating and cooling loads are calculated with the above equation using Duhamel's integration method. Temperatures and heat loads are calculated using the calculated temperatures in other rooms and the calculated ventilation rates as values Δt before. In the following case studies, the interval time Δt was decided to be 5 minutes. The values are calculated using the standard weather data from Society of Heating, Air-conditioning and Sanitary Engineers of Japan and the rates of solar radiation through windows are calculated considering the effect of shades. The thermal loads by human behaviors such as cooking, watching television and cleaning rooms, are calculated from the daily schedule model of a family. The air-conditioners and the windows are operated to make the indoor climate comfortable considering the daily schedule of a family.

Calculation of air flow rate in the multi-cell system using the equation of power at the openings: the airflow rates are calculated using the following equations which are led by the balances of power at openings.

$$[D] \cdot \{q^n\} + [K]\{\int q dt\} = \{F_{wind}\} + \{F_{temp}\} + \{F_{fan}\} \tag{5}$$

where q the airflow rate, n the exponent of airflow friction, $[D]$ the matrix of airflow friction, $[K]$ the matrix of room air elasticity, $\{F_{wind}\}$ the power of wind, $\{F_{temp}\}$ the power by room air density $\{F_{fan}\}$ the power of fan.

The equations can be solved using Newmark's numerical integration method. The ventilation rates are calculated considering stack effect, wind pressure and mechanical power using the standard weather data, the ratios of wind pressure, the ratio of wind speed considering the circumstances and the performance of fans. In the case of the following studies, the ratio of wind speed at the town to the speed at the plain flat ground was 0.4.

Dynamic calculation of concentrations of pollutants using the equation of the amount of pollutants: the concentrations of pollutants in each room are calculated using the following equation which is led by the balance of the volume of pollutants.

$$[Q] \cdot \{C(t)\} + [V] \cdot \{C'(t)\} = \{M(t)\} \tag{6}$$

where $[Q]$ the matrix of airflow rate $Q(i,j)$: the airflow rate from room-i to room-j, $Q(k,k) = -\Sigma_{k \neq i} Q(k,i)$, $C(t)$ the concentration of a pollutant, $[V]$ the volume of rooms, $\{M\}$ the emission rates of a pollutant in each room.

The equations can be solved using Newmark's numerical integration method. The emission rates of CO_2 are calculated using the average Japanese daily schedule and the data on the emission rates caused by people's behavior in houses shown in Table1. The daily schedule of each family in a house is calculated considering the plan of the house using the results of the survey on the Japanese daily schedule by NHK. Figure 11 shows the calculated emission rates of CO_2 on a holiday and a weekday in the house model. The emission rates of CO_2 change with the behavior of the family and the emission rates are high in the bedrooms on the second floor at night and the emission rates are high in the living room on the first floor at daytime. This is a typical pattern of emission rate of CO_2 in a general Japanese detached house.

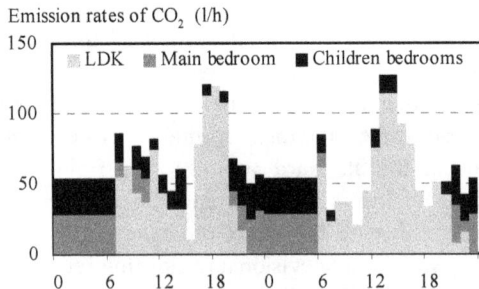

Figure 11. Calculated emission rates of CO2

The emission rates of formaldehyde were calculated using the following equation. The influences of temperature and sink were considered in the equation as in [6].

$$E = E25 \cdot a^{(T-25)} - \beta \cdot C(t) \tag{7}$$

Where E: emission rate ($\mu g/h.m^2$), E_{25} (=100 $\mu g/h.m^2$): emission rate measured in small chamber when temperature is 25deg.C, T: temperature, a=1.11: measured in small chamber, β (= 0.06): ratio of sink measured in small chamber, C(t): concentration ($\mu g/m^3$)

In these studies, in order to make clear the influence of the infiltration of chemical compounds from beam spaces and a crawl space to indoor space upon the indoor air quality, SF6 was emitted at the beam spaces and R22 was emitted at the crawl space in the simulation models shown in Figure 12. Formaldehyde emission rates in the concealed spaces are set to be 100$\mu g/h.m^2$ considering the surface area of emission sources like plywood and the emission rates. This model was designed considering the shape of a low two-storied house. The emission rates of SF6 and R22 were 300 ml/h. The movements and the concentrations of each gas were calculated. The outdoor concentrations were set as follows. The concentration of carbon dioxide is 400 ppm, that of formaldehyde 0 $\mu g /m^3$, that of SF6 0 ppm and that of R22 0ppm.

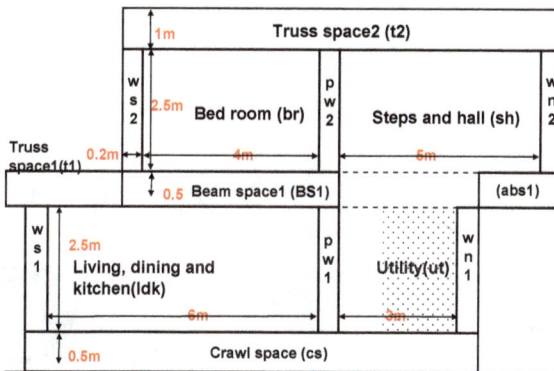

Figure 12. Building model for simulations

4. Adaptation of the measurement results of dwellers' behaviour on opening and closing windows

A monitor instrument to measure the open width at a sliding window was devised. The instrument consists of pulleys, a piece of string, a spring, a potential meter and a recorder as shown in Figure 13. When a dweller opens the sliding window, the sucking disk on the glass pulls the string and the pulley and the potential meter rotates. When a dweller closes the sliding window, the thread is rewound by the spring. The recorder records the resistance of the potential meter was recovered every 10 minutes. The open width was calculated using the recorded resistance. Not only the open width but the temperature and the humidity both in the room and outside were measured successively for almost a year in thirteen detached houses from the temperate zone to the sub frigid zone in Japan. The living habits and the building types were investigated using a questionnaire form.

Figure14 shows the indoor temperature and the width of window opening. The width shows the president's behavior corresponding to weather condition. Figure 15 shows the change of the open width and the room temperature and the annual cycle of the relation between ambient temperature and open width.

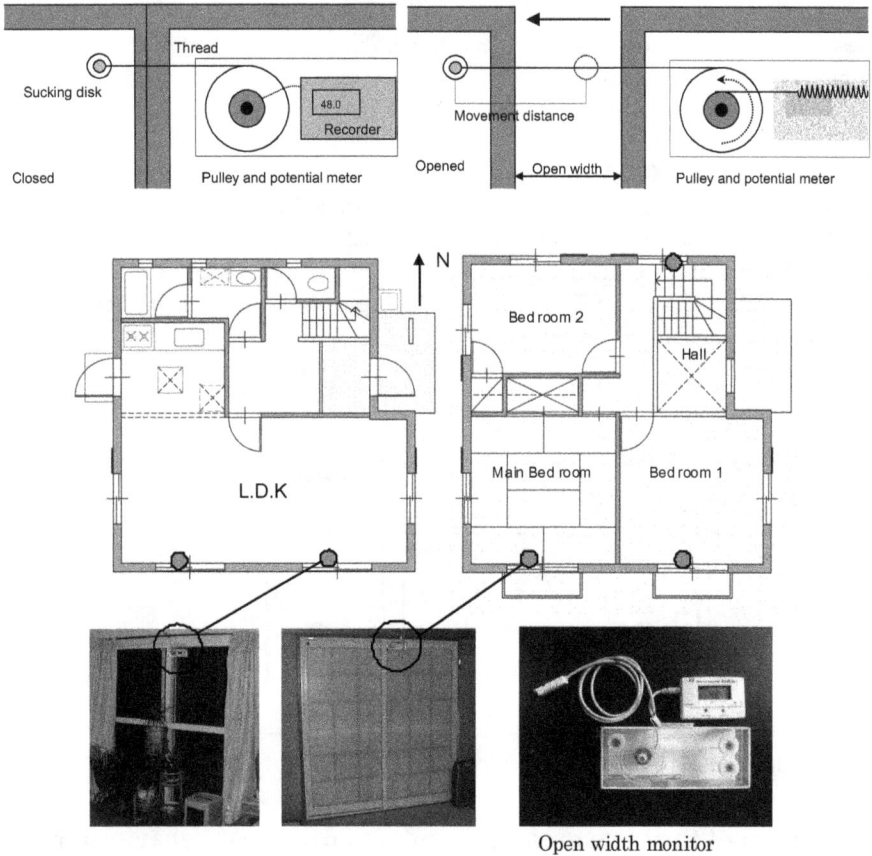

Figure 13. Open width monitor at a sliding window

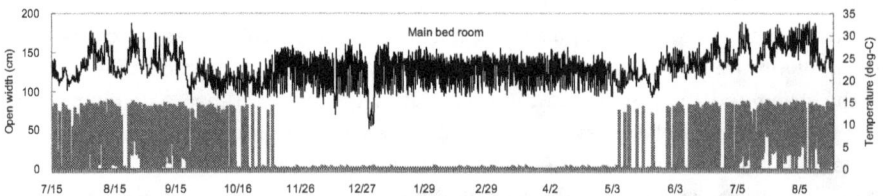

Figure 14. Measured indoor temperature and open width in a bedroom

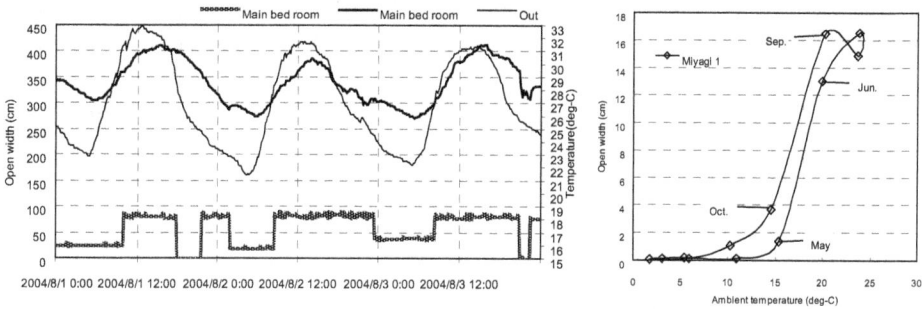

Figure 15. Open width and temperatures in a house

The results of the measurement showed that the averages of open widths were very small even in hot season and mild season. The open width was large when they open window, but windows were open for a very short in hot season. The results of questionnaire survey showed that the reasons: windows were closed were insects, noise and the prevention of criminals. The closing habit was detected clearly in the case of the measured open width in the room on the first floor.

The open width changes with the indoor temperature and the living style. When dwellers go into the room and feel hot, they open windows. In many cases, they close windows when they leave the room. Therefore the open width changes with the dweller's living schedule. The open width is influenced by the indoor and outside temperature. Dwellers keep open width a few centimeters to ten centimeters when they feel a little hot. Such small open widths were shown at night in many cases. When they feel very hot, they close windows and turn on the air conditioner.

The simulations were made on the following conditions based on the results mentioned above. The dwellers operate air-conditioner and windows when they use the room. When they are sleeping, they don't operate windows. The grades of open width were 1: closed (0cm), 2: slit open (10cm) and 3: full open (80cm). The indoor temperature is higher than 26 deg-C, the grade was made up and when the indoor temperature is lower, the grade was made down. If the temperature became higher than 28 deg-C, the grade was set to 1 and the room was cooled by air conditioner. If the temperature became lower than 22 deg-C, the room was heated by air conditioner. These operations were made at the interval of one hour. The simulation was made using HASP weather data of Tokyo.

5. Results of simulations

The equivalent leakage areas of the three structures were measured using this simulation program. A large fan was set and the inside air was exhausted. The airflow rates are controlled to meet five ranks of airflow rate and the pressure differences were calculated. The equivalent leakage areas were also calculated. The equivalent leakage area of the model with a common wooden structure was 5.0 cm^2/m^2, that of an improved structure was 2.8 cm^2/m^2 and that of a wooden (2 inch x 4 inch) stud structure was 0.3cm^2/m^2.

Figure 16 and Figure 17 show the hourly change of formaldehyde concentrations in winter. The concentrations in the beam space (bs1) in the case of exhaust and supply ventilation in Figure 17 is higher than those in the case of exhaust ventilation in Figure 16. The formaldehyde stays longer in the beam space in the case of an exhaust and supply ventilation. But the indoor concentrations are lower in the case of exhaust and supply ventilation. These results show that the influences of the concealed spaces upon the indoor air concentrations depend on the ventilation method.

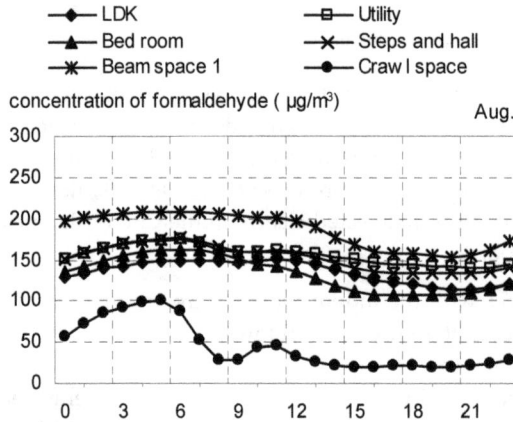

Figure 16. Hourly change of formaldehyde in a house with an improved structure and an exhaust ventilation system

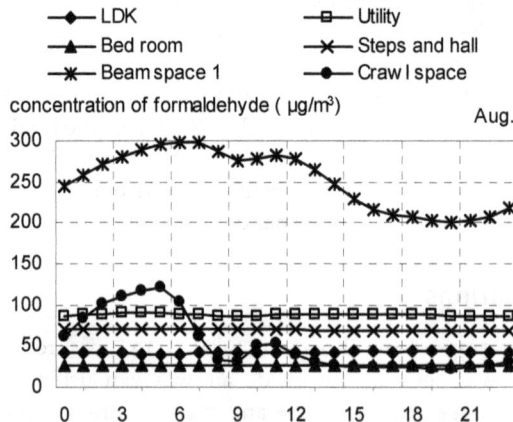

Figure 17. Hourly change of formaldehyde in a house with an improved structure and an exhaust-and-supply ventilation system

Figure 18 and figure 19 show the airflows in a common structure. In the case of exhaust ventilation in figure 18, the air goes from concealed spaces to indoor spaces. The air goes from the crawl space to the indoor space on the second floor through the wall and truss space (t1). Several routes from concealed spaces to indoor spaces are shown in these figures. In the case of exhaust and supply ventilation in Figure19, these routes are also recognized.

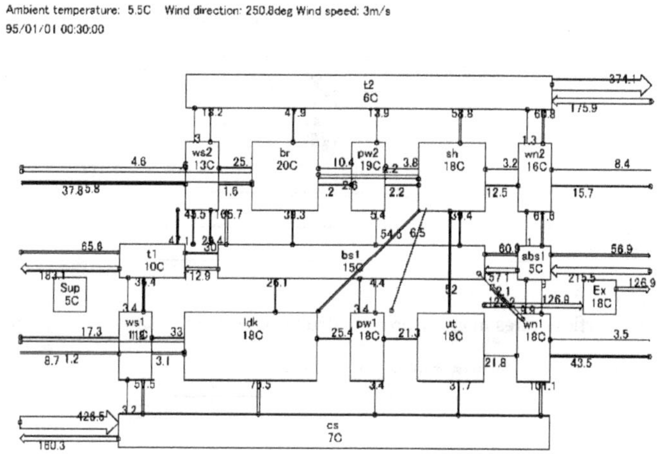

Figure 18. Calculated airflow rates in a common structure with an exhaust ventilation system

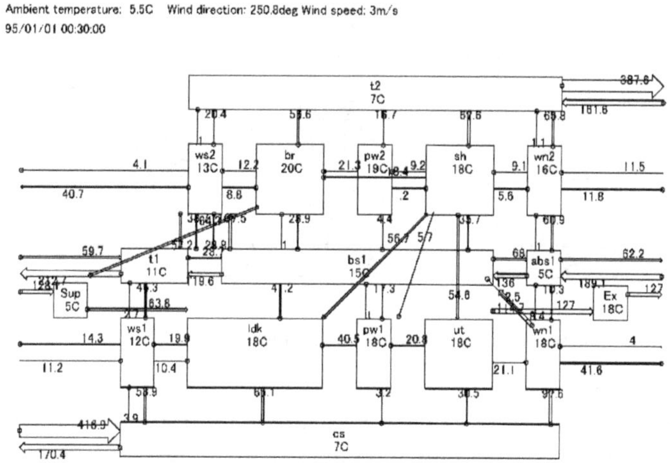

Figure 19. Calculated airflow rates in a common structure with an exhaust and supply ventilation system

Figure 20 and Figure 21 show the airflows in the case of exhaust ventilation. The airflow rates through the above routes became lower in these airtight structures: an improved structure and a wooden (2 inch x 4 inch) stud structure. The airflow rates through these routes become very low in the cases of airtight structures.

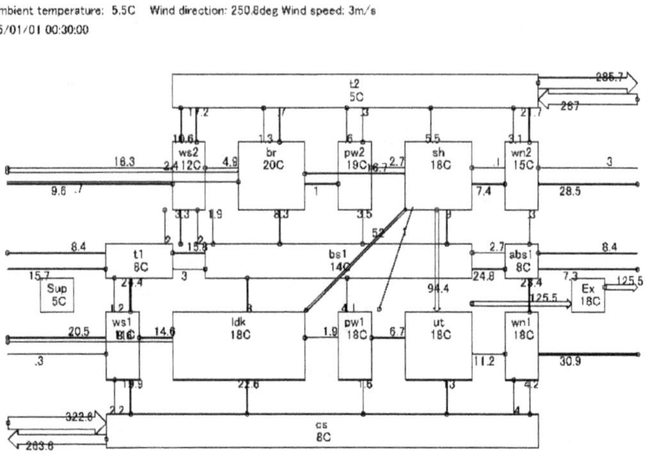

Figure 20. Calculated airflow rates in an improved structure with an exhaust ventilation system

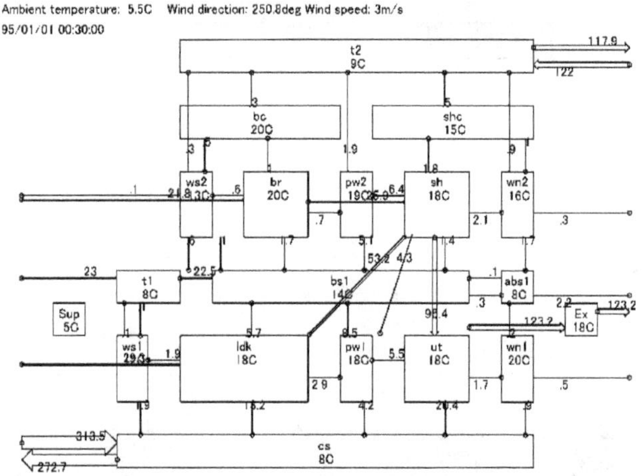

Figure 21. Calculated airflow rates in a wooden (2 inch x 4 inch) structure with an exhaust ventilation system

Figure 22 shows the annual change of temperature difference and wind speed in Tokyo. Figure 23 shows the calculated airflow rates directly from the outside to indoor spaces. In mild seasons like June and September, some airflow rates became higher than in winter, because the windows were opened to make indoor climate comfortable. In mid-summer, the indoor spaces were cooled by air conditioner and these airflow rates became lower than those in mild seasons.

Figure 24 shows the annual change of calculated concentrations. These concentrations are monthly averages. The concentrations of carbon dioxide were steady and low. The

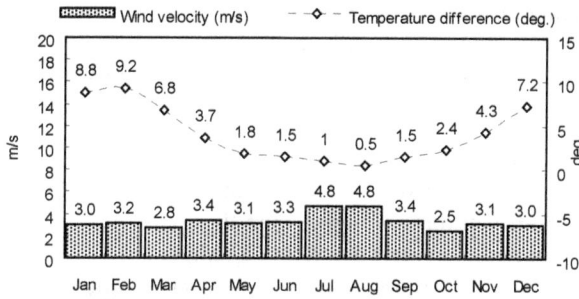

Figure 22. Annual change of temperature difference and wind speed

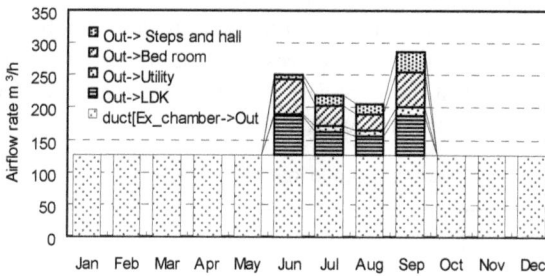

Figure 23. Airflow rates in an improved structure with an exhaust and supply ventilation system

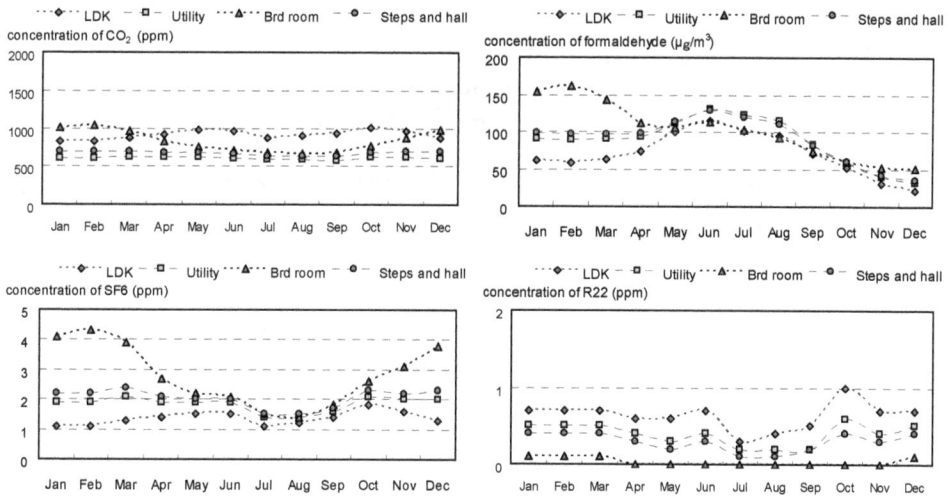

Figure 24. Calculated concentrations in an improved structure with an exhaust ventilation system

concentrations in the bedroom (br) were the highest because the air supply through the ventilator was not high enough. These characteristics were shown in common structures with many air leaks. The annual change of formaldehyde concentrations shows interesting characteristics. Generally, the concentrations decrease but the concentrations were high in summer except for the concentrations in bedroom that were high in early months of the year. These characteristics were based on the following mechanisms. The emission rate of formaldehyde increases with temperature. The emission ability declines with the integral volume of emission.

SF6 is emitted in the beam space (bs1). The concentrations of SF6 are low in summer and the concentrations in the bedroom (br) are higher than those in other spaces. In winter, the air supply to bedroom is not high enough due to large temperature differences but it becomes sufficient in mild seasons and summer. In mild seasons windows were opened. These changes of conditions influenced the changes of concentrations.

R22 is emitted in the crawl spaces (cs). The concentrations of R22 are lower than SF6 concentrations. The crawl spaces are connected to the outside through openings according to the building code to keep wooden structures. But the indoor concentrations were not zero. This result shows that it is necessary to keep the emission rates of chemical compounds like organic phosphorus insecticide low even in the crawl space.

Infiltration ratios from beam spaces to indoor spaces

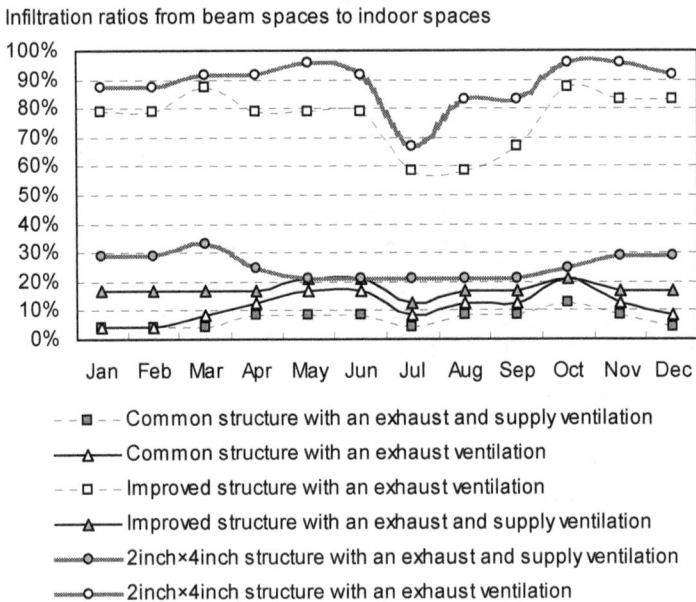

- - ■ - - Common structure with an exhaust and supply ventilation
——△—— Common structure with an exhaust ventilation
- - □ - - Improved structure with an exhaust ventilation
——▲—— Improved structure with an exhaust and supply ventilation
——○—— 2inch×4inch structure with an exhaust and supply ventilation
——○—— 2inch×4inch structure with an exhaust ventilation

Figure 25. Annual changes of infiltration ratios from the concealed spaces to indoor spaces

Figure 25 shows the infiltration ratios from beam space to indoor space. The infiltration ratio accords to a ratio: κ of the infiltration rate of gas to indoor spaces to the emission rate in concealed spaces. The ratios are high in the cases of exhaust ventilation system and airtight structures. In these cases, indoor spaces are decompressed and pollutants are pulled inside from the concealed spaces. Therefore when windows are open, the ratios become lower. In the case of crawl spaces, the ratios are lower than those in the case of beam space.

Figure 26 shows the comparison of concentrations. In the case of carbon dioxide, the concentrations are higher with an exhaust ventilation system and an airtight structure. This tendency is much stronger in the case of formaldehyde, SF6 and R22.

Figure 27 shows the infiltration ratios from concealed spaces to indoor spaces. The figure also shows the tendency mentioned above and shows that it is necessary to design ventilation routes considering air leaks and the emission of pollutants in the concealed spaces for better indoor air quality.

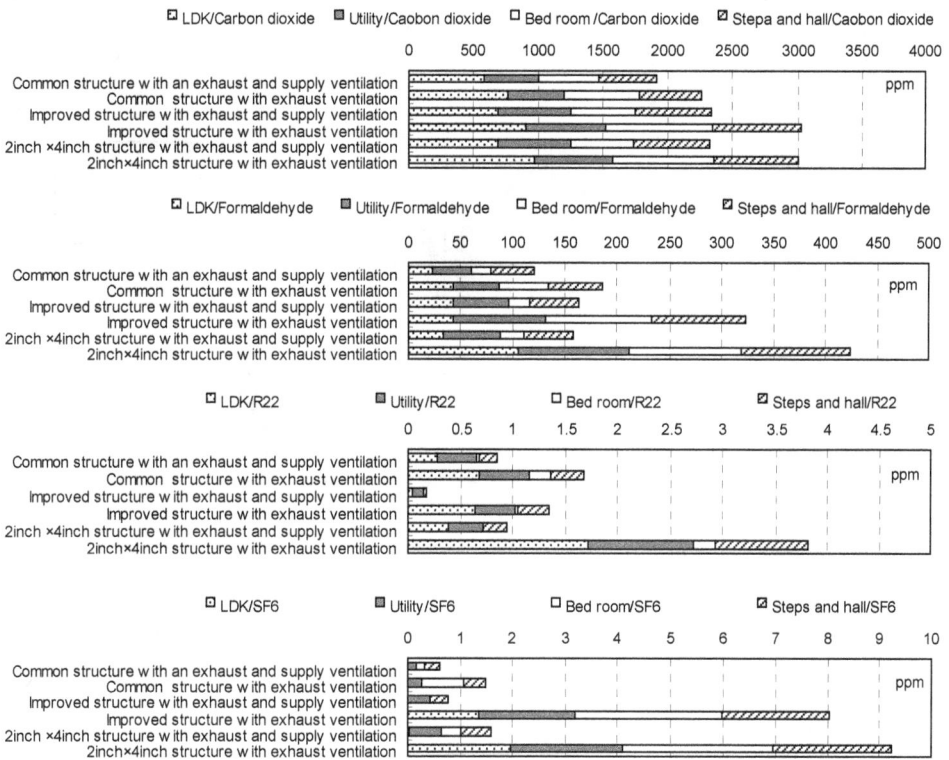

Figure 26. Calculated indoor concentrations of each case study

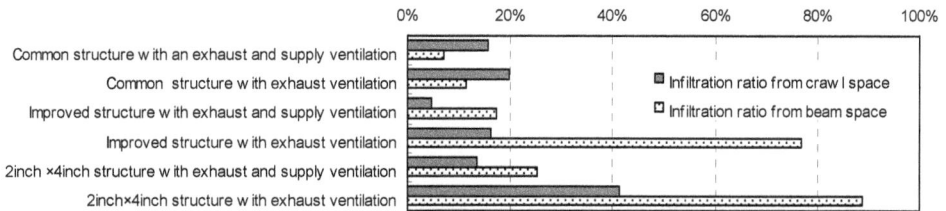

Figure 27. Infiltration ratios from the concealed spaces to indoor spaces in each case study

The calculated infiltration ratios were compared with the measured ratios in real houses: common wooden houses, prefabricated houses and a wooden (2 inch x 4 inch) stud houses as shown in Figure 28. The ratios are measured using tracer gases: SF6 and R22. The tracer gases are emitted constantly in the concealed spaces and in the indoor spaces and the concentrations were measured indoors. The ratios were calculated from the concentrations and the emission rates.

The calculated infiltration ratio from beam space increases with the decompression level. The calculated ratio in the case of a wooden (2 inch x 4 inch) stud structure is the highest and the level of decompression is also the highest. In the case of a common structure, the calculated ratio and the decompression level are both the lowest. The measured ratios have the same tendency. In the case of the ratio from crawl spaces, the measured ratios are lower but the above-mentioned tendency remains.

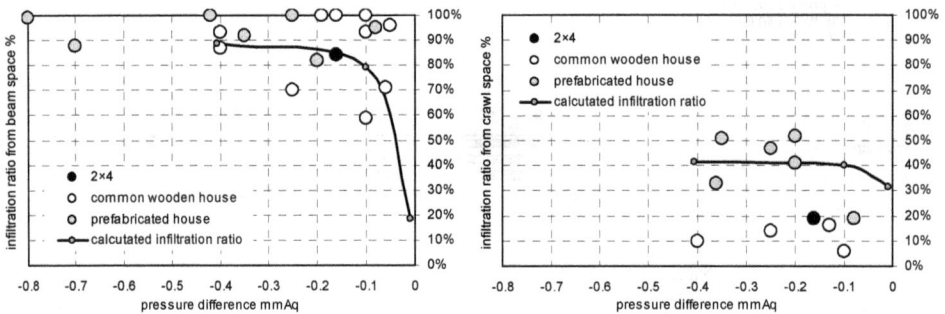

Figure 28. Comparison between the calculated infiltration ratios and the measured ratios in real houses

6. Conclusions

The concealed infiltration routes were shown by the measurements of equivalent leakage areas using cut models of Japanese houses and the simulation considering the weather and Japanese living habit. The indoor concentrations of the chemical compounds which volatilized in concealed spaces changed with the weather and the behaviors of the residents. The infiltration ratios from the concealed spaces to indoor spaces were influenced by mechanical ventilation. The influence of the infiltration upon the indoor air quality was larger in the house with an exhaust ventilation system than with any other ventilation system. These results show that it is necessary to consider the materials and the leakages in the concealed spaces for a countermeasure against sick house syndrome especially in the case of exhaust ventilation. The results will show a guide line for designing a house with better indoor air quality.

Nomenclature

B_0: the steady value of thermal-flow rate
$C(t)$: the concentration of a pollutant

[D]: the matrix of airflow friction
{F$_{temp}$}: the power by the room air density
{F$_{wind}$}: the power of wind
h(t): the initial response of thermal-flow rate
[K]: the matrix of room air elasticity
{M}: the emission rate of a pollutant in each room.
n: the exponent of airflow friction
q : the airflow rate
[Q]: the matrix of airflow rate
Q(i,j): the airflow rate from room-i to room-j
[V]: the volume of a room
δ(t): Delta function

Author details

Motoya Hayashi
Miyagigakuin Women's University, Japan

Yishinori Honma
Iwate Prefectural University, Japan

Haruki Osawa
National Institute of Public Health, Japan

Acknowledgement

The study was a part of a national project "Development of Countermeasure Technology on Residential Indoor Air Quality" by National Institute for Land and Infrastructure Management under the Japanese government. The study was carried out by Grant-in-Aid Scientific Research of Japan Society for the Promotion of Science. The investigations were made with the cooperation of Center for Housing Renovation and Dispute Settlement Support, The Center for Better Living and the students of Miyagigakuin Women's University. The authors express their gratitude to Dr.NoboruAratani, Prof. MasamichiEnai, Prof. HiroshiYoshino, Dr. TakaoSawachi, and Prof. AtsuoNozaki.

7. References

[1] Motoya Hayashi, Haruki Osawa: The influence of the concealed pollution sources upon the indoor air quality in houses, The International Journal of Building Science and its Applications · BUILDING AND ENVIRONMENT,Vol.43,pp.329-336,2008

[2] Hiroshi Yoshino, Kentaro Amano, Mari Matsumoto, Koji Netsu, Koichi Ikeda, Atsuo Nozaki, Kazuhiko Kakuta, Sachiko Hojo and Satoshi Ishikawa, 2004. Long-Termed Field Survey of Indoor Air Quality and Health Hazards in Sick House, Journal of Asian Architecture and Building Engineering, Vol. 3 (2004) No. 2 pp.297-303

[3] Haruki Osawa, Motoya Hayashi: Status of the indoor air chemical pollution in Japanese houses based on the nationwide field survey from 2000 to 2005, The International Journal of Building Science and its Applications BUILDING AND ENVIRONMENT, Vol.44, pp. 1330-1336, 2009

[4] M.Hayashi, M.Enai and Y.Hirokawa, 2001. "Annual characteristics of ventilation and indoor air quality in detached houses using a simulation method with Japanese daily schedule model", The International Journal of Building Science and its Applications 'BUILDING AND ENVIRONMENT' Vol.36, No.6 July 2001,721-731

[5] NHK. 1990. "The survey on the Japanese daily schedule 1990"

[6] Rika Funaki and Shin-ichi Tanabe, 2002. Chemical Emission Rates from Building Materials Measured by a Small Chamber, Journal of Asian Architecture and Building Engineering, Vol. 1 (2002) No. 2 pp.2_93-100

Chapter 4

Information Systems for Air Quality Monitoring

Silvije Davila, Ivan Bešlić and Krešimir Šega

Additional information is available at the end of the chapter

1. Introduction

Today, for the successful performance of the work of monitoring air quality information system is essential. Before some 30-odd years ago when it was beginning to monitor air quality measurement results were written on paper in the table. Increasing use of computer data has begun to digitize. "Database" or tables were first written on the magnetic tape, then on floppy disks, CDs and DVDs. Today, air quality monitoring is impossible to perform without the use of computers and information systems.

2. Information systems

In developing (design) information system for monitoring air quality you must look primarily to the needs of users and data collection methods. Will the information system only collect data and store them in a database, or would have done, and other functions to the data (e.g. calculating AQI, making DEM files, perform data validation, forecasts data, etc.). In addition you have to look at the methods of collecting. If you can only use so-called "Manual methods" data must be manually entered into the system, however if you are using automatic methods data are automatically entered into the system. Very often, information systems for monitoring air quality are made because of changes in the law on air protection, or the suggestions of the competent institutions. If an information system developed for the needs of state institutions or the information that is obtained through an information system is to be forward to the government institutions or later perhaps even to international institutions (e.g. European Environment Agency (EEA)) then the information systems must operate according to certain standards. EEA has issued standards for the development of information systems [1]. Such a rule is a well-respected if the information system developed for the needs of the agency that will directly communicate with the EEA.

When designing information systems people often uses three-tier system architecture (Figure 1). This architecture consists of three layers. In the first layer there are databases, on second layer other helper applications, services and systems, and the third layer there are applications that use the data, other information systems or end users.

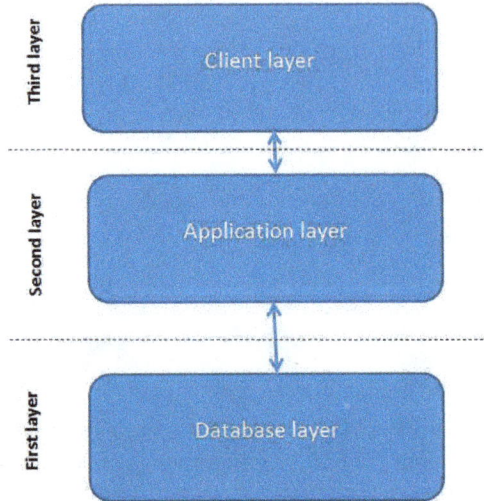

Figure 1. Three-tier architecture of the information system

The preparation of an information system can be divided into software and hardware part. When we talk about the software part we think of the database, various applications to perform various tasks in information systems, services that are needed to operate the information system. While the hardware components are talking about devices for the analysis of the air and on the communication devices that are connected to the information system. First we handle the software and then the hardware part of the information system.

3. The software part of the information system

3.1. Databases

Nowadays the amount of data that has to do with the information system for air quality is enormous. Several years ago, data on air quality were recorded in the tables in different programs (e.g. MS Excel, Quattro, etc.), but such programs are unsafe and not used for storing data, but for their treatment.

These types of programs are called flat files (executable files). The main problem with such files (or such an organization) is to write to physical media (hard disk, CD, DVD, etc.) depends on the media, also with them we cannot talk about any kind of organization the additional data. Other problems that exist among them are:

- competitiveness - at one point the data can access more than one person or one application;

- Integrity - when multiple applications are using the same information (i.e. information), there may be a problem or an error in the data because there is no control of access to data;
- Relationships - set a relation or connection between the data is impossible because there is no predefined structures;
- Re-utilization - direct files are designed for a specific system and cannot work on another (e.g. Doc files do not work on Mac OS or Unix operating system because they are designed to operate on the Windows operating system);
- Security - when it comes to security files, we cannot talk about a common method of providing access to data and are therefore vulnerable to unauthorized access to data.

When all this is taken into account, the question is what would be the definition of a database? The database is a collection of interrelated data stored in external memory and simultaneously available to various users and applications.

The data of a database are stored as files on your computer. In the previous section we talked about random files are also stored on your computer. So the question is how is it different than the database files directly? If we think, we realize that actually the database itself and not much more than what we used to direct the file. That of which we have great benefit is the DBMS (DataBase Management System) system or database management. DBMS server is actually located between the database and its users. He actually performed for us all database operations

Today there are several freeware and commercial DBMS's (MySQL, PostgreSQL, Microsoft Access, SQL Server, Oracle, etc.). DBMS not only retrieves data from the database but is also responsible for adding them, changing and deleting, for storing backups, keeping the integrity of the database, authenticate users who access the database and watch the simultaneous database access by multiple users. The most important advantage that distinguishes a database from a direct file is independent of data provided by the DBMS.

The data in the database are logically organized in accordance with some of the data model. The data model is a set of rules that determine how it may appear logical structure of the database. Models of data from the oldest to the newest are listed below.

- **Hierarchical Model** - in a hierarchical database model is presented to a tree or a set of trees. Nodes represent the types of records, while the hierarchical relationships expressed attitude "superior - subordinate".
- Network Model - when the network database model is presented directed graph. Nodes are the types of records, and the links between them are defined arches. The previously mentioned hierarchical model is actually a special case of network models.
- The relational model - it is based on the mathematical notion of relation. When it presents the data in tables. These tables are connected by links to avoid redundancy (repetition), integrity (wholeness) of the data and to maximize the speed of searching the database.

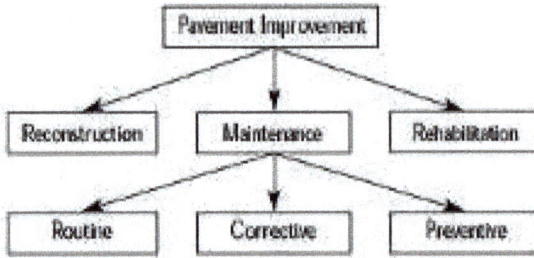

Figure 2. An example of hierarchical models

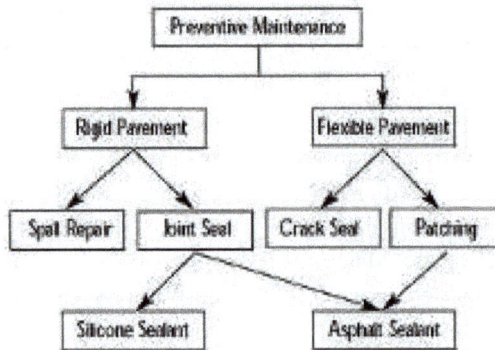

Figure 3. Example of network model

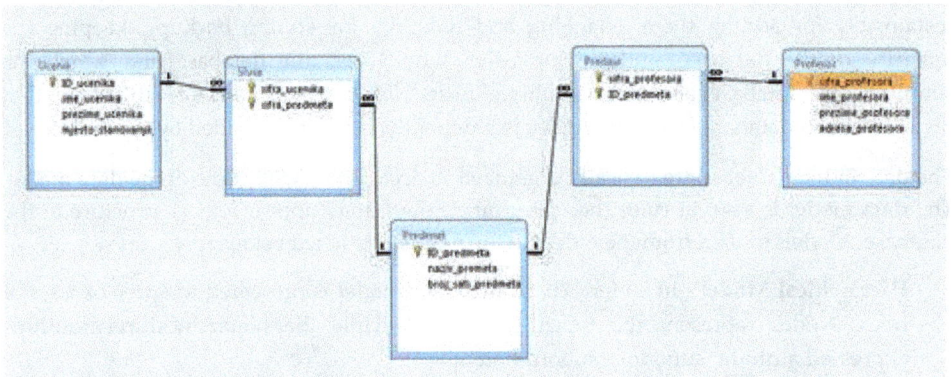

Figure 4. An example of the relational model

- **Object model** - the object model is inspired by object-oriented programming languages. The database is a collection of permanently stored objects that consist of its internal data and methods for working with these data. Each object belongs to a class and the connection is established between classes inheritance, aggregation, or mutual use of operations.

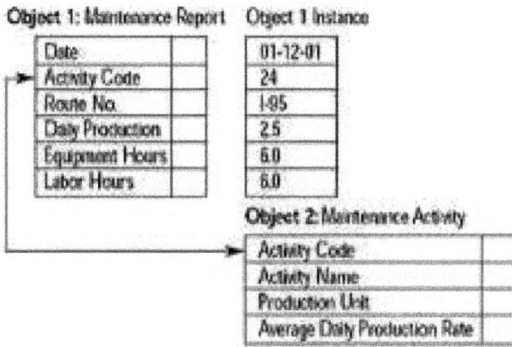

Figure 5. Example of object model

- **Semi structured Model** - in such models the information that would otherwise be normally connected to the scheme are within the data. The advantages are that it can be used to display information about the data that other models do not support, and providing flexible format for exchanging data between different types of databases

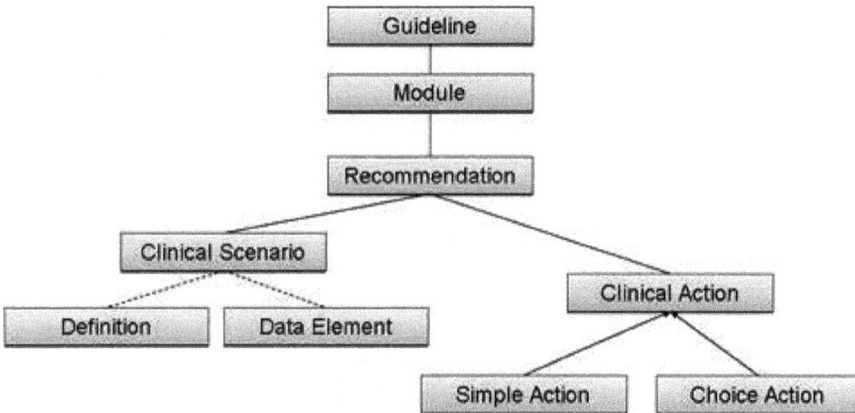

Figure 6. An example of semi-structured models

- Semantic (RDF) Model - allows the organization and management of unstructured data. Mostly to be encountered when connecting XML documents in database.
- Geospatial Model - At this time more and more data related to the Earth's surface and to the need for analysis, visualization and distribution of such data types. In principle, it is a combination of relational and object models (because they did not support relational spatial and temporal component object model and has not had a good enough standard).

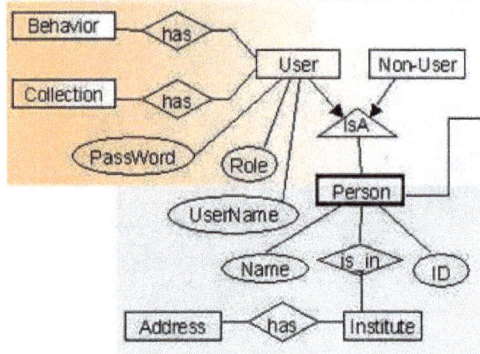

Figure 7. An example of RDF models

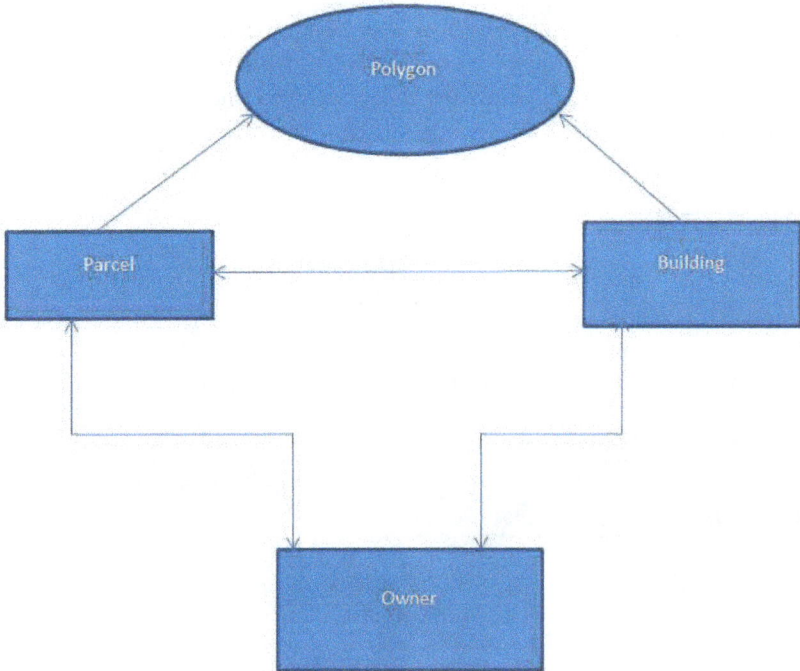

Figure 8. An example of geospatial models

Each model has its bad and good sides. Somewhere in the 80-ies of the relational model is becoming increasingly popular due to its simplicity and short time to overcome. In contrast, the object model is a lot more powerful than all other models, but also more complicated and there are still no generally accepted standards. Geospatial model began to develop rapidly in the last 10-odd years along with a comprehensive application of the GPS system and use geographic coordinates as in science and research segment, as well as economic, military, technological and political environment.

When we look only information systems for monitoring air quality the most common model used is relational, however, at the Institute for Medical Research and Occupational Health (IMROH), we started with the experimental use of geospatial models. Why geospatial model for monitoring air quality? If you look at the data that is handled first, we can see that they are dependent on the temporal and spatial components. Data obtained from the monitoring stations are different for each time of day and also demonstrated their dependence on weather conditions [2]. Because of these characteristics of air pollution, and also because of air pollution were never a point in the air but the best I can describe irregular three-dimensional geometric figure, we believe that it is the easiest and best described in a geospatial model. Tests of such models are only just beginning and it will take some time if and when this model began to be used in larger information systems.

What is for sure the database is the foundation of information system and therefore in its design must be given enough attention. If in the early design phase of information system in preparing the database we detected possible problems later, they can be corrected or avoided.

3.2. Applications

After the databases themselves, very important element of information systems for monitoring air quality are applications that are required for monitoring air quality such as application for DEM files and applications for the AQI.

3.3. DEM files

Within the European Union adopted method for transferring data between institutions within countries and within countries. The Data Exchange Module (DEM) can be used to exchange of information on air quality data (i.e. raw data, statistics and ozone exceedances) and meta information on operational air quality networks, stations and measurement configuration [3,4]. The file format for Air Quality DEM was developed by the European Topic Centre on Air and Climate Change (ETC / ACC), and every institution that sends data to the EEA must send DEM files prepared in that format [4].

There are many free or commercial applications such as Data Communication Server or DEM for creation DEM files.

Application Data Communication Server 2008 is created by the GEMI GmbH. The application is designed in a way that takes data directly from the database converts them into DEM files and saves them on hard disk. Application DEM v14.0 is the latest version that was created in 2011 by the European Topic Centre for Air Pollution and Climate

Change (ETC-ACC) (Figure 9). The application is mainly intended for working with large networks of stations for monitoring air quality, such as the National Network.

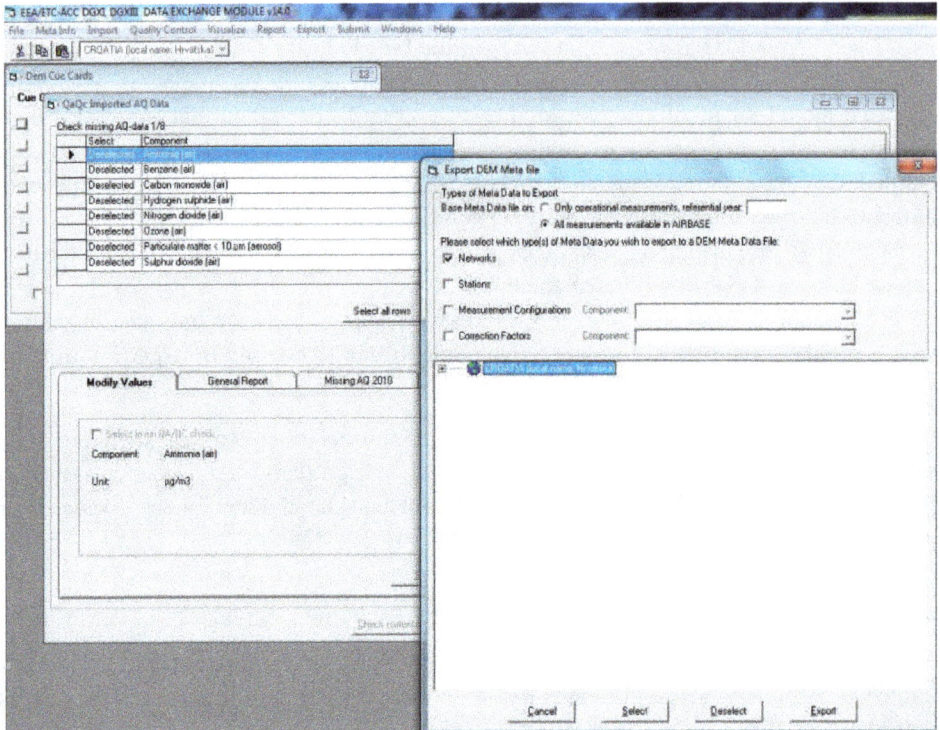

Figure 9. DEM v14.0

Each application has its advantages and disadvantages, so if you cannot find an application that will meet the specific requirements for each information system you can always make your own applications. Of course it is necessary to follow the instructions that gave ETC / ACC regarding looks DEM files [5].

3.4. AQI

Air Quality Index (AQI) is also one more very common application that meets the information systems for monitoring air quality. AQI serves as an indicator of reporting daily air quality. It shows how clean or polluted the air and what the possible health problems could be caused by inhaling air. AQI focuses on health effects that may be caused by inhaling polluted air during a few hours or days. The European Union uses mostly CAQI (Common Air Quality Index) air quality model calculations [6]. In CAQI air quality index can have values from 0 to 100. What is the index value lower, the better air quality is. For this measurement number 75 represents the presence of low levels of each pollutant, the level below the legally prescribed limit value. Therefore, the air quality index below 75 represents good air quality and little possibility of causing health problems of the population. Also the air quality index scale is divided into five different colored categories, from green to red. For example, the index value of 27 drops into the green-yellow zone, this indicates good air quality. On the other hand, values above 101 belong to the red zone, which points to the danger.

Index class	Grid	Traffic				City Background					
		Mandatory pollutant			Auxiliary pollutant	Mandatory pollutant				Auxiliary pollutant	
		NO_2	PM_{10}		CO	NO_2	PM_{10}		O_3	CO	SO_2
			1-hour	24-hours			1-hour	24-hours			
Very low	0	0	0	0	0	0	0	0	0	0	0
	25	50	25	12	5000	50	25	12	60	5000	50
Low	26	51	26	13	5001	51	26	13	61	5001	51
	50	100	50	25	7500	100	50	25	120	7500	100
Medium	51	101	51	26	7501	101	51	26	121	7501	101
	75	200	90	50	10000	200	90	50	180	10000	300
High	76	201	91	51	10001	201	91	51	181	10001	301
	100	400	180	100	20000	400	180	100	240	20000	500
Very High*	> 100	> 400	>180	>100	>20000	> 400	>180	>100	>240	>20000	>500
NO_2, O_3, SO_2:		hourly value / maximum hourly value in $\mu g/m^3$									
CO		8 hours moving average / maximum 8 hours moving average in $\mu g/m^3$									
PM_{10}		hourly value / daily value in $\mu g/m^3$									

* An index value above 100 is not calculated but reported as "> 100"

Figure 10. Table for CAQI model (taken from http://citeair.rec.org)

Another very popular Air Quality Index was developed by the United States Environmental Protection Agency (EPA) [7]. In their index value is 0-500. The higher the AQI value, the greater the level of air pollution and the greater the health concern is. For example, an AQI value of 50 represents good air quality with little potential to affect public health, while an AQI value over 300 represents hazardous air quality.

An AQI value of 100 generally corresponds to the national air quality standard for the pollutant, which is the level EPA, has set to protect public health. AQI values below 100 are generally thought of as satisfactory. When AQI values are above 100, air quality is considered to be unhealthy-at first for certain sensitive groups of people, then for everyone as AQI values get higher.

Each category corresponds to a different level of health concern. The six levels of health concern and what they mean are:

- "Good" AQI is 0 - 50. Air quality is considered satisfactory, and air pollution poses little or no risk.
- "Moderate" AQI is 51 - 100. Air quality is acceptable; however, for some pollutants there may be a moderate health concern for a very small number of people. For example, people who are unusually sensitive to ozone may experience respiratory symptoms.

Air Quality Index (AQI) Values	Levels of Health Concern	Colors
When the AQI is in this range:	..air quality conditions are:	...as symbolized by this color:
0-50	Good	Green
51-100	Moderate	Yellow
101-150	Unhealthy for Sensitive Groups	Orange
151 to 200	Unhealthy	Red
201 to 300	Very Unhealthy	Purple
301 to 500	Hazardous	Maroon

Figure 11. Example table AQI developed by EPA (taken from the site http://www.airnow.gov/index.cfm?action=aqibasics.aqi)

- "Unhealthy for Sensitive Groups" AQI is 101 - 150. Although general public is not likely to be affected at this AQI range, people with lung disease, older adults and children are at a greater risk from exposure to ozone, whereas persons with heart and lung disease, older adults and children are at greater risk from the presence of particles in the air.
- "Unhealthy" AQI is 151 - 200. Everyone may begin to experience some adverse health effects, and members of the sensitive groups may experience more serious effects.
- "Very Unhealthy" AQI is 201 - 300. This would trigger a health alert signifying that everyone may experience more serious health effects.
- "Hazardous" AQI greater than 300. This would trigger a health warning of emergency conditions. The entire population is more likely to be affected.

Applications for calculating the AQI rule does not exist. There are several variants of a program for calculating the AQI in MS Excel. So those responsible for the information system monitoring the air quality must themselves make the necessary application to display the AQI based on a model they choose.

3.5. Website

Another very important part of information system for monitoring air quality is a web page or web site. In it you can show some important information related to the quality of the air, warning to the citizens about pollution and archive data for previous days, months and years. What information will be displayed on the website depends on the interest of citizens for air quality and capabilities of the institution itself.

Figure 12. Examples of web pages to display the air quality (http://kvaliteta-zraka.imi.hr/ ; http://uk-air.defra.gov.uk/ ; http://airnow.gov/ ; http://www.airqualitynow.eu/)

4. The hardware part of the information system

4.1. Measuring devices

The most important part in making information systems are measuring instruments for analyzing air quality. There are various companies that produce them, which will be chosen depends on the person who purchased them.

Automatic analyzers used by IMROH for monitoring gaseous pollutants are trademarks Horiba. We pick them because of good service and the existence of accredited calibration laboratory in Croatia. All Horiba devices have the ability to display data every 3 minutes, half hour, hour or every 3 hours. Three minute value is obtained by taking the average value of the seconds for 3 minutes. Half-hour value is the average value of three-minute value in half an hour. 3-hour value is the average value of the half-hour values in these three hours. Hourly value of the integration of information that is obtained by dividing the measured value (current value) measured every second with 3600 and adding these data for a particular accounting period. Maximum capacity of memory measured data for 3-minute value, the value of half-hour and hour value is 1000 data; a three-hour value is 100 data. The temperature of the room in which the devices can be found in the range of 5 ° C to 40 ° C, although the device works best at a temperature of 20 ° C. Automatic methods for monitoring the mass concentrations of gases at reference methods.

Figure 13. APOA 370 [8]

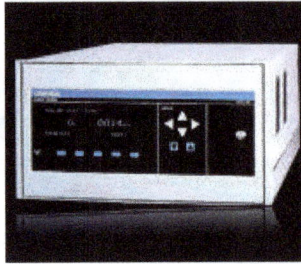

Figure 14. APOA 360 [8]

Automatic analyzer Horiba APOA 370/360 is used for measuring the concentration of ozone in the air. The analyzer uses a cross-modulation UV absorption method. The range of detection of mass concentrations that can be measured ranges from 0 to 1.0 ppm. The minimum detection sensitivity is 0.5 ppb for the range of 0 to 0.2 ppm, and for a range of up to 1 ppm sensitivity is 5 ppb. The analyzer automatically changes the range of detection. The analyzer has the option of displaying the mass concentration in ppm or mg/m^3 (with a correction factor of 2.143). Sample collection rate is 0.7 l / min.

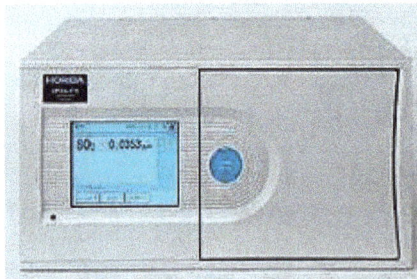

Figure 15. APSA 370 [9]

Automatic analyzer Horiba APSA 370 is for measuring concentrations of sulfur dioxide in the air. The analyzer uses ultraviolet fluorescence method. The range of detection of mass concentrations that can be measured ranges from 0 to 10 ppm. The minimum detection sensitivity is 0.5 ppb for the range of 0 to 0.2 ppm, and for a range of up to 10 ppm sensitivity is 5 ppb. The analyzer automatically changes the range of detection. The analyzer has the option of displaying the mass concentration in ppm or mg/m³ (with a correction factor of 2.86). Sample collection rate is 0.7 l / min.

Figure 16. APNA 370 [10]

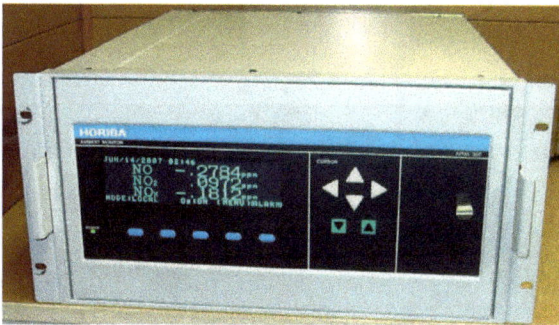

Figure 17. APNA 360 [10]

Automatic analyzer Horiba APNA 370/360 is used for measuring concentration of nitrogen oxide in the air. The analyzer uses a cross-modulation chemiluminescence method. The range of detection of mass concentrations that can be measured ranges from 0 to 10 ppm. The minimum detection sensitivity is 0.5 ppb for the range of 0 to 0.2 ppm, and for a range of up to 10 ppm sensitivity is 5 ppb. The analyzer automatically changes the range of detection. The analyzer has the option of displaying the mass concentration in ppm or mg/m³ (with a correction factor of 1.34 for nitrogen oxide (NO) and 2.054 for nitrogen dioxide (NO2)). Sample collection rate is 0.8 l / min.

Figure 18. APMA 370 [11]

Automatic analyzer Horiba APMA 370 is used to measure the concentration of carbon monoxide in the air. Cross-modulation analyzer uses non-dispersive infrared spectroscopic method. The range of detection of mass concentrations that can be measured ranges from 0 to 100 ppm. The minimum detection sensitivity is 0.5 ppm for the range of 0-100 ppm. The analyzer automatically changes the range of detection. The analyzer has the option of displaying the mass concentration in ppm or mg/m^3 (with a correction factor of 1.25). Sample collection rate is 1.5 l / min.

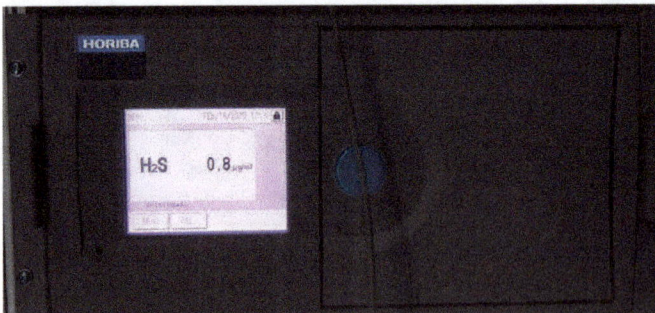

Figure 19. APSA-H370 [12]

Automatic analyzer Horiba APSA H370 is used to measure the concentration of hydrogen sulphide in the air. The analyzer uses the method of oxidation catalysts in combination with UV fluorescence. The range of detection of mass concentrations that can be measured is in the range from 0 to 1 ppm. The minimum detection sensitivity is 1 ppb for the range of 0-1 ppm. The analyzer automatically changes the range of detection. The analyzer has the option of displaying the mass concentration in ppm or mg/m³ (with a correction factor of 2.86 for sulfur dioxide and hydrogen sulfide to 1.521). Sample collection rate is 0.7 l / min.

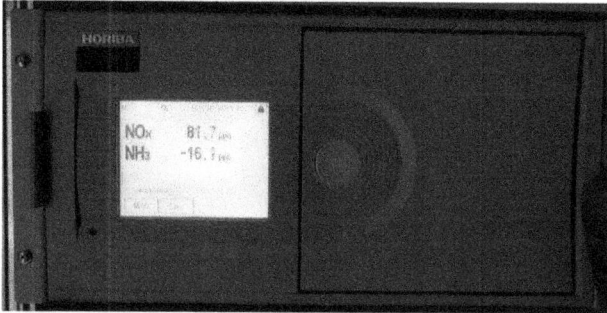

Figure 20. APNA 370 [13]

Automatic analyzer Horiba APNA 370 is used to measure the concentration of ammonia in the air. The analyzer uses the method of oxidation catalysts in combination with UV fluorescence. The range of detection of mass concentrations that can be measured ranges from 0 to 0.5 ppm. The analyzer automatically changes the range of detection. The analyzer has the option of displaying the mass concentration in ppm or mg/m³ (with a correction factor of 0.76 for ammonia). Sample collection rate is 2 l / min.

Particles that are considered to be pollutants in the air and for which medical and scientific evidence that environmental impact on air quality are particles PM10 (diameter less than 10μm), PM2.5 (diameter less than 2.5μm) and PM1 (diameter less than 1μm).

5. Automatic methods

An automatic method used at the IMROH is reduced to an automatic device Verewa Beta-Dust Monitor F-701-20 [14]. The device shows the mass concentration of particles in mg/m³. Mass concentration of particles is determined by radiometric method. Radiometric measurement is done using the Beta transmitter (C-14) and Geiger-Müller counters. The range of detection is in ranges from 0 to 10 mg/m³. The minimum sensitivity of detection is 0.001 mg/m³. The device has a heated tube so that at low temperatures there is no formation of ice inside the pipe. Will it be collecting particles of PM10, PM2.5 or PM1 only depends on the head with nozzles of different diameters. Sampling rate is 1000 l/h. Sample collection time is one hour, but the device has the ability to display the current value of the mass concentration. Automatic methods for collecting airborne particles are not the reference methods.

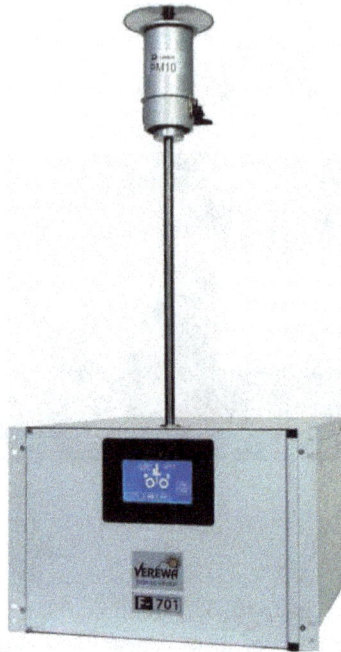

Figure 21. Verewa Beta-Dust Monitor F-701-20

6. Manual methods

Manual method for measuring the mass concentration of particles in the air is gravimetric. Gravimetric method is very strictly defined in the standard 12 341 [15]. That can be used gravimetric method laboratory must have a special room to hold the filter and weighing them. In gravimetric methods filters that are used for collecting samples of air must before and after sampling be specially treated. Filters before sampling must be in a desiccator at a constant temperature of 20 ° C ± 1 ° C and at a relative humidity of 50% ± 5%, for 48 hrs. After 48 hours the filters are weighed on vase sensitivity 1µg, once again returned to the desiccator for 24 hours. After 24 hours, repeat the procedure for weighing the filters and then they are ready for sampling. After re-sampling filters are placed in a desiccator for 48 hours, and weighed after 48 hours. They return back to the desiccator for 24 hours, and after 24 hours they are weighed. Only then the mass concentration of particles for the sampling day can be calculated. Gravimetric method is the reference method for determining the mass concentration of particles in the air. Mass concentrations of particles are shown in mg/m^3. The IMROH for the gravimetric method uses two types of devices. Both devices in a small flow and meet the requirements to be a reference.

Sven Leckel Low Volume Sampler LVS3 [16]

The machine uses air flow of 2.3 m³/h. Samples are taken at intervals of 24 hours ± 1 hour. Filters used for sampling are the quartz filters 47 mm in diameter. The device has the option of taking the sample ranged from 1h to 999h; however, norms and laws stipulate that the sample must be taken 24 hours. The device allows you to adjust the flow of 1 m³/h, 1.6 m³/h, 2 m³/h or 2.3 m³/h. Also the device has a microprocessor to maintain the flow deviation <2%. Filters can be a diameter of 47mm to 50mm. What will be collected particulate fractions (PM10, PM2, 5 or PM1) depends on the head and the nozzle of the device.

Figure 22. Sven Leckel Low Volume Sampler LVS3

Sven Leckel Sequential Sampler SEQ47/50 [17]

Unlike Sven Leckel Low Volume Sampler LVS3 SEQ47/50 device does not require every day manually changing the filters but it contains a repository of multiple filters (maximum 17). Air flow that supports the 1m³/h, 1.6 m³/h, 2m³/h and 2.3 m³/h. Sampling time can be a between 1h and 168h. Filters can be a diameter of 47 - 50mm. Also the device has separate compartments for storing the sampled filter that is regulating to maintain a constant temperature of 20 ° C and relative humidity at 50%. What fraction of particles will also be collected as in previous devices depends on the head and nozzle. By their specifications of the device also meets the standard to be a reference device.

Figure 23. Sven Leckel Sequential Sampler SEQ47/50

7. Communication

Communication between devices (analyzers), in most cases is done by using some of the company's application that exist in the market. As each program and they have their advantages and disadvantages.

In most cases, information systems are not located in the same position where there are stations for air quality control so that all data must be sent from the station into the information system if possible using a network cable, or in some cases using the mobile Internet or WIFI.

8. Conclusion

The section shows the general items of an information system of monitoring air quality. A large part of the components of the information system for monitoring air quality described are at the Institute for Medical Research and Occupational Health in Zagreb. Monitoring of air quality today is unthinkable without the use of good and quality information systems.

The future of information systems at the IMROH will be improved and pollution forecasts depicting pollution using GIS technology.

Author details

Silvije Davila*, Ivan Bešlić and Krešimir Šega
Institute for Medical Research and Occupational Health,
Environmental Hygiene Unit, Zagreb, Croatia

9. References

[1] EEA (2010) Software standards of EEA and Eionet Available: http://www.eionet.europa.eu/software/swstandards

[2] Ivan Bešlić (2007) Concentrations of particulate fractions of PM2.5 and PM10 in the atmosphere in Zagreb and their dependence on meteorological parameters. Zagreb : Faculty of Science

[3] W. Garber, J. Colosio, S. Grittner, S. Larssen, D. Rasse, J. Schneider, M. Houssiau (202) Guidance on the Annexes to Decision 97/101/EC on Exchange of Information as revised by Decision 2001/752/EC, for the European Commission, DG Environment, pp.71.

[4] G. Endregard, K. Karatzas, K.-H. Carlsen, I. Floisand, B. I. Skaanes, S. Larssen (2007) EEA air quality web dissemination solution - Recommendations for further development, The European Topic Centre on Air and Climate Change (ETC/ACC), Technical Paper 2006/9, pp. 79.

[5] W. Garber, J. Colosio, S. Grittner, S. Larssen, D. Rasse, J. Schneider, M. Houssiau (2002) Guidance on the Annexes to Decision 97/101/EC on Exchange on Information as revised by Decision 2001/752/EC, for the European Commission, DG Environment

[6] S. Elshout, K. Leger (2007) Comparing urban air quality across borders. Environmental Protection Agency Rijnmond, Netherland

[7] EPA (2011) Air Quality Index (AQI) – A guide to air quality and your health, Available: http://www.airnow.gov/index.cfm?action=aqibasics.aqi

[8] Horiba (2008), "Ambient O3 monitor APOA-370 Operation Manual"

[9] Horiba (2009) "Ambient SO2 monitor APSA-370 Operation Manual"

[10] Horiba (2009) "Ambient NOx monitor APNA-370 Operation Manual"

[11] Horiba (2008) "Ambient CO monitor APMA-370 Operation Manual"

[12] Horiba (2005) "Ambient SO2/H2S monitor APSA-H370 Instruction Manual"

[13] Horiba (2007) "Ambient NH3 monitor APNA-370 Instruction Manual"

[14] Verewa, "Verewa Beta-Dust Monitor F-701-20 Technical Manual"

[15] Croatian Standards Institute (2006), "Croatian Standard HRN EN 12341"

[16] Ingenieurbüro Sven Leckel "Low Vloume Sampler LVS3 Instruction Manual"

---

* Corresponding Author

[17] Ingenieurbüro Sven Leckel "Sequential Sampler SEQ47/50 Instruction Manual"

Chapter 5

Development of Real-Time Sensing Network and Simulation on Air Pollution and Transport

Yasuko Yamada Maruo and Akira Sugiyama

Additional information is available at the end of the chapter

1. Introduction

This chapter describes an air quality-monitoring project carried out in Metro Manila in the Philippines in 2006. We developed a real-time environmental and atmospheric sensing system and used it to measure air pollution and weather parameters remotely in order to simulate the gathered data with a 3D fluid simulator and another simulation model. Traffic count surveys were also conducted to acquire road traffic data to allow us simulate traffic density and traffic flow.

In urban areas of the Philippines, industrialization and the rapid growth of motorization has caused serious environmental problems to local residents (Lodge, 1992; Isidrovaleroso et al., 1992). In this study we established a system and methodology for monitoring air pollution in real time and to evaluate related environmental aspects by using the latest sensing network technology. For this project, we used a small nitrogen dioxide (NO_2) sensor and a suspended particulate matter (SPM) sensor, both of which we developed. These air pollutant sensors can perform real-time monitoring at several monitoring points along a road network. SPM has recently been recognized as an air pollutant that has a causal correlation with the human fatality rate (Sabin et al., 2005), and NO_2 is recognized as an air pollutant that causes respiratory illness (Smith et al., 2000). In conventional studies, air pollutant amounts are measured with passive samplers that require very longer measurement periods of up to a month, and no dynamic validation has been conducted in real time. The air pollutant concentration will fluctuate widely over short intervals and the weekly/monthly average concentration could be far lower than the actual daily peak. Macroscopic and microscopic transportation and environment models have been developed to predict the pollutant concentration from the aggregate transportation demand and traffic flow simulations, but there has been no dynamic validation of the pollutant concentrations. By measuring pollutant concentrations and weather parameters at ten- or sixty-minute intervals, the daily variation can be measured and the correlation between pollutant concentration, weather parameters and traffic volume can be simulated.

The results obtained with the simulation system will be useful for the evaluation and implementation of urban transportation policies, pollution countermeasures, guidelines and human resource programs. The environmental impact of policies and social experiments can be directly with an SPM and NO_2 measurement network system, which will be very useful for policymakers and decision-makers in national and local government agencies.

2. Development of real-time sensing network

2.1. System outline

A real-time sensing network system was established with the following process:

- The observation data are periodically transferred over a cell-phone (GPRS) from the sensors installed along an arterial road in Metro-Manila to Server I (PH-SV) in which they are stored.
- The data stored in PH-SV are periodically transmitted to Server II (JP-SV) over the Internet, and stored.
- In JP-SV, raw SPM data are converted into mass density ($\mu g/m^3$), and NO_2 data are converted into concentration (ppb) using the above-mentioned arithmetic expression.

In JP-SV, the calculated data are stored in RDBMS (PostqreSQL) and the WEB application provides some functions (time-series graphs of SPM and NO_2, data search and data download, etc.). By employing these functions, these data can be accessed from anywhere and at any time via a PC that is connected to the Internet. The data measured and output by a weather sensor are understandable as they are, so they do not need to be processed with an arithmetic expression. The outline of our project is shown in Fig. 1.

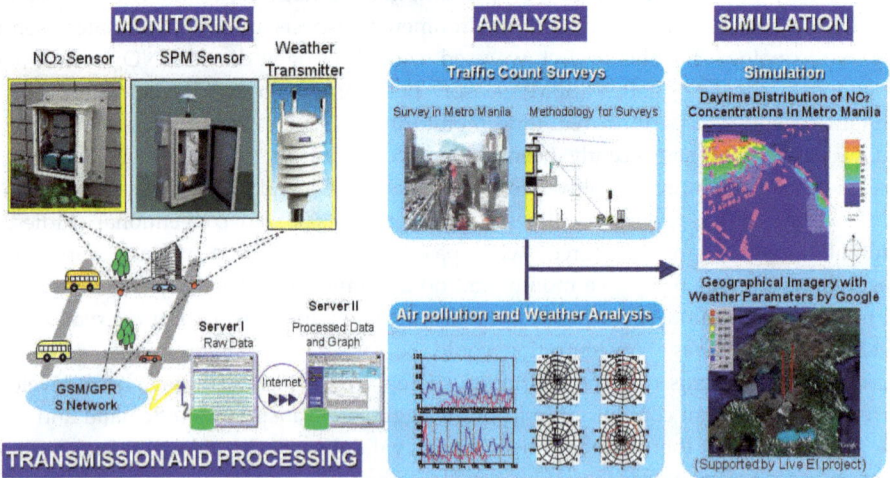

Figure 1. Outline of this research

2.2. System diagram

A chart showing the composition of the entire system is shown in Fig. 2. In this section, we describe each mechanism of the system separately. The system is divided into the following four parts:

1. Sensor to PH-SV: Transfer each sensor's measurement data (raw data) to PH-SV over the GSM/GPRS wireless with a controller programming method.
2. PH-SV to JP-SV: Transfer the raw data from PH-SV to JP-SV over the Internet with a UNIX secure protocol.
3. Calculation application: On JP-SV, calculate the raw data into concentrations by using each expression and store them in a database.
4. Web application: Provide certain functions (e.g. graph view, data selection and data download) for web application.

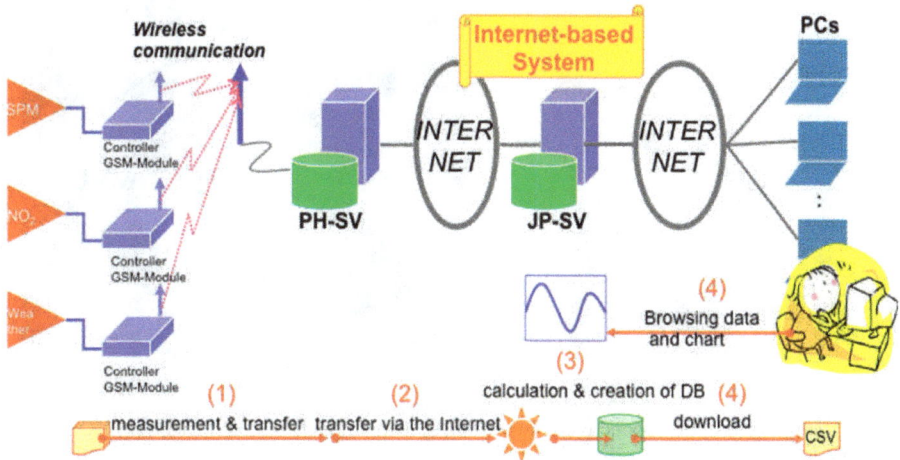

Figure 2. System diagram of this research

2.3. Principles and performance of environmental sensors

2.3.1. NO₂ sensor

The NO₂ sensor is an accumulation type sensor (Tanaka et al. 1999). The sensor element that detects NO₂ has a reagent contained in porous glass that reacts chemically only with a particular gas (diazo coupling reaction) as shown in Fig. 3 (a). This sensor consists of 1-mm thick porous glass sheets with an average pore diameter of 4.6 nm, a vacancy ratio of 28%, and a specific surface area of 195 m²/g, cut into 8 x 8 mm chips. After exposure to ambient NO₂, the chips turn pink. A specific single absorption peak appears at 525nm and it is clear that the absorbance increases as the exposure time increases. This pink material is stable and there is no change in the absorbance after the preservation of the exposed samples in a dry nitrogen atmosphere. The NO₂ concentration was calculated from the absorbance difference at 525 nm and the exposure time.

2.3.2. NO₂ monitoring device

A device has been developed that can detect low levels of NO₂ gas (Maruo et al. 2003). The device consists of a sensor element, an absorbance meter with a single wavelength (525 nm) light source, a controller containing a data converter and a mobile modem to make it possible to transfer data from the device to a central PC through a wireless network. The device is outlined in Fig. 3 (b). It is very compact with dimensions of 200 mm W x 100 mm D x 85 mm H. Another feature of the device is that it has no pumping unit for replacing the air above the sensor. Instead, the body includes an electric heater (electric board) to provide a convection flow and several slits for ventilation. It is possible to calculate the ventilation speed using a simple structural model and factors related to the energy consumption of the electric board and the difference between the temperatures outside and inside the device. This gives a value of about 1l/min indicating that the inside air is replaced in <1.5 min.

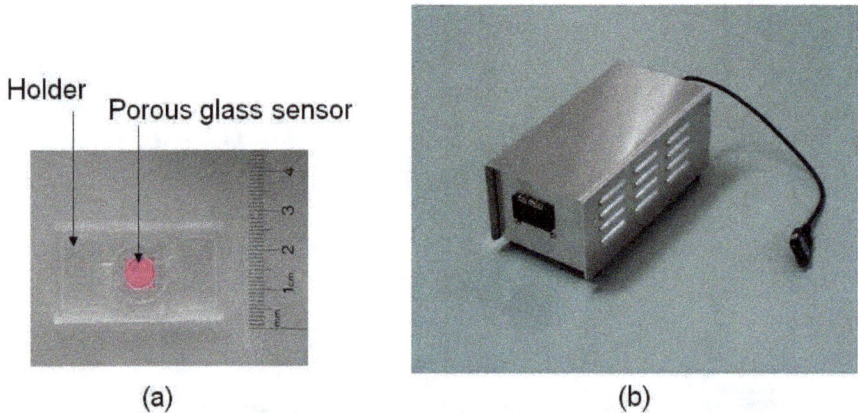

(a)   (b)

Figure 3. Photographs of developed NO₂ monitoring system. (a) NO₂ sensor element, (b) NO₂ monitoring device.

2.3.3. SPM sensing device

A suspended particulate matter (SPM) sensing device has been developed (Tanaka, et al., 2004) and used to measure the relative SPM concentration in ambient air with the right-angle light scattering method according to the particle diameter at the time the particles pass through it. SPM is thought to be a cause of respiratory ailments such as asthma, because it can easily reach the respiratory organs. In this study, we set up this compact SPM sensor with a GSM/GPRS network module along with an NO₂ sensor and a weather transmitter (WXT520, VAISALA) to develop a real-time sensing network.

As illustrated in the Fig. 4, this SPM sensor employs an optical particle counter based on the right-angle light scattering caused by suspended particles, and determines the number of particles and their diameter. By combining a dilutor with a particle counter, high SPM concentrations can be detected up to a maximum of 12,000 particles/cc. Since the amount of light that is scattered depends on the particle size, a pulse-height analyzer makes it possible to count particles classified into five diameter ranges, namely 0.3-0.5, 0.5-1.0, 1.0-2.5, 2.5-7 and 7-10 µm, according to the scattered light pulse intensity.

The SPM sensor used here is less than one tenth the size of conventional SPM monitoring instruments.

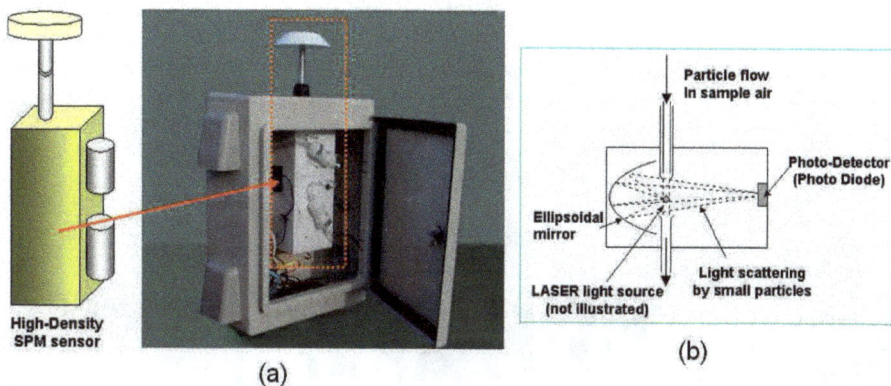

Figure 4. Photographs of developed SPM monitoring system. (a) SPM sensing device, (b) particle detection method.

3. Experiment in Metro Manila

3.1. Outline of experiment

In this study, we established three monitoring. Of these monitoring points, two were beside a busy thoroughfare within the Makati Commercial Business District (Makati CBD) to study the variation in air pollution effected by street canyons and traffic on the road. While the third monitoring point focused on the background conditions used for running the simulation study.

The equipment employed for the study consisted of two NO_2 monitoring devices, one SPM sensing device, and two weather stations. One NO_2 sensor, one SPM sensor and one weather station were prepared for roadside installation in the Makati CBD. The remaining NO_2 sensor and weather station were used to collect background data.

The weather stations should be more than 5 meters above the roof deck. This is because, in this study, the representative wind velocity of the monitoring point is measured by the weather station and the measurement should be 5 meters above the roof deck or side wall of a building to avoid the influence of the boundary layer from those objects.

The Intellectual Property Office (IPO) Building and the Board of Investments (BOI) Building, both on Senator Gil J.Puyat Avenue in the Makati CBD in Metro Manila were selected as the locations of the first two monitoring points for the study. And both have canopies over their front entrances that jut out towards the roadside and over which the SPM and NO_2 sensors were easily mounted.

Both buildings also have roof decks that have no street canyon effect. Therefore, portable weather stations can be installed on those roof decks to provide wind speed, wind direction, air temperature, air pressure, relative humidity, and precipitation data. The third monitoring point is located inside the National Mapping and Resource Information Authority (NAMRIA) compound inside the Fort Bonifacio military reservation less than 6 km south east of the first two monitoring points.

NO_2 monitoring device   SPM sensing device   Weather Transmitter

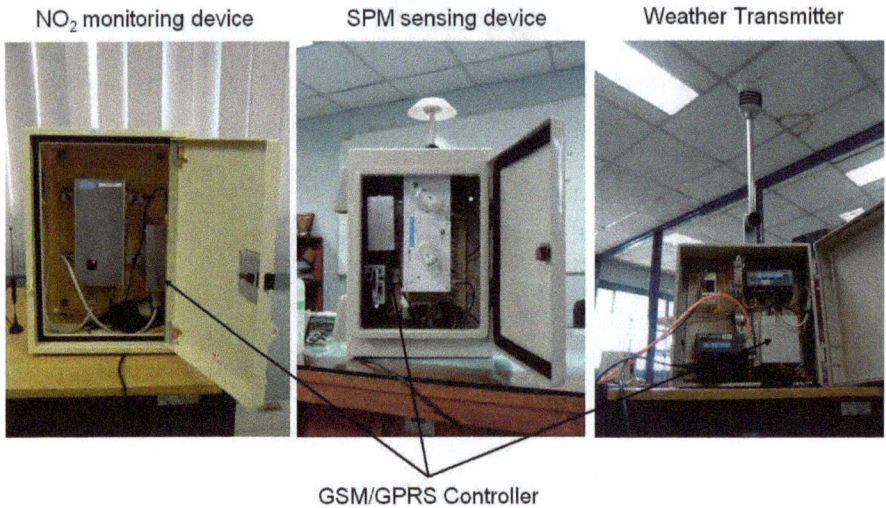

GSM/GPRS Controller

Figure 5. Real time monitoring devices

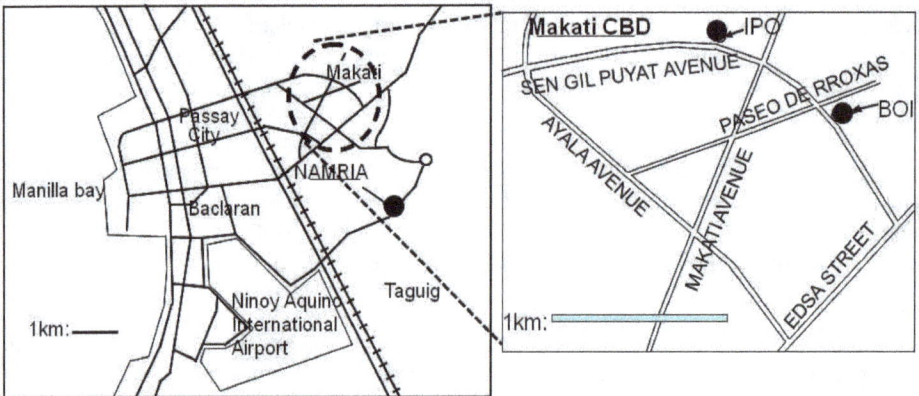

Figure 6. Air monitoring locations in the Manila and in Makati CBD.

3.2. Monitoring point; IPO

The IPO building is located on Senator Gil J.Puyat Avenue about 300 meters far from its corner with Makati Avenue. The road in front of the IPO has six lanes and is about 22 meters wide, and the IPO is in a street canyon. The IPO building also has a canopy that induces a street canyon effect. NO_2 and SPM sensors were mounted on the canopy because it is 3 meters above the ground. The IPO building also has an accessible roof deck, but there is a wall on the roof deck. To avoid the boundary layer and turbulences caused by this wall, the weather station was mounted on the top of a pole that was fixed to handrails.

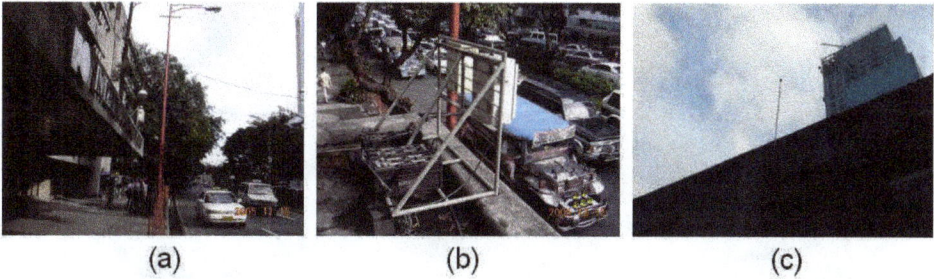

(a)   (b)   (c)

Figure 7. Photographs of roadside monitoring points; IPO. (a) The Senator Gil J.Puyat Avenue and IPO building, (b) NO_2 monitoring device and SPM sensing device, (c) weather station.

3.3. Monitoring point; BOI

The BOI building is also located along Senator Gil J.Puyat Avenue. The road in front of the BOI also has six lanes and is about 22 meters wide. The BOI building is not in a street canyon but in an open space. The NO_2 sensor and the SPM sensor were installed at the top of the canopy at a height of 3meters.The canopy and therefore the sensors are 3 meters away from the vehicular road. The BOI building also has a roof deck that we can access, but there is a structure on the roof deck. To avoid the boundary layer and turbulences induced by this structure, the weather station was set on the top of a pole that was fixed to the handrails of the structure.

(a)   (b)   (c)

Figure 8. Photographs of roadside monitoring point; BOI. (a) The Senator Gil J.Puyat Avenue and BOI building, (b) NO_2 monitoring device and SPM sensing device, (c) weather station.

3.4. Monitoring point; NAMRIA

NAMRIA is located along McKinley Road. The road has four lanes and is 12 meters wide. To avoid the air pollution from McKinley Road, the NO_2 sensor was mounted on the roof of a security office more than 150 meters from the road. The roof is only 3 meters above the ground but there is adequate security so the NO_2 sensor was set on the roof of the security office. The weather station was installed on the top of a pole, which was raised on the roof deck of the NAMRIA building.

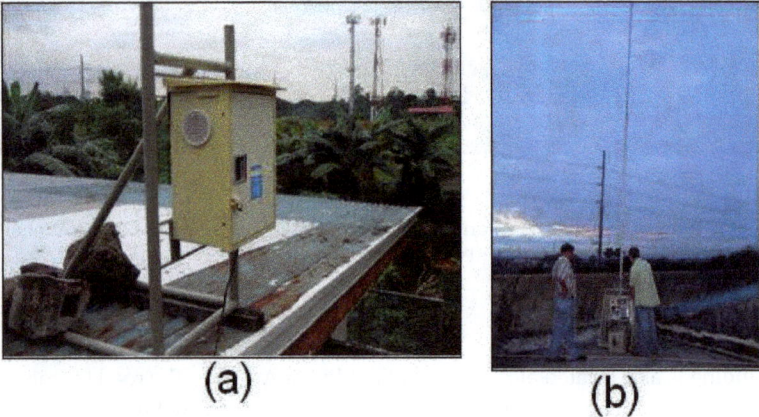

Figure 9. Photographs of background monitoring point; NAMRIA. (a) NO_2 monitoring device, (b) weather station.

3.5. Measurement schedule

In this study, measurements were gathered for over 2 weeks at each monitoring site, IPO and BOI. Two NO_2 sensors, one SPM sensor and two weather stations were mounted along the time schedule indicated in the table 1.

Month/Year	12/2005				1/2006					2/2006	
Day	5	12	19	26	2	9	16	23	30	6	13
SPM			IPO			BOI					
NO₂-1			IPO			BOI					
NO₂-2					NAMRIA						
Wet.-1			IPO			BOI					
Wet.-2					NAMRIA						

Table 1. Time schedule of measurement

4. Methodology for traffic surveys

4.1. Traffic counts

Traffic volume, defined as the number of vehicles passing a predetermined observation line during a certain time period, is a very important variable with regard to obtaining the relationship and the degree of traffic involvement with in-situ concentrations of particulate and gaseous pollutants in a roadside environment.

In this project, traffic volume was determined at a midblock section and an intersection. Monitoring at the first location involved establishing an observation line across the road of interest and locating it directly in front of the sensors. We then counted the number of vehicles that crossed the line during a pre-set time period of 3-4 hours for peak observations (4 hours in the morning and another 3 hours in the afternoon) 2 hours at night and 2 hours for off-peak observations. A video camera was mounted on top of the IPO and BOI buildings to record the traffic and counts were measured later (not in real time). The typical set-up is shown in Fig. 10.

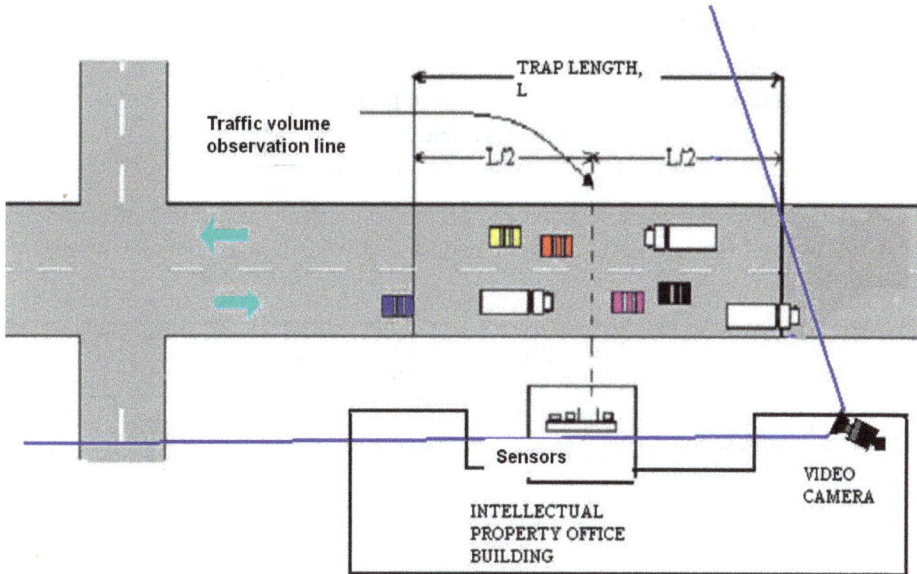

Figure 10. Plan view of traffic surveys set-up.

The survey was scheduled to be undertaken on at least one weekday and one weekend (Sunday) during the following time periods:

- 6:00 AM – 10 AM (AM peak)
- 11:00 AM – 2 PM (daytime off-peak)
- 4:00 PM – 7 PM (PM peak)
- 2:00 AM – 4:00 AM (nighttime off-peak)

Vehicle classifications for the traffic counts include the following:

(1) passenger car, (2) light truck, (3) private jeepney, (4) van, (5) medium truck, (6) minibus, (7) taxi, (8) heavy truck, (9) bus, (10) FX taxi, (11) public utility jeepney, and (12) motorcycle/tricycle

4.2. Vehicle speed surveys

Vehicle speed, generally referred to as the distance in kilometers that a point on a vehicle travels per hour of elapsed time, is as important as traffic volume because different speeds will yield different emission levels, resulting in varying amounts of pollutant deposition. It is for this purpose that the speeds of vehicles traversing the road were obtained for later association with traffic volume and pollutant emissions for modeling studies.

5. Results of experiment

5.1. Results of traffic count surveys

Traffic surveys were conducted on four days:

- January 5, 2006 [along Sen. Gil Puyat Avenue at IPO Building]
- January 8. 2006 [along Sen. Gil Puyat Avenue at IPO Building]
- January 18, 2006 [along Sen. Gil Puyat Avenue at BOI Building]
- January 22, 2006 [along Sen. Gil Puyat Avenue at BOI Building]

The raw data for each day were encoded and processed for the morning and afternoon peak periods as well as the noontime off-peak period. Sample outputs from the surveys are shown in Fig. 11.

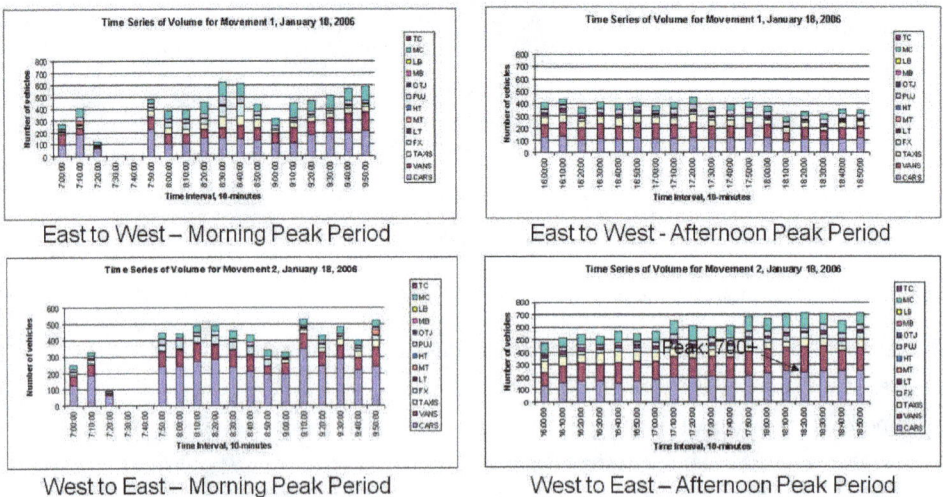

East to West – Morning Peak Period

East to West - Afternoon Peak Period

West to East – Morning Peak Period

West to East – Afternoon Peak Period

Figure 11. Variations in traffic volume along the Sen. Gil Puyat Avenue on 18 Jan (Wed), 2006.

Speed samples were also derived from the field surveys. Example measurements are shown in Table 2.

Time Period			Speed	
			meter/sec	km/hr
6:10:00	-	6:15:00	7.6	27.5
6:15:00	-	6:20:00	8.8	31.7
6:20:00	-	6:25:00	9.6	34.6
6:25:00	-	6:30:00	9.1	32.6
6:30:00	-	6:35:00	11.8	42.6
6:35:00	-	6:40:00	10.8	39.0
6:40:00	-	6:45:00	9.3	33.3
6:45:00	-	6:50:00	9.1	32.7
6:50:00	-	6:55:00	9.6	34.7
6:55:00	-	7:00:00	11.1	39.8
7:00:00	-	7:05:00	11.0	39.7
7:05:00	-	7:10:00	11.6	41.6
7:10:00	-	7:15:00	9.5	34.3
7:15:00	-	7:20:00	8.3	29.7
7:20:00	-	7:25:00	10.4	37.6
7:25:00	-	7:30:00	10.0	35.8
7:30:00	-	7:35:00	9.6	34.6
7:35:00	-	7:40:00	11.0	39.4
7:40:00	-	7:45:00	10.2	36.6
7:45:00	-	7:50:00	9.7	34.9
7:50:00	-	7:55:00	8.1	29.0
7:55:00	-	8:00:00	8.9	32.1

Table 2. Sample speed measurements (Jan. 22, 2006/AM Peak/Direction towards Ayala Avenue).

5.2. NO₂ measurement results

Figure 12 shows examples of time-series plots of NO_2 concentrations obtained at two different points (IPO and NAMRIA). The experiment at IPO was performed between Dec. 16, 2005 and Jan. 9, 2006, and that at NAMRIA from Dec. 22, 2005 to Jan. 23, 2006. We used two NO_2 sensors; one was used for the measurement at NAMRIA and the other for the measurements at IPO and subsequently at BOI.

Data were acquired by a sensor and transferred to a server at one-hour intervals. Although some data were lost when the data collection failed, we were nevertheless able to measure the accumulated NO_2 concentration. This is because we used an accumulation type sensor as described above. The one-hour NO_2 concentration value for a period where data collection failed was calculated as a mean value. For example, we successfully collected data at 15:00 after failing for four hours (from 11:00 to 14:00 on Dec. 16, 2005). The accumulative NO_2

amount for five hours (from 11:00 to 15:00) was calculated at 242 ppb, giving a one-hour
mean value of 48 ppb. In Fig. 12, the parts showing the mean values are indicated by A.

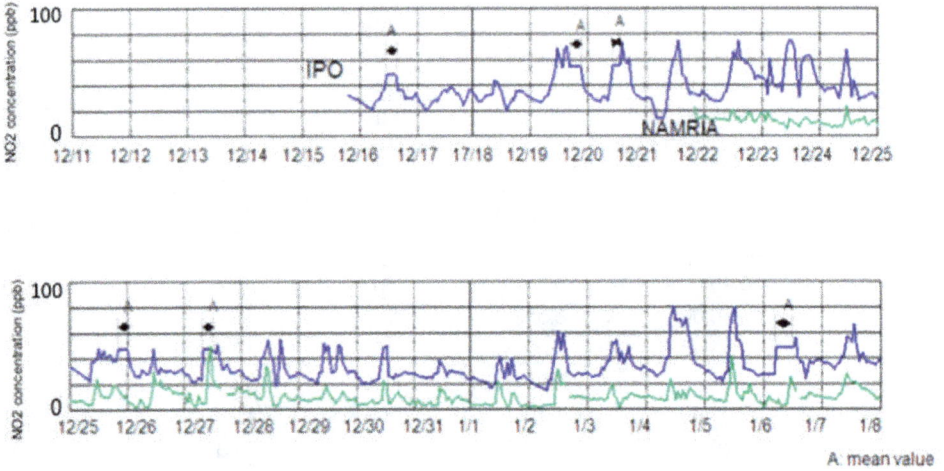

Figure 12. Time series plots of NO₂ concentrations at two different points (IPO and NAMRIA).

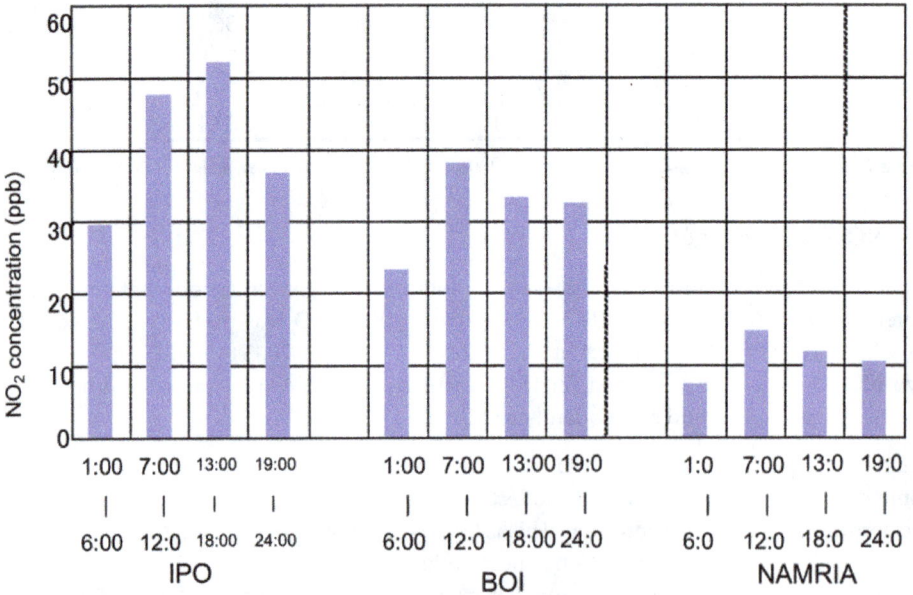

Figure 13. Calculated six-hour average NO₂ concentrations for weekdays at three points.

Figure 13 shows the six-hour average NO₂ concentration for weekdays at the three points. For both BOI and NAMRIA, the peak concentration hour appears during the second period (from 7:00 to 12:00), which corresponds to peak traffic hours. On the other hand, for IPO, the peak concentration hour appears during the third period (from 13:00 to 18:00), which does not correspond to peak traffic hours. The peak concentration hours for IPO appeared after those of both BOI and NAMRIA. We assume this to be because IPO is in a canyon. It has been reported that NO₂ becomes concentrated in canyon areas after peak traffic hours (Maruo et al. 2003).

5.3. SPM measurement results

Figure 14 shows time-series plots of SPM concentrations at a monitoring point (IPO). The experiment at IPO ran from Dec. 16, 2005 to Jan. 5, 2006. Since we had one SPM sensor set, we used it first for the measurement at IPO and then moved it to BOI. Data were acquired by the sensor and transferred to a server at ten-minute intervals.

The one-hour average SPM mass concentration for weekdays was about 25 % higher than the weekend concentration. The mass concentration was considerably smaller than the already reported value (Santos et al. 2006, Oanh et al., 2006). Figure 15 shows the six-hour average SPM mass concentration for weekdays at IPO. The peak concentration hours appear during the second period (from 7:00 to 12:00), which corresponds to peak traffic hours.

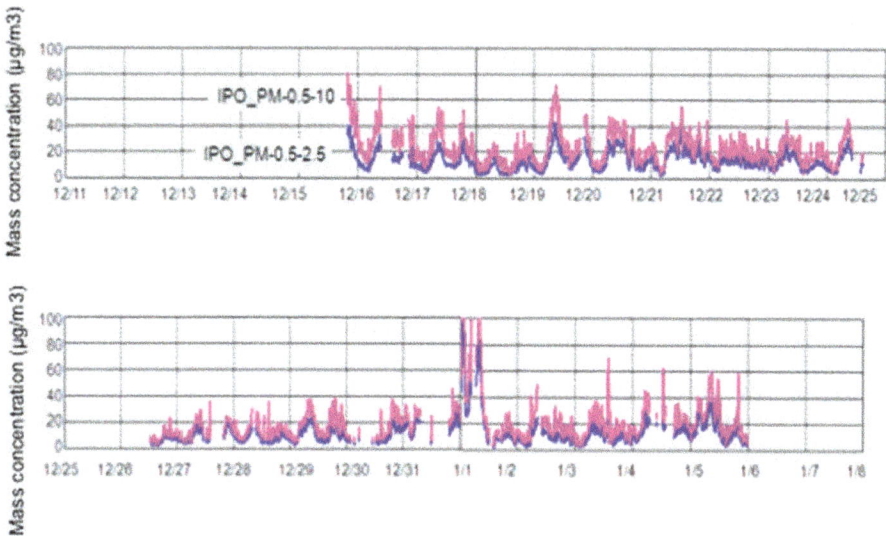

Figure 14. Time series plots of SPM concentrations at IPO. The high concentration on Jan. 1 was due to firecrackers being let off on the New Year's holiday.

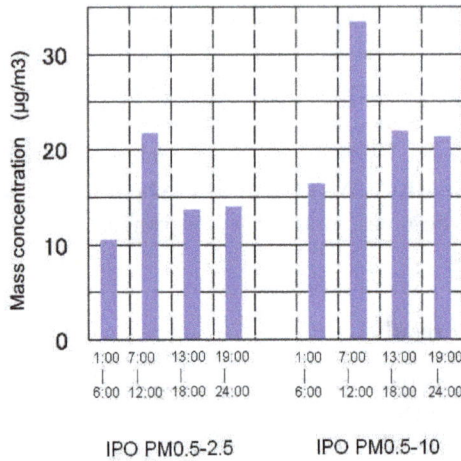

Figure 15. Calculated six-hour average SPM concentrations for weekdays at IPO.

6. Simulation and consideration

6.1. Objective of simulation

An analysis of the data we obtained from the sensors located on the IPO and the BOI buildings provided a lot of information about the SPM and NO₂ concentrations as described in section 5. Although these sensors gave us detailed information, it was only about the areas around their locations. It is difficult to infer what is happening in the target area around the Makati CBD solely from the sensor data.

A 3-D fluid simulator can show us what is happening in a target area better than sensors. We simulated pollutant concentrations using the data obtained from the sensors to understand which areas have high concentrations of pollutants in the Makati CBD and how wide these areas are. The results will be useful for evaluating and implementing urban transportation policies, pollution countermeasures, guidelines, and human resource programs. These results will also be very useful for policymakers and decision makers in national and local government agencies.

An analysis of the data obtained from the NO₂ sensor located on the IPO building showed that the NO₂ concentrations exceeded the Japanese environmental quality standard levels of 40 and 60 ppb (EQS, 2012). For public health reasons, it is preferable that NO₂ concentrations not exceed that standard. The distribution of the NO₂ concentration was calculated around the Makati CBD in this study, and regions where the NO₂ concentration exceeded 40 or 60 ppb were identified.

6.2. Model description

The 3-D localized atmospheric environment simulator is an integrated system that can consistently perform all the required processes ranging from collecting data, such as a map model and initial conditions, to confirming the calculation results. The simulator is based on an air flow simulator developed by Kozo Keikaku Engineering, Inc. (KKE, 2012). A hydrodynamic method is used to calculate the advection and diffusion of pollutants in the atmosphere in three dimensions. This allows us to take account of the shapes and heights of buildings, which was not possible with conventional two-dimensional calculations. A 3-D computational domain can be easily made from a digital map that includes the shape and height of buildings and road networks. As a result, the simulator allows the calculation of the distributions of temperature, humidity, car exhaust emissions, heat radiated by buildings, etc. in an urban environment where wind flows are very complex. Numerical analysis by the finite volume method is used to solve the continuity equation that is derived from the assumption that the flow is a three-dimensional incompressible viscous flow, and from the Navier-Stokes, energy, and advective-diffusion equations.

The main characteristic of this 3-D fluid simulator (Tanaka et al., 2004) is that it calculates the photochemical reactions of pollutants and air flow simultaneously. Motor vehicle exhaust gas contains many air pollutants including NO and NO_2. NO reacts quickly with ozone to form NO_2. However, sunlight breaks down NO_2 via a photochemical reaction, so the simultaneous calculation of photochemical reactions and air flow makes it possible to simulate the NO_2 concentrations more precisely. We simulated the NO_2 concentration around the Makati CBD, and analyzed the simulation results to identify areas of high concentration.

6.3. Conditions and input parameters

Various parameters including wind conditions and the amount of air pollutants from motor vehicles should be established to simulate air pollutant concentrations by numerical analysis. The settings of these parameters are described in this section.

The NO_2 concentrations in the air near the IPO and BOI buildings fluctuated widely throughout the day as discussed in section 5. Although these fluctuations were more prominent because the sensors were located near the main road, the pollutant concentrations over the whole area fluctuated like those near the roadside. Since the purpose of the simulation was to identify the areas where the NO_2 concentration was high, and these areas varied based on the time of day, we selected three characteristic time periods when calculating the distribution of the NO_2 concentration:

1.   low concentration: 2:00 – 3:00 AM,
2.   high concentration: 11:00 AM – 12:00 PM,
3.   many people are outdoors: 5:00 – 6:00 PM.

We calculated the most frequent wind direction and the average wind velocity and background concentration for each time period and we used all the data obtained in this experiment to calculate these average values and thus simulate the distribution of the NO_2 concentration under typical conditions in the Makati CBD in winter. Data gathered on weekends were excluded from the calculation of the average background concentrations because of the difference between weekends and weekdays. We simulated weekday conditions because the pollutant concentrations on weekdays were higher than those at weekends.

6.3.1. Computational domain

3-D and 2-D representations of the computational domain are shown in Fig. 16(a) and (b), respectively. This computational domain was made from a digital map that includes road networks, road widths, building shapes, and building heights. The size of the target area was 2.2 square kilometers, and the center was around the intersection of Makati and Sen. Gil Puyat Avenues. Sen. Gil Puyat Avenue is shown as a blue line and the IPO and BOI buildings are shown as green blocks.

An orthogonal grid system was applied. Its specifications are as follows.

1. Number of grids: 110 (x) X 110 (y) X 53 (z).
2. Width of grids: regular horizontal intervals at 20 m, regular vertical intervals at 1 m (up to 10 m high), irregular vertical intervals between 1 and 50 m (higher than 10 m).

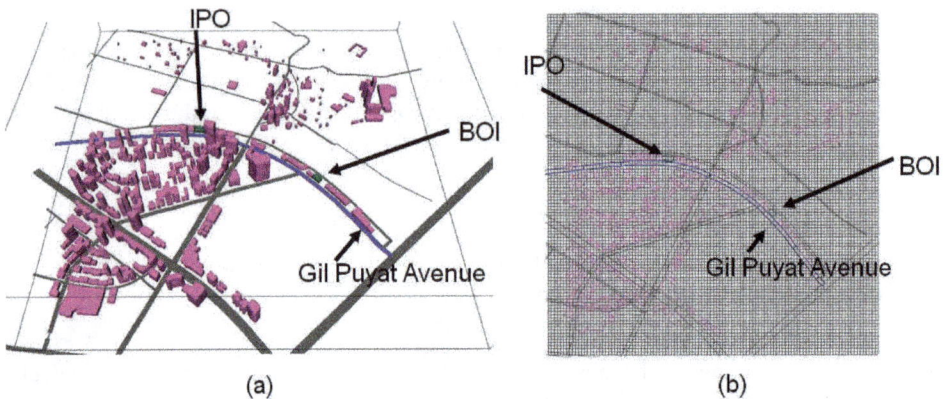

(a)   (b)

Figure 16. Calculation description of the Makati CBD, (a)3-D representation, (b)2-D representation and grid description

6.3.2. Sources

Motor vehicles that drive down Sen. Gil Puyat Avenue were considered to be sources of air pollution. Although the simulated NO_2 concentration distribution, which assumes that air pollutants are emitted from only one road, may differ from the actual distribution in the Makati CBD, we can identify the effect of air pollutants emitted from vehicles on the road. The volumes of emitted air pollutants were estimated from traffic volumes and vehicle speeds measured in this experiment. We used the emission factor proposed by the Asian Development Bank in 1992 (ADB, 1992). Although the 1992 emission factor is different from the current one, we assumed that the ADB emission factor is the most reliable one available in the Philippines. The vehicle classifications used with the ADB emission factor are different from those for the traffic counts in our experiment. The correspondence between the ADB vehicle classification and ours is shown in Table 3.

Classification in ADB	Classification in our experiment
Gas Car	Passenger car ,Van ,Taxi
Gas Jeep	Private Jeepney
Diesel Car	FX Taxi
Diesel Jeepny	Light Truck, Medium Truck, Public Utility Jeepney, Minibus
Diesel Bus	Heavy Truck ,Bus
Gas Tricycle	Motor Cycle, Tricycle

Table 3. Correspondence between ADB vehicle classification and our experimental classification

We calculated the NOx emission volume using the ADB emission factor. The NOx emission volume is the sum of the NO and NO_2 emission volumes. The NO and NO_2 emission volumes are needed to simulate the photochemical reaction process. Emission volumes of pollutants other than NOx, such as NMHC, are also needed. These emission volumes were estimated using the ratio of these emissions compared with the NOx emissions reported by Tange (Tange, H. et al. 1987). We used the traffic volume measured on January 5, 2006. Table 4 lists the traffic and the NOx emission volumes.

Time of day	Traffic Volume(10^{-4}vehicle s^{-1})						NOx emission (g km^{-1} s^{-1})
	Gas car	Gas jeep	Diesel car	Diesel jeepney	Diesel bus	Gas tricycle	
2:00-3:00	1515	6	30	166	48	109	0.6
11:00-12:00	4896	48	183	1100	51	951	2.27
17:00-18:00	4109	34	222	868	40	962	1.9

Table 4. Traffic volume and NOx emission volume

6.3.3. Boundary conditions

Wind flowing into the target area and the concentrations of air pollutants in the inflowing wind were established as boundary conditions. Values measured by sensors located near the NAMRIA building were used to calculate these conditions. The average values for each period were also calculated as typical background conditions in winter from the data obtained in this experiment.

The average direction and velocity of the wind were calculated using the data collected for days between December 19 and January 9. The most frequent wind direction was chosen as the average wind direction for each period. The average wind velocity was calculated using the data when the wind direction was the same as the average wind direction. Table 5 lists the average wind direction and velocity for each period. The 1/4-power law (Peterson and Hennessey, 1978) was applied to obtain the vertical profile of the wind velocity.

Time of day	Wind Direction	Wind Velocity(ms^{-1})	Concentration(ppb)			
			NO	NO$_2$	O$_3$	NMHC
2:00-3:00	North east	0.7	15.1	9.5	4.3	407
11:00-12:00	East-northeast	4	1.2	16.9	37.2	272.3
17:00-18:00	East-northeast	2.7	15.4	19.9	11.8	407.4

Table 5. Wind direction, wind velocity and background concentration of chemical species

Average NO, NO$_2$, O$_3$, and NMHC concentrations were established as background concentrations. These chemical substances were incorporated in a photochemical reaction model that was used in this study. Their concentrations were calculated using the data for days between January 4 and 25. Data gathered at weekends were excluded from the calculation of the average background concentrations because of the difference between weekends and weekdays. Table 5 also lists the background concentrations of each pollutant for each period.

6.3.4. Photochemical reaction model

Various photochemical reaction models that consist of hundreds of elementary reactions have been previously been proposed for precisely simulating the complex chemical reaction process (Finardi et al., 2008, Kamin et al., 2008, Vuilleumier et al., 2001, Collins et al., 1997). Although these models can simulate the chemical reaction process precisely, they require long calculation times. In this study, we used the RS32 model, which consists of only 12 elementary reactions and that can be used to simulate a photochemical reaction (Tange et al., 1987). Table 6 lists the characteristics of RS32 model. Some chemical reaction velocities are proportional to light intensity; therefore, the light intensity should be estimated in order to simulate a photochemical reaction. This simulator has a function for calculating the light intensity of each grid taking the shade provided by buildings into account. The light intensity was calculated based on the altitude of the sun in Manila on January 1.

No.	Reaction	Reaction Velocity
1	NO$_2$ + hv → NO + O$_3$	k1 min^{-1}
2	NO + O$_3$ → NO$_2$	20.8 ppm^{-1}min^{-1}
3	HC1 + O$_3$ → 2RO$_2$ + 2HC4	0.016 ppm^{-1}min^{-1}
4	HC1 + OH → 1.3RO$_2$	25000 ppm^{-1}min^{-1}
5	HC2 + OH → 2RO$_2$	3800 ppm^{-1}min^{-1}
6	HC4 + OH → RO$_2$	23000 ppm^{-1}min^{-1}
7	HC4 + hv → 2RO$_2$	0.01k1min^{-1}
8	RO$_2$ + NO → NO$_2$ + 0.4OH + 0.4RO$_2$ + 0.6HC4	2500 ppm-1min-1
9	2RO$_2$ → PROD1	5300 min^{-1}
10	NO$_2$ + OH → PROD2	15000 ppm^{-1}min^{-1}
11	O$_3$ + OH$_2$ → PROD3	0.046 ppm^{-1}min^{-1}
12	HNO$_2$ + hv → OH + NO	0.05 k1 min^{-1}

k1:ratio of the light intensity

Table 6. RS32 photochemical reaction model

6.4. Results and discussion

Figure 17 (a), (b) and (c) show the distributions of the NO₂ concentration and wind vectors at a height of 1.5 m in the target area that were calculated based on the conditions at each period of time. The wind vectors are plotted at 40-m intervals. The NO₂ concentrations above Sen. Gil Puyat Avenue were high and gradually decreased at locations away from the avenue. The NO₂ concentration decreased at various gradients with distance from the avenue and was strongly influenced by the winds around areas near the avenue. For example, in the area south of the BOI building, where there are no buildings, the NO₂ concentration decreased appreciably with increased distance from the avenue. Meanwhile, in an area with high density of buildings, such as south of the IPO building, the NO₂ concentration decreased slightly. This indicates clearly that 3-D fluid simulators are suitable for calculating the distributions of air pollutant concentrations in an urban environment where wind flows are very complex.

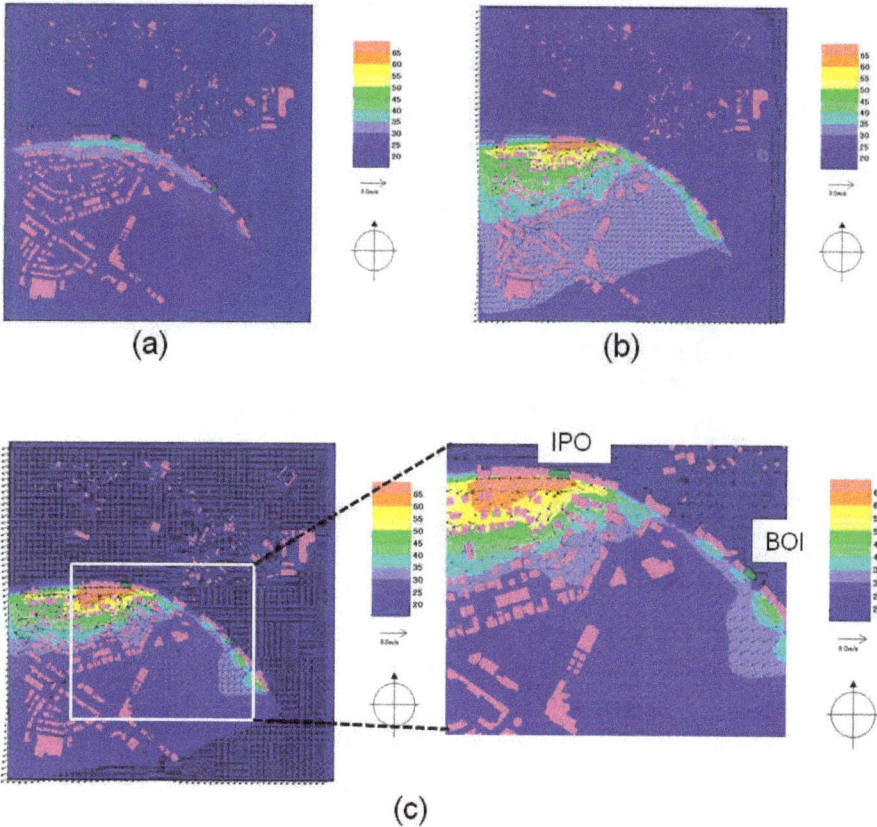

Figure 17. Distribution of NO₂ concentration (ppb) and wind vector at a height of 1.5 m in the Makati CBD. (a) Calculated results based on conditions at 2:00-3:00, (b) calculated results based on conditions at 17:00-18:00 and (c) calculated results based on conditions at 11:00-12:00

A comparison of the average NO_2 concentration of data measured on weekdays and simulated NO_2 concentrations at a height of 3.5 m in front of the IPO and BOI buildings is shown in Fig. 18 These bar charts indicate that the simulated NO_2 concentrations in front of the IPO building were close to the averages measured NO_2 concentrations. On the other hand, the simulated NO_2 concentrations in front of the BOI building based on conditions at 11:00 – 12:00 and 17:00 – 18:00 were lower than the measured averages. We infer that there are two main reasons for these differences. One is that air pollutants emitted from motor vehicles on Paseo De Roxas, which runs near the BOI building, were not considered in this study. The other reason is that the BOI building is located about 90 m from the nearest intersection. Most of us would accept that the emissions from vehicles driven in the vicinity of an intersection would exceed those of other locations on the avenues because of acceleration and idling. The simulated NO_2 concentrations were lower than the actual values because our calculation were based on the assumption that the same emission volumes were emitted regardless of the distance from the intersections. We assume that the simulation results are reliable for analyzing the effect of air pollutants emitted from vehicles on Sen. Gil Puyat Avenue. The simulated NO_2 concentrations in front of the IPO building are similar to the measured values. The simulated and measured NO_2 concentrations obtained in front of the BOI building are somewhat different as explained above.

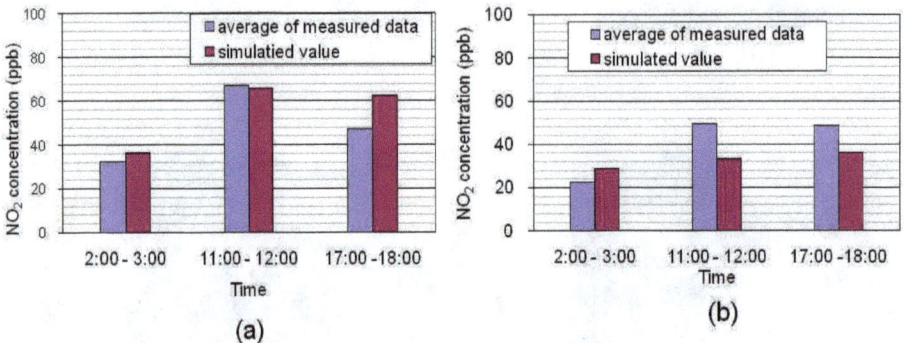

Figure 18. Comparison between average NO_2 concentration of data measured on weekdays and simulated NO_2 concentration at a height of 3.5 m in front of the building. (a) IPO, (b) BOI.

The distributions of the NO_2 concentrations around the IPO and BOI buildings that were calculated based on conditions at 11:00 – 12:00 are shown in Fig. 17(c). The NO_2 concentration in front of the IPO building was higher than that in front of the BOI building. There are high buildings on both sides of the street in front of the IPO building, thus forming a street canyon; it is difficult for air pollutants to escape from such canyons and so they tend to have high concentrations.

We discuss the simulation results for each time period to identify areas where the NO_2 concentration exceeds the Japanese environmental quality standard levels of 40 and 60 ppb at a height of 1.5 m. The distribution of the NO_2 concentration, which was calculated based on the condition at 2:00 – 3:00, had no areas where the NO_2 concentration exceeded 40 ppb. In the distribution of the NO_2 concentration that was calculated based on the condition at 11:00 – 12:00, a high NO_2 concentration area extended to the area southwest of the IPO building. The area where the NO_2 concentration exceeded 40 ppb extended about 200 m away from the street. The area where the NO_2 concentration exceeded 60 ppb also extended about 80 m away from the street. In the distribution of the NO_2 concentrations calculated based on the condition at 17:00 – 18:00, a high NO_2 concentration area extended to the area southwest of the IPO building similar to the condition before noon. Although the amount of emission during 17:00 – 18:00 was lower than that from 11:00 – 12:00, the area where the NO_2 concentration exceeded 40 ppb was larger than the area at 11:00 – 12:00. This area extended about 300 m away from the street. On the other hand, the area where the NO_2 concentration exceeded 60 ppb was small and limited to the roadside.

An analysis of the simulation results indicates that the areas where the NO_2 concentration exceeded the standard level extended much further at noon and in the evening. We conjectured that the high-concentration areas should be larger than those in the calculated simulation results. We expect that an effort will be made to reduce air pollution by adopting such measures as improving traffic management and reinforcing the regulation of motor vehicle exhaust emissions. Meanwhile, as mentioned above, the 3-D localized fluid simulator is useful in relation to reducing the air pollutant problem because air pollutant dispersion is strongly influenced by winds around the local area.

7. Conclusion

In this project, we measured the NO_2 concentration, the SPM concentration, and the traffic volume at Makati in Manila. We also simulated the NO_2 concentration, and estimated the area with a high NO_2 concentration.

The results of our measurements and measures for reducing air pollutants are summarized below.

1. The NO_2 concentration has a tendency to increase, especially in canyon areas. One effective measure for reducing the NO_2 concentration in a canyon area is to reduce the traffic volume. For example, a 35% reduction in traffic volume can lead to a 12% reduction in NO_2 concentration.

2. The NO_2 concentration has a tendency to increase, especially in canyon areas, as already mentioned. Therefore, if it is difficult to reduce traffic volume, a city plan with open spaces instead of canyons is effective for reducing air pollutants.

3. Recently, it has been reported that certain plants have the ability to reduce air pollutants such as NO_2 (Takahashi et al., 2005). It is therefore recommended that city planners plant trees with this capability beside the road.
4. The SPM concentration depends more on traffic volume than on land use. Therefore, traffic regulation in open spaces will function successfully in reducing the mass concentration of SPM.

Author details

Yasuko Yamada Maruo and Akira Sugiyama
NTT Energy and Environment Systems Laboratories, Japan

Acknowledgement

We thank the members of this project, Mr. P. A. Varilla, Mr. R. V. Cabana and Mr. H. C. Paquiz (Commission on Information and Communications Technology), Mr. D. F. Villorente, Mr. J. C. Manio, Mr. J. Bolo, Mr. I. C. A. Lomugdang, Mr. J. S. G. Sicam, Mr. J. R. D. Ty, Mr. G. P. Guba and Mr. M. T. Patula (Advanced Science and Technology Institute), Dr. K. B. N. Vergel and Dr. J. R. F. Regidor (Affiliate National Center for Transportation Studies), Dr. J. Nakamura, Dr. S. Ogawa and Mr. S. Sakata (NTT Energy and Environment Systems Laboratories), Mr. K. Matsunaga, Mr. T. Murai and Mr. W. Ishida (Nippon Telegraph and Telephone East Corporation), Mr. S. Morimoto and Mr. M. Osaka (Nippon Telegraph and Telephone West Corporation). We appreciate the financial support provided by the APT HRD Program for Exchange of ICT Researchers/Engineers through Collaborative Research of the Ministry of Internal Affairs and Communications, Japan. We also appreciate the Live E! project in Japan for lending us the weather stations.

8. References

ADB (1992). Asian Development Bank, Vehicular emission control planning in Metro Manila Final Report.

Collins, W.J., Stevenson, D.S., Johnson, C.E. & Derwent, R.G. (1997). Tropospheric ozone in a global-scale three-dimensional Lagrangian model and its response to NOX emission controls, Journal of atmospheric chemistry, 26, 223-274.

EQS (2012) Environmental Quality Standards in Japan, 2012/02/20, Available from: http://www.env.go.jp/en/air/aq/aq.html

Finardi, S., De Maria, R., D'Allura, A., Cascone, C., Calori, G. & Lollobrigida, F. (2008). A deterministic air quality forecasting system for Torino urban area, Italy, *Environmental modeling & software*, 23, 344-355.

Isidrovaleroso, I., Monteverde, C.A. & Estoque, M.A. (1992). Diurnal-variations of air-pollution over metropolitan Manila, *Atmosfera*, 5, 142-257.

Kaminski, J. W., Neary, L., Struzewska, J., McConnell, J. C., Lupu, A., Jarosz, J., Toyota, K., Gong, S. L., Cote, J., Liu, X., Chance, K. & Richter, A. (2008). GEM-AQ, an on-line global multiscale chemical weather modelling system: model description and evaluation of gas phase chemistry processes, *Atmospheric chemistry and physics*, 8, 3255-3281.

KKE (2012) Kozo Keikaku Engineering, 2012/02/20, Available from: http://www4.kke.co.jp/kaiseki/software/air-design.html

Lodge, J.P. (1992). Air-quality in metropolitan Manila-inferences from a questionable data set, *Atmospheric Environment*, 26, 2673-2677.

Maruo, Y.Y., Ogawa, S., Ichino, T., Murao, N. & Uchiyama, M. (2003). Measurement of local variations in atmospheric nitrogen dioxide levels in Sapporo, Japan, using a new method with high spatial and high temporal resolution, *Atmospheric Environment* 37,1065-1074.

Oanh, N.T.K., Upadhyaya, N., Zhuang, Y.H., Hao, Z.P., Murthy, D.V.S., Lestari, P., Villarin, J.T., Chengchua, K., Co, H.X., Dung, N.T. & Lihdgren, E.S. (2006). Particulate air pollution in six Asian cities: Spatial and temporal distributions, and associated sources, *Atmospheric Environment*, 40, 3367-3380.

Sabin, L.D., Kozawa, K., Behrentz, E., Winer, A.M., Fitz, D.R., Pankratz, D.V., Colome, S.D., & Fruin, S.A. (2005). Analysis of real-time variable affecting children's exposure to diesel-related pollutants during school bus commutes in Los Angeles, *Atmospheric Environment*, 39, 5243-5254.

Santos, F.L., Pabroa, P.C.B., Esguerra, L.V., Racho, J.M., Almoneda, R.V. & Sucgang, R. (2006). Roadside air particulate monitoring in the PM10 range at the Poveda Learning Center, EDSA, Metro Manila, *Proceedings of the FNCA 2004 Workshop on the Utilization of Research Reactors*, 17-22.

Smith, B.J., Nitschke, M., Pilotto, L.S., Ruffin, R.E., Pisaniello, D.L. & Willson, K.J. (2000). Health effects of daily indoor nitrogen dioxide exposure in people with asthma, *European Respiratory Journal*, 16, 879-885.

Takahashi, M., Higaki, A., Nohno, M., Kamada, M., Okamura, Y., Matsui, K., Kitani, S. & Morikawa, H. (2005). Differential assimilation of nitrogen dioxide by 70 taxa of roadside trees at an urban pollution level, *Chemosphere*, 61, 633-639.

Tanaka, T., Ban., H., Ogawa, S., Maruo., Y.Y., Delaunay, J.J. & Ieyasu, T. (2004). Environment Assessment technology, *NTT Technical Review* 2, 1-5.

Tanaka., T., Gulleux, A., Ohyama, T. Maruo, Y.Y. & Hayashi, T. (1999). A ppb level NO2 gas sensor using coloration reactions in porous glass, *Sensors and Actuators* B56,247-253(1999).

Tange, H., Kataya, K., Okamoto, S. Kobayashi, K. & Yoshida, Y. (1987). Simplified photochemical reaction model for forecasting photochemical air pollution, *Industrial Pollution*, 23, 12-18(in Japanese).

Vuilleumier, L., Bamer, J.T., Harley, R.A. & Brown, N.J. (2001). Evaluation of nitrogen dioxide photolysis rates in an urban area using data from the 1997 Southern California Ozone Study, *Atmospheric Environment* 35,6525-6537.

Chapter 6

Atmospheric Concentration of Trace Elements, Dry Deposition Fluxes and Source Apportionment Study in Mumbai

D.G. Gajghate, P. Pipalatkar and V.V. Khaparde

Additional information is available at the end of the chapter

1. Introduction

Presence of toxic elements in the atmosphere is of a great concern due to their adverse affect on human health and ecosystem. Despite the requirement of some of the elements for all living organisms, certain elements cause various toxic effects on accumulation in animal tissues (Yasutake and Hirayama 1997). There are various types of sources emitting these elements into the atmosphere, e.g. fossil fuel combustion contributes Al, Fe, Ca, Mg, K, Na, As, Pb, Cd, Sc, Hg elements (Furimsky 2000), elements like Pb and Zn are contributed by wood combustion (Mohn *et al.*, 2002), vehicular traffic contributes Cd, Cr, Cu, Ni, Pb, Zn (Westerlund 2001), electroplating contributes Cr (Flower *et al.*1994) metal alloy industries emit different elements like Cd, Cr, Al, Fe, Ni, Zn, Pb, Cu, etc. in the air (Harrison *et al.*, 1986). Dry deposition of trace elements has gained the importance due to its environmental significance and concerns. There have been many studies undertaken over the past decade to estimate the dry deposition (Injuk *et al.*, 1998; Jiries *et al.*, 2002; Fang *et al.*, 2004; Sakata *et al.*, 2006; Herut 2001). Recently, measurement on dry deposition fluxes of elements and their velocities studies were also reported at Brazil, Taiwan and Tokyo (Fang *et al.*, 2007; Pedro *et al.*, 2007; Sakata *et al.*, 2008). In India, status of airborne toxic elements at different land used pattern locations in major urban cities has been reported (Tripathi *et al.*, 1993; Gajghate and Hasan, 1999; Gajghate and Bhanarkar 2005a, b; Gajghate *et al.*, 2005). Bhanarkar *et al.*, (2005) have been carried out inventory of toxic elements for industrial sources in Greater Mumbai. Atmospheric deposition of trace elements like Pb, Cd, Cu and Zn was studied at Deonar, Bombay, India (Tripathi et al., 1993). They indicated that dry deposition flux of these trace elements was in the range of 0.2 to – 104.6 kg km^{-2} yr^{-1}. Mudri *et al.*, (1986) reported the dry deposition of dust fall on monthly basis for whole year at the Visakhapatnam, which amounted to 12.35 MT km^{-2} month^{-1}. However, in comparison, there is still relatively little known about atmospheric fluxes in India. This paper presents concentration and estimated dry deposition flux of selected trace elements (Zn, Pb, Al, Cd, Ni, Sr, Mn, Fe, Cr, Cu, Mg, V and Ca) in the atmosphere of Mumbai. Estimates the elements loading from the atmospheric deposition over the Mumbai are also presented. Enrichment factors (EFs) of elements with respect to crustal composition and source apportionment are also discussed.

1.1. Study area

Greater Mumbai is located on the western coast of India and above 11 m mean sea level. The city is spread over an area of 603 km^2 extending from Island city in the south to Borivali and Mulund in the north. It is one of the largest metropolises of the world with a population of 12 million in 2001 which is expected to touch 15 million in 2011. There are a large number of small and big industries include thermal power plant, refineries, petrochemicals, fertilizer, textiles mills, and engineering and foundry units in the city. Mumbai has a tropical savanna climate with relative humidity ranging between 57% and 87%, and annual average temperature of 25.3°C, with a maximum of 34.5° C in June and minimum of 14.3° C in January. The average annual precipitation is 2078 mm, with 34% of total rainfall occurring in July. The prevailing wind directions are from West and northwest, with West and southwest shifts during monsoon. Some easterly component is observed during winter.

2. Materials and methods

2.1. Sample collection and analysis

Four sites were selected to represent different segments of the region and activities, which include one control/reference (C), one industrial (I), and two kerb sites (K-1 and K-2) for sampling of PM$_{10}$ in ambient air of Mumbai. Details of sampling location are presented in Table 1 and also shown in Figure 1.

Site	Classification of area	Description
1	Control (C)	Protected area under Indian Navy, minimum traffic
2	Kerbside (K-1)	Commercial area, only four wheelers and buses are plying, 6 roads junction, cinema theatre, educational institutions
3	Kerbside (K-2)	Busy Swami Vivekanand Marg, Pawan Hans Civil air port, Santacruz BEST Bus Depot
4	Industrial (I)	Heavy traffic of trucks, tankers, multi wheeler dockyard activity

Table 1. Description of sampling sites

24-hrly sampling was carried out round the clock during the winter season using respirable high volume sampler with a flow rate of 1 m^3 min^{-1}. PM$_{10}$ collected on glass fibre filter paper were determined by gravimetric method (Katz, 1997). PM$_{10}$ samples collected from various sites were analyzed for selected trace elements (Zn, Pb, Al, Cd, Ni, Sr, Mn, Fe, Cr, Cu, Mg, V and Ca). A known portion of the filter paper is taken and digested with high purity HNO_3 in clean Teflon bombs using microwave digester. Then the digested solution was filtered through glass fibre filter and make up to a final volume of 100 mL with double distilled water. Blank samples were extracted and concentrated with the same method described. Trace elements analysis was performed with Inductively Coupled Plasma - Atomic Emission Spectroscopy (ICP-AES, Model: JY-24, Jobin Yvan, France).

2.2 QA/QC protocol

QA/QC protocol is undertaken for precise measurement of particulate matter by calibrating the air flow measurement device using top load calibrator prior to sampling and air flow is corrected at STP. Blank test background contamination was monitored by using operational blanks (unexposed filter) which were processed simultaneously with field samples. The field blanks were exposed in the field when the field sampling box was opened to remove and replace field samples. Blanks were cleaned and prepared with the same procedures applied to the actual samples. Concentration of particulate matter was determined by gravimetrically using calibrated balance (Mettler AE 163). Element analysis was carried out using Inductively Coupled Plasma - Atomic Emission Spectroscopy. High quality glass wares were preferred, throughout the sampling, digression and analysis steps, to prevent any metal contamination. The filter paper were digested with concentrated nitric acid (Merck) in Teflon vessel in a microwave digestion chamber (ETHOS make-milestone, Italy).The sample was digested for twenty minutes and then filtered through Whatman 42 (Ashless filter papers 125mm, cat No 1442 125) filter paper into properly cleaned volumetric flask. Calibration standards of 0.5ppm, 1ppm and 2ppm were prepared through serial dilution of standard stock solution of multi element having concentration of 1000 mg/lit (Merck, Cat No.1.11355.0100) and used for the calibration of the instrument. Samples are analyzed by spiking with a known amount of elements to calculate recovery efficiencies. The analysis procedure for the recovery test is the same as that described for the field samples. The recovery tests of elements were 102%, 95.5%, 106.5%, 94%, 98.0 114%, 103.5%, 96.5 %, 94.5%, 98.0%, 90.0%, 93.0% and 95.5 % for Al, Zn, Fe, V, Pb, Ca, Mg, Cd, Ni, Mn, Cr, Cu and Sr respectively. The reproducibility test indicates the stability of the instruments. Analysis of elements of the same concentration standard solution is repeated for many times. The standard solution of 0.5 mg/lit was repeated and reproducibility of results indicates that 103.9%, 96.65%, 96.61%, 97.5%, 99.40%, 96.02%, 99.2%, 99 %, 97.24%, 98.62%, 94.4%, 96.6% and 98.02 % for Al, Zn, Fe, V, Pb, Ca, Mg, Cd, Ni, Mn, Cr, Cu and Sr respectively was observed.

3. Results and discussions

3.1. Concentration of PM10

The air quality was determined in the activity zones namely control, industrial zone and high traffic area (Kerbside). The average PM10 concentration for control site was observed to be 78 $\mu g/m^3$ with minimum and maximum of 29 $\mu g/m^3$, 146 $\mu g/m^3$ respectively. Ambient air quality in industrial site showed the average PM_{10} was 135 $\mu g/m^3$. The average concentration of PM_{10} were maximum at all kerbsides. Average level of PM_{10} was 186 $\mu g/m^3$ and 315 $\mu g/m^3$ at K-1 and K-2 respectively. Concentration of PM_{10} has been exceeded CPCB standards for mixed area at all the sampling location except at control site. Kerbside K-2 showed the 315$\mu g/m^3$ as higher average concentration of PM_{10} with 450 $\mu g/m^3$ and 166 $\mu g/m^3$ as maximum and minimum concentration respectively.

3.2. Concentration of toxic elements

The measured average elemental concentrations at various sites were presented in Fig 1 (a & b). Average individual trace element concentrations fluctuated between 0.003 $\mu g/m^3$ (Cr) and 3.432 $\mu g/m^3$ (Zn). Average concentration of Zn was found to be ranged from 2.59 $\mu g/m^3$ to 5.45 $\mu g/m^3$. A minimum average Zn concentration of 2.59 $\mu g/m^3$ at control site and maximum concentration of 5.45 $\mu g/m^3$ at industrial location. Average concentration of Fe varied from 0.23 $\mu g/m^3$ to 3.5 $\mu g/m^3$ with average mean concentration of 1.69 $\mu g/m^3$. Similarly average minimum Ca concentration of 0.61 $\mu g/m^3$ observed at control site and maximum concentration of 3.07 $\mu g/m^3$ at kerbside (K-1) with average mean concentration of 2.28 $\mu g/m^3$. Average mean V concentration of 0.52 $\mu g/m^3$ was found with maximum of 1.2 $\mu g/m^3$ at kerbside (K-2). As regards to Al, average concentration was found to be 0.60 $\mu g/m^3$ with minimum of 0.01 $\mu g/m^3$ and maximum of 1.22 $\mu g/m^3$ at industrial site. The highest mean concentration are found for Zn (3.43 $\mu g/m^3$), following Ca (2.28 $\mu g/m^3$), Fe (1.69 $\mu g/m^3$), Al (0.60$\mu g/m^3$), V (0.52$\mu g/m^3$), Mg (0.19 $\mu g/m^3$) and Pb (0.13 $\mu g/m^3$). The remaining elements (Ni, Mn, Cd, Cu, and Cr) were found to follow a sequence of decreasing concentration starting from Mg (0.10 $\mu g/m^3$) and terminating at Cr (0.003 $\mu g/m^3$). Overall average basis decreasing element concentration trend was Zn>Ca>Fe>Al>V>Mg> Pb>Ni>Mg>Cd>Cu>Cr. However, overall average basis decreasing anthropogenic element concentration trend was Zn>V>Pb>Ni>Cd>Cr and crustal elements trends as Ca>Fe> Al >Mg>Mg>Cu respectively.

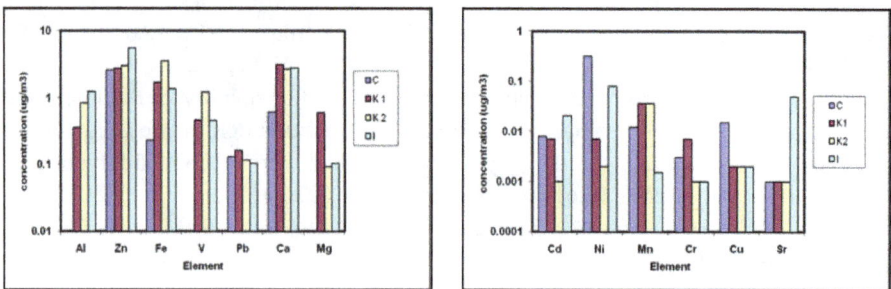

Figure 1. (a & b) Average trace element concentrations at various sites

3.3. Dry deposition estimation

Dry depositions flux for the various elements are estimated as the product of the atmospheric concentration and a suitable deposition velocity. Dry deposition fluxes were calculated using the equation:

$$F = C_i \times V_d \qquad\qquad (1)$$

Where, F is the dry deposition flux, C_i trace element mean concentration and V_d is the elemental settling velocity.

GESAMP,(1989) suggested that the best values for the dry deposition velocities of Zn, Pb and Cd should be 0.1 cm s^{-1} and those for Al, Fe and Mn should be 1.0 cm s^{-1}. Dry deposition velocities reported by Wu et al., (1994) for Cr and Cd are 0.26 cm s^{-1} and 0.47 cm s^{-1} respectively. In our study, dry deposition estimate for Al, Mn, Fe, Zn, Pb have been undertaken using dry deposition velocity values adopted from GESAMP (1989) and for Cr and Cd dry deposition velocity values adopted from Wu et al., (1994). Deposition velocity values for V, Ca, Sr, Cu and Ni have been adopted from Injuk et al., (1998) and for Mg from Wu et al., (2006).

Site wise variation in dry deposition fluxes of selected trace elements is shown in Fig 2 (a & b). Dry deposition fluxes of elements at industrial site showed the dominance of Ca (1401.63 µg/m²/d) followed by Fe (1162.94 µg/m²/d), Al (1054 µg/m²/d), Zn (470.88 µg/m²/d) and Mg (124.58 µg/m²/d) was observed. At control site, prominent elemental fluxes were Ca (305.68 µg/ m² /d), Zn (223.77 µg/m²/d), V (207.36 µg / m² /d), Fe (198. 72 µg / m² /d) and Ni 168.65µg / m² /d). Whereas at kerb site, the highest dry deposition fluxes were observed for Fe (3024 µg/m²/d), Ca (1538.43 µg/m²/d), Mg (725.76 µg/m²/d), Al (717.12 µg/ m²/d), Zn (258.33 µg/m²/d) and V (248.83 µg/m²/d).

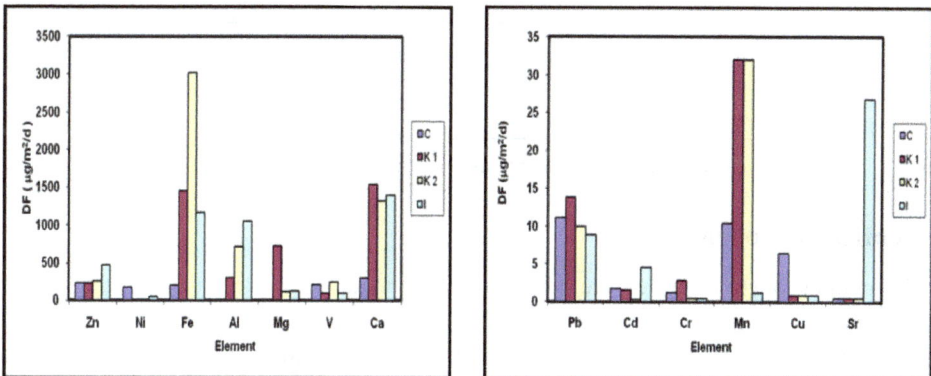

Figure 2. (a & b) Average dry deposition fluxes of trace elements

The average estimated trace element fluxes ranged between 1.218 (Cr) and 1461 (Fe) µg m^{-2} d^{-1}. The dry deposition fluxes for dominant elements were Fe (1461.45µg/m²/d), Ca (1143.43 µg/m²/d), Al (521.64 µg/m²/d), Zn (296.56 µg/m²/d) and Mg 241.014 µg/m²/d) were observed. The deposition fluxes for remaining metal (V, Pb, Cd, Ni, Mn, Cr, Cu, Mg and Sr) ranged from highest of V (161.27 µg/m² /d) to minimum of Cr (1.21 µg / m² /d).

Average deposition fluxes of preliminary crustal elements were significantly higher than the anthropogenic elements. Fe, Mg and Ca were the most significant crustal elements followed by abundant anthropogenic elements (Zn, V and Ni). However, anthropogenic element Cd and Pb were comparatively small. Al was most abundant crustal elements.

Dry deposition loading (Mg yr^{-1}) of trace elements to the Greater Mumbai was calculated using the elemental dry deposition fluxes from Table 2. Loading of Al, Mn, Fe, Zn, Pb, Cd, V, Ca, Ni, Mg, Sr, Cu and Cr to the entire area of Greater Mumbai ranged from 0.27 Mg yr^{-1} for Cr to 321.7 Mg yr^{-1} for Fe. Fe, Ca, Al, Zn Mg and V were found the highest loading to the Greater Mumbai, which may support the atmospheric pathway as an important source of this metal.

Elements	Estimated deposition rates (Mg yr^{-1})
Al	114.8
Mn	4.2
Fe	321.7
Zn	65.3
Pb	2.4
Cd	0.45
Cr	0.27
Cu	0.50
Ni	11.9
Mg	53.0
Ca	251.7
Sr	1.56
V	35.5

Table 2. Estimate of yearly element deposition rates to Mumbai

3.4. Enrichment of elements

Enrichment factor (EF) is a measure to which trace elements are enriched or reduced relative to a specific source. The enrichment factor (EF) is the ratio of the concentration of any trace metal or ion (X) to Fe in the sample divided by the corresponding ratio in crustal material (Weisel *et al.*, 1984). If the EF is less than 10, trace metal in aerosol has a significant crustal source but elements with enrichment factors greater than 10 are assumed to be due to the other sources rather than background contribution in that sampling region.

Results of the EF computation are plotted in Fig 3. Iron is the common reference element for crustal particles in EF calculations. The order of EF values were Cd>Zn>Pb>V>Ni > Cu> Ca >Sr>Mn>Cr >Mg>Al. Crustal elements such as Al, Ca, Mg, Mn, Cr, Cu and Sr were not enriched to higher than 10 (log EF < 1) indicating that they have originated from local soil dust. Elements such as Cd, V, Ni, Zn and Pb have enrichment factors higher than 10, indicating these elements had significant anthropogenic sources.

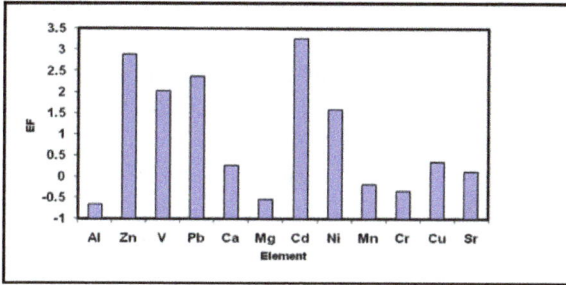

Figure 3. Enrichment of elements in the study area

3.5. Source apportionment

The varimax rotated factor analysis technique based on the principal components (Zhou *et al.*, 2004) has been used in the determination of the source contribution of PM$_{10}$ pollution in Mumbai city. The components or factors rotated had eigen values greater than one after rotation. Based on the results, contribution of ambient PM$_{10}$ by each source at different sites is presented in Fig 4.Overall, the study indicates the diesel and gasoline vehicle exhaust emissions contribute to the extent of 6% to 14% ambient PM$_{10}$ levels. The contribution of vehicular exhaust emissions at kerb side however, varies between 33% and 54%. Resuspended dust due to movement of traffic also contributes significantly (10% to 20% in ambient air and 9% to 17% to kerbside). Marine sources contribute 13 %to 14% to ambient PM$_{10}$ and upto 12% to kerb side PM$_{10}$ levels.

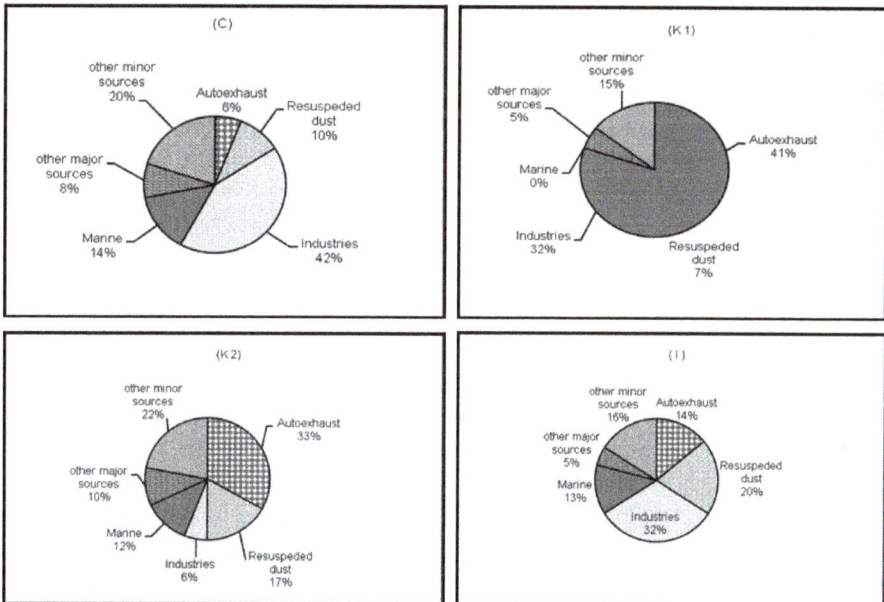

Figure 4. Source apportionment study for PM$_{10}$ using factor analysis

4. Conclusions

Airborne PM samples collected from Greater Mumbai area was used for examine the concentrations of selected trace elements (Zn, Pb, Al, Cd, Ni, Sr, Mn, Fe, Cr, Cu, Mg, V and Ca) and to perform dry deposition flux estimates. Average individual trace element concentrations fluctuated between 0.003 µg/m^3 (Cr) and 3.43 µg/m^3 (Zn). The primarily crustal elemental concentrations were higher than the primarily anthropogenic elemental concentrations. Zn, Fe, Ca, Al, V, Mg and Pb had the highest measured concentration followed by Ni, Mn, Cd, Cu, Cr and Sr. The estimated trace element flux values ranged between 1.2 (Cr) and 1461 (Fe) µg m-2 d-1. The study indicates that crustal elemental fluxes were significantly higher than anthropogenic elemental fluxes. Estimated loading of trace elements to Mumbai region ranged from 0.27 Mg yr-1 for Cr to 321.7 Mg yr-1 for Fe. Higher crustal trace element concentrations and fluxes in this study may be due to re-suspension of dust released during traffic activities and soil erosion. However, the concentration and fluxes of Zn and V (anthropogenic elements) may be attributed to industrial emission. Source apportionment study reveals that PM10 is contributed by vehicles exhaust, followed by resuspension of dust and marine sources in Mumbai city.

Author details

D.G. Gajghate, P.Pipalatkar and V.V. Khaparde
Air Pollution Control Division, National Environmental Engineering Research Institute (NEERI), Nehru Marg, Nagpur, India

Acknowledgement

We thank Dr S.R.Wate, Director, NEERI, Nagpur for encouragement and permission to publish the paper. We deeply acknowledge to other colleagues who participated and provide supports in the conduct of the study

5. References

Bhanarkar A.D., Rao P.S., Gajghate D.G., Nema P. (2005). Inventory of SO₂, PM and toxic metals emissions from industrial sources in Greater Mumbai, India. Atmos Environ, 39, 3851-3864.

Fang G.C., Wu Y.S., Huang S.H., Rau J.Y. (2004). Dry deposition (downward, upward) concentration study of particulates and heavy metals during daytime night time period at the traffic-sampling site of Sha-Lu, Taiwan. Chemosphere, 56, 509-518.

Fang G.C., Wu Y.S., Chang S.Y., Lin J., Lin J.G. (2007). Overall dry deposition velocities of trace elements measured at harbor and traffic site in central Taiwan. Chemosphere, 67, 966–974.

Flower W.L., Peng L.W., Bonin M.P., French N.B., Johnsen H.A., Ottesen D.K., Renzi, R.F., Westbrook, L.V.A.(1994). Laser-based technique to continuously monitor metal aerosol emissions. Fuel process technol., 39, 277–284.

Furimsky E. (2000). Characterization of trace element emissions from coal combustion by equilibrium calculations. Fuel process technol., 63: 29–44.

Gajghate D.G., Hasan M.Z. (1999). Ambient lead levels in urban areas. Bull. Environ. Contam. Toxico., 62, 403-408.

Gajghate D.G., Bhanarkar A.D. (2005a). Tracking toxic metals in ambient air of Agra city, India. Bull. Environ. Contam. Toxico, 72,806-812.

Gajghate D.G., Bhanarkar A.D. (2005b). Characterization of particulate matter for toxic metals in ambient air of Kochi city, India. Environ. Monit. Assess., 102, 119-129.

Gajghate D.G., Thawale P.R., Vaidya M.V., Nema P., (2005). Ambient respirable particulate matter and toxic metals in Kolkata city. Bull. Environ. Contam. Toxico., 75, 608-614.

GESAMP (1989). Group of Experts on Scientific Aspects of Marine Pollution Working Group 14, The atmospheric input of trace species to the world ocean. Rep. Stud. 38, World Meteorol Urgan, Geneva, 106-106.

Harrison R.M. (1986). Handbook of Air Pollution Analysis In: Harrison, RM Perry R (eds) Champman and Hall, London, p 215.

Herut B., Nimmo M., Medway A., Chester R., Krom, M.D. (2001). Dry atmospheric inputs of trace metals at the Mediterranean coast or Israel (SE Mediterranean): sources and fluxes. Atmos. Environ., 35, 803-813.

Injuk J., Van Grieken R., de Leeuw G. (1998). Deposition of atmospheric trace elements into the North Sea: coastal, ship, platform measurements and model predictions. Atmos. Environ., 32, 3011-3025.

Jiries A., EI-Hasan T., Manasrah W. (2002). Qualitative evaluation of the mineralogical and chemical composition of dry deposition in the central and southern highlands of Jordan. Chemosphere, 48, 933-938.

Katz M. (1977). Methods for air sampling and analysis, 2nd edn. APHA Press Inc.

Mohn J., Figi R., Graf P., Gujer E., Haag R., Honegger P., Mattrel P., Nagel O., Schmid P., Seiler C., Schreiner C., Steinhauser E., Zennegg M., Emmenegger L. (2002). Wood Combustion-Clean Energy In: proceedings of 5th International Conference on Emission Monitoring, Odense, Denmark

Mudri S.S., Vankatrao D., Ramarao K.G., Ramaprasad R.V., Ravishankar V., Pamattawar V.I., Aggrawal A.L., Murty Y.S. (1986). Ambient air quality at Visakhapatnam-A case study. Indian J. Environ. Hlth., 28, 284-295.

Pedro de A., Pereira P., Lopes W.A., Carvalho L.S., da Rocha G.O., Bahia N. de C., Loyola J., Quiterio S.L., Escaleira V., Arbilla G., de Andrade J.B. (2007).Atmospheric concentrations and dry deposition fluxes of particulate trace metals in Salvador, Bahia, Brazil. Atmos. Environ., 41, 7837–7850.

Sakata M., Marumoto K., Narukawa M., Asakura K. (2006). Regional variation in wet and dry deposition fluxes of trace elements in Japan. Atmos. Environ., 40, 521-531.

Sakata M., Tani Y., Takagi T. (2008). Wet and dry deposition fluxes of trace elements in Tokyo Bay. Atmos. Environ., 42, 5913– 5922.

Tripathi R.M., Ashawa S.C., Khandekar R.N. (1993). Atmospheric deposition of Pb, Cd, Cu and Zn in Bombay, India. Atmos. Environ., 27, 269-273.

Weisel C.P., Duce R.A., Fasching J.L., Heaton R.W. (1984). Estimates of the transport of trace metals from the oceans to the atmosphere. J. Geophy. Res., 89, 11607-11618.

Westerlund K.G. (2001). Metal Emissions from Stockholm Traffics- Wear of Brake Linings, Reports from SLB-analysis, Environment and Health Protection Administration in Stockholm, Stockholm.

Wu Z.Y., Han M., Lin Z.C., Ondov J.M. (1994). Chesapeake Bay atmospheric deposition study year 1; sources and dry deposition of selected elements in aerosol particles. Atmos. Environ., 28, 1471-1486.

Wu Y.S., Fang G.C., Chen J.C., Lin C.P., Huang S.H., Rau J.H., Lin J.G. (2006). Ambient air particulate dry deposition, concentration and metallic elements at Taichung Harbor near Taiwan Strait. Atmos. Res., 79, 52-66.

Yasutake K. (1997). Hirayama Animal models. In: Massaro EJ (ed) Handbook of Human Toxicology, CRC Press, Boca Raton, New York.

Zhou L., Kim E., Hopke P.K., Stanier C.O. (2004). Advanced Factor Analysis on Pittsburgh particle size distribution data. Aero. Sci. Tech., 38, 118-132.

Chapter 7

Emission Inventory of Air Pollutants and Trend Analysis Based on Various Regulatory Measures Over Megacity Delhi

Manju Mohan, Shweta Bhati, Preeti Gunwani and Pallavi Marappu

Additional information is available at the end of the chapter

1. Introduction

An emission inventory is defined as an accounting of all air pollution emissions and associated data from sources within a specified area and over a specific time interval [1]. Air emissions inventory information assists in planning to reduce emissions to meet air quality goals and tracking progress of control initiatives towards pollution mitigation. The development of information for an emissions inventory can be carried out in one of two methods. One method is often referred to as the top-down approach. In the case of a top-down approach, generalized factors such as total fuel use, total population, total activity data, for example, are used as indicators of emissions. Emission factors are developed that predict emissions per unit of a process or fuel mass or per person or such. The product of the emission factor with the relevant emissions indicator provides an estimate of emissions. These emissions can be disaggregated sector wise. In other method, known as a bottom-up approach, the region of interest is divided into sectors of interest and specific information is developed for each sector. This information is then used to estimate the emissions that will occur in each sector [2].

Fast developing economy of Delhi, the capital city of India, has led to the rapid increase in its population. Consequently, urbanization and increasing numbers of vehicles in the city are causing high levels of air pollution. In response to the growing environmental concerns, various regulatory actions have also been initiated in Delhi over the past decade such as introduction of CNG, stringent emission norms, development of public transport etc. Construction of emission inventories is significant to understand sources of pollution so as to define priorities and set objectives for pollution management. Different groups have developed emission inventories for different parts of India such as Nagpur [3], Jamshedpur [4] and Delhi [5,6,7]. Similarly mathematical modeling tools aid in understanding the fate of various pollutants emitted into ambient air and estimations of their levels. In this context, emission inventories are gridded on urban scale [8] and applied for air quality modelling for varied purposes [9,10]

With this context, the objective of the present study is to develop an annual emission inventory of some selected air pollutants of Delhi and analyse the trends in light of various control measures enacted by the regulatory authorities. Emission data for different source categories has been calculated annually for Delhi for the year 2001-2008 using top to bottom approach. Estimated emissions have been used to compute ambient concentrations by air quality modeling and further compared with observed concentrations. In addition efforts are being made by various organizations to construct emission inventories for a particular sector, geographical location, specific pollutants and time span etc. This study also examines the various other emission inventories that are available in part or full over Delhi.

2. Study area and sources of air pollution

Delhi, the capital city of India, is located at 28.6 N and 77.2 E and geologically, this region is bounded by the Indo-Gangetic alluvial plains in the North and East, by Thar Desert in the West and by old Aravalli hill ranges in South. The city is the highest populated megacity of India with 16.3 million inhabitants belonging to different economic strata. The city also boasts of the highest number of registered motor vehicles in the nation [11]. The number of vehicles has increased by over 80 % from year 2000 to about 6 million in year 2008-09 [12] which can be sees in Figure 1. Consequently, vehicle fuel consumption has also increased. Major pollutants emitted from vehicles are CO, NOx, particulate matter and hydrocarbons. Upto year 2009, the city's power demand were met by four thermal power stations, three of which were coal based. The remaining one is gas based. Coal based thermal power stations are major sources of particulate pollution and some gases such as SO_2.

Delhi is the largest municipal solid waste producer in the country [13]. Treatment of waste and enteric fermentation in animals is source of emissions like methane while animal manure contributes towards gases like NH_3 and CH_4.

3. Emissions sectors

Emissions for the city have been calculated for transport sector, power plants, domestic sector and from animals using the emission factors approach for different pollutants. The pollutants considered for the assessment of emission inventory are criteria pollutants like SO_2, CO, NOx, TSP and some other pollutants such as hydrocarbons (HC), methane (CH_4) and Carbon-di-oxide (CO_2).

3.1. Transport

Delhi city is home to the largest number of registered motor vehicles in India. The city has witnessed consistent increase in number of registered motor vehicles. Figure 1 shows the rise in increase in total number of registered motor vehicles in Delhi from 2001-2008 [12]. A consistent increase is observed in the number of vehicles in past years. The ever-rising number of vehicles is one of the major concerns for the city in terms of air pollution control.

The calculation of emission in Delhi from vehicles require the data on emission factor for the specific vehicle type, the distance traveled by a particular vehicle type and number of vehicles and their distribution in the type of the fuel used. The emission from vehicles is calculated using:

$$E_i = \sum (Veh_j \times D_j) \times E_{i:j:km}$$

where, E_i: emission of compound (i) ; Veh_j: number of vehicles per type (j) ;D_j: distance traveled in a year per different vehicle type (j) and $E_{i:j:km}$: emission of compound (i), vehicle type (j) per driven kilometer.

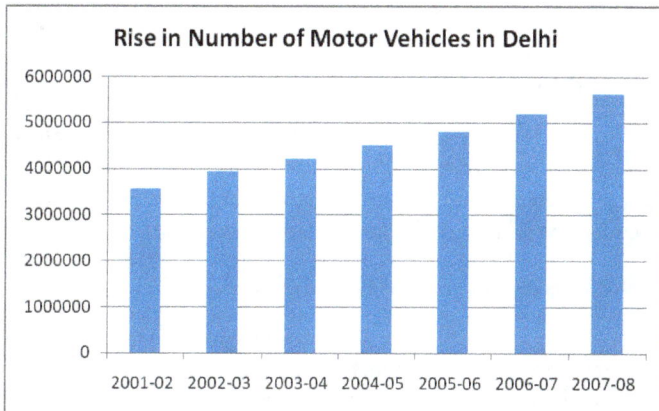

Figure 1. Vehicular population of Delhi

The emission factors are derived from Gurjar et al [6] which are based on sources such as EEA [14], Foell et al.[15] , Bouwman et al. [16] and Reddy and Venkataraman [17]. Cars in Delhi run on petrol, diesel and CNG. Phasing out of diesel based buses and petrol based autorickshaw (three wheeler) was initiated in year 2000 and by year 2002 all diesel based buses were phased out. Hence for buses and autorickshaw CNG consumption emission factors are considered. Two wheelers use gasoline while goods vehicles use diesel. The distance travelled by each vehicle is derived from surveys carried out in past studies [18,19].

Emissions have been estimated for four pollutants viz Carbon monoxide (CO), Hydrocarbons (HC), nitrogen oxides (NOx) and total suspended particulate matter (TSP). Figure 2 displays total annual emissions from vehicles for these four pollutants. The dominant pollutant is CO and among these four pollutants the least emitted pollutant is TSP. CO emissions have shown an increase till 2005 and then a decrease while other pollutants are showing an overall increase over the time period of 2001-2008.

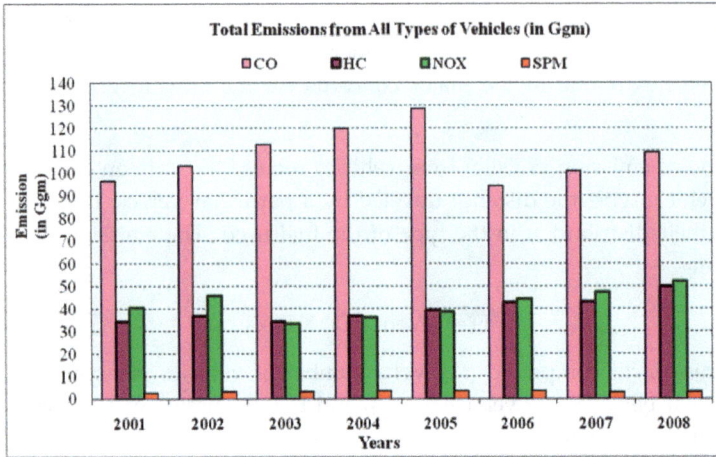

Figure 2. Emissions from transport sector

3.2. Power plants and fuel consumption in industries

Three coal-based Power Plants of Delhi (Badarpur, Indraprastha and Rajghat) have been considered in Delhi. The coal consumption for these Power Plants was collected from annual performance review reports of thermal power sector of India [20] . Using relevant emission factors, emissions from power plants are calculated as:

Total Emission= coal consumption (in 000′ ton) X emission factor for that power plant

Table 1 displays the annual coal consumption in the city from these power plants. As can be seen the total coal consumption has been increasing steadily from year 2001 to 2008. Figure 3 displays total emissions of NOX, SO_2, CO and TSP from power plants which are the dominant pollutants from this sector.

Sl. No.	Thermal Power Plants	Consumption in 000T							
		2001	2002	2003	2004	2005	2006	2007	2008
1.	Badarpur	3767	3818	3554	3605	3732	3768	3739	4104
2.	I.P. Stn.	695	650	495	639	789	934	946	982
3.	Rajghat	612	542	671	629	541	503	529	736

Table 1. Coal Consumption for year 2001-2008 for all the three thermal power plants [20]

Use of beneficiated coal was implemented in late nineties and early 2000s in Delhi for lower sulphur content to control TSP emissions [21]. There isn't a consistent increase in emissions although overall increase in CO emissions has been observed.

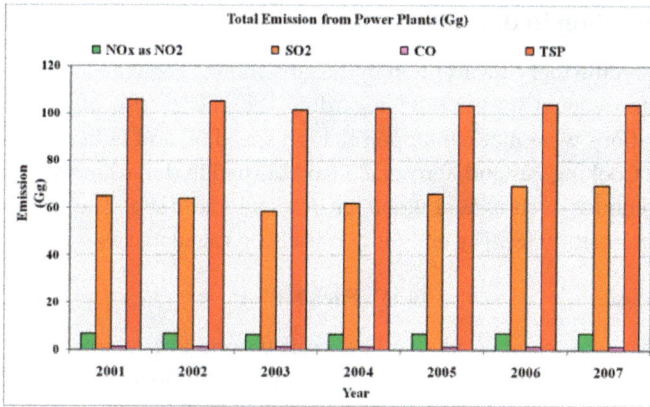

Figure 3. Emissions from power plants

Through a government regulation, a total of 2210 polluting industries were closed/relocated between 1998-2001 in Delhi [21]. Thus emissions from the industrial sector are mainly due to consumption of industrial fuels such as HSD and LDO. The emissions from industrial fuel consumption are calculated as:

$$E_i = \sum (Fuel_j \times EF_{ij}) \tag{1}$$

where, E_i: emission per compound (i) ; $Fuel_j$: consumption of fuel per fuel type (j) ; $EF_{i,j}$: emissions of compound (i) per unit of fuel (j) consumed.

The statistical handbook of Delhi [12] gives information on consumption of the HSD and LDO fuel. Emission factors for these fuels are as used in Delhi inventory preparation by Central Pollution Control Board [5] which are based on revised AP-42 emissions [5]. Figure 4 displays emissions from fuel consumption in industries for TSP, SO_2, NO_x and CO. Although, there is no consistent trend of increase or decrease, there is an overall decrease in emissions from 2001 to 2008.

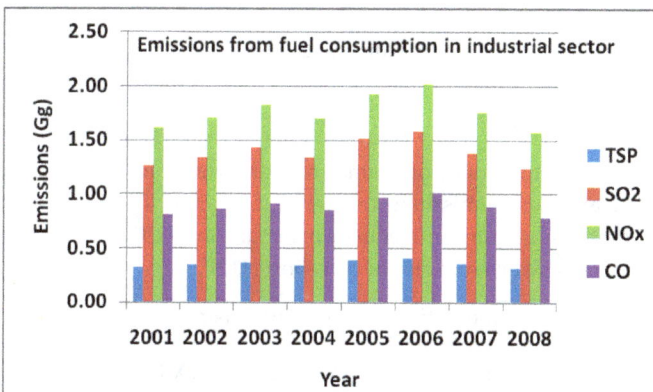

Figure 4. Emissions from industrial sector

3.3. Fuel consumption in domestic sector

The information on the fuel consumption in the government regulated statistical reports [12] does not provide detailed information regarding the energy use in power or industrial sector so assumptions were made that: petrol, high speed oil and light diesel oil are used in transport sector; Cooking gas and kerosene oil are burned in domestic sector other energy is assumed to be biomass such as fuel wood, crop waste and dung. The similar assumptions were used by earlier study [6]. The emissions have been calculated as

$$E_i = \sum (Fuel_j \times EF_{ij}) \tag{2}$$

where, E_i: emission per compound (i) ; $Fuel_j$: consumption of fuel per fuel type (j) ; $EF_{i,j}$: emissions of compound (i) per unit of fuel (j) consumed. Emission factors of as used in [6] have been used. Figure 5 displays emissions for CO, NOx, SO2 and TSP from fuel consumption in domestic sector. Since fuel consumption in domestic sector is directly associated with population growth, a steady increase of emissions is observed in domestic sector.

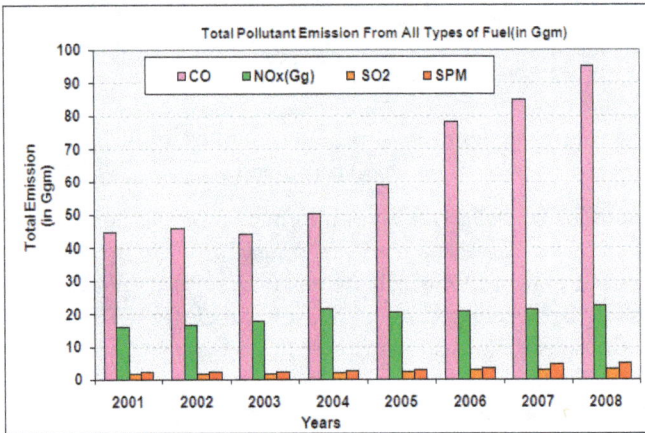

Figure 5. Emissions from fuel consumption in domestic sector

3.4. Emissions from animals

Enteric fermentation and treatment of animal manure are main processes contributing to methane, N_2O and ammonia emissions from animal sector. These emissions have been calculated using the formulation and emission factors in IPCC tier I methodology [22]. Livestock count has been taken from statistical reports [12] .

CH_4 emissions from enteric fermentation are calculated using

$$Emission_{CH4} = \sum (Number\ of\ animals_j \times EF_{CH4\ j}) \tag{3}$$

Here, j denotes the animal type.

N_2O and NH_3 emissions are calculated using equation given below, assuming unmanaged livestock keeping and non-treatment of manure.

$$\text{Emission}_{N2O/NH3} = \sum (\text{Number of animals}_j \text{ X Nitrogen-excretion}_j \text{ X } EF_{N2O/NH3j}) \qquad (4)$$

Figure 6 shows annual emissions of CH_4, NH_3 and N_2O from animal sector. With increasing urbanization, the agricultural activities in Delhi are on decline. The agricultural land has decreased by 27 % in the decadal period of 2001-2010 [23]. This has had an influence on animal population as well which has declined by 18 % from 2001 to 2008. Thus the emissions also show a gradual decline especially in case of ammonia.

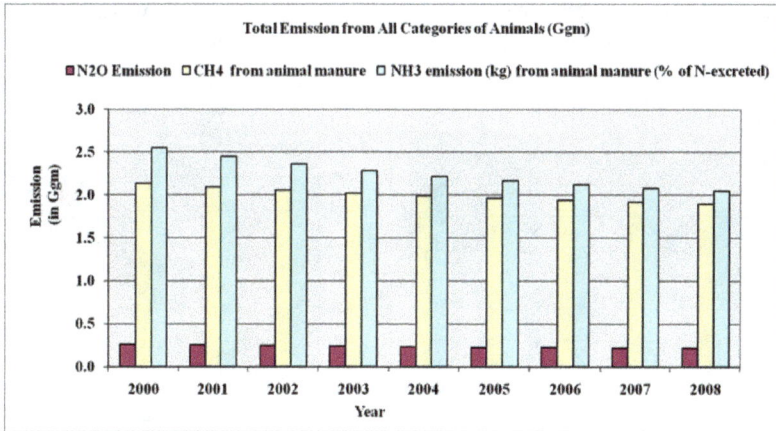

Figure 6. Emissions from animals

3.5. Emissions from waste sector

About 6500 - 7000 tons of municipal solid waste (MSW) is generated each day in Delhi with per capita generation rate of 0.47 kg day-1. MSW mainly comprises of biodegradable materials, which undergo anaerobic decomposition in landfills generating landfill gas (LFG) consisting of about 60% methane (CH_4) together with small quantities of non-methane organic compounds and other trace gases [24] . Major processes in the waste treatment sector are treatment of municipal solid waste and wastewater which leads to leads to CH_4 emission from landfills. NH_3 emissions are resulted from waste composting. Further some waste is left out on streets for open burning which also leads to emission of several pollutants.

CH_4 emissions from solid waste disposal are calculated using the following formula [25,6].

$$\text{Total Emission } CH_4 = [(MSW \text{ X } MCF \text{ X } DOC \text{ X } DOC_f \text{ X } F \text{ X } (16/12) - R] \text{ X } [1 - OF] \qquad (5)$$

where, MSW = Municipal Solid Waste [per capita waste produced X Delhi population] ; MCF = Methane Correction Factor ; DOC = Degradable Organic Carbon ; DOC_f = Fraction of DOC dissimilated ; F = Fraction of CH_4 in landfill gas and OF = Oxidation factor

Figure 7 shows annual emissions of three major pollutants from waste treatment sector viz. CH_4, NH_3 and N_2O. Contribution of waste sector for NH_3 and N_2O emissions is very small in comparison to methane emissions. Again, as in the case of fuel consumption in domestic sector, waste generation is also related to population and thus emissions from waste sector have also shown a consistent increase during 2001-2008.

Methane estimations can have different methodologies yielding different estimations. Chakraborty et al [24] estimated methane emission estimations using different methods viz. the in-situ CH_4 measurements, IPCC 1996 default methodology (DM), Modified Triangular Method (MTM) and First Order Decay (FOD) method based on data collected between 2008-2009. The annual average methane emission rates from three landfills were 45.7 Gg by IPCC method, 31.1 Gg y^{-1} by the FOD; 41.1 Gg y-1 by the MTM respectively.

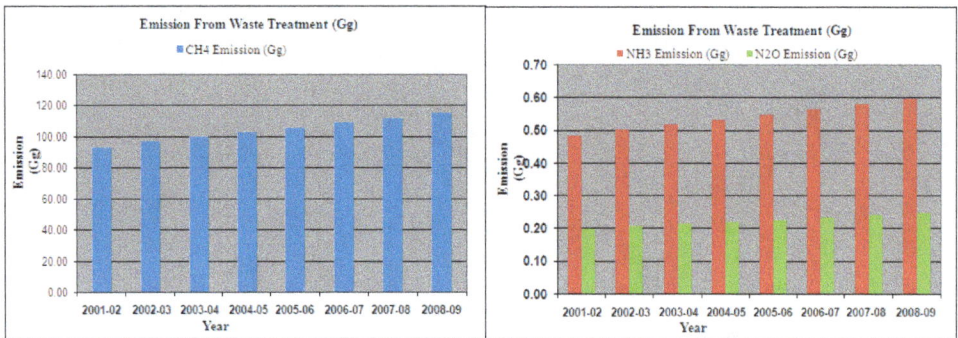

Figure 7. Annual emissions from waste treatment

4. Sectoral contribution towards emissions

Different sources have varying contribution towards different pollutants. Figures 8-11 display sector wise contribution and total emissions for CO, NOx SO_2 and TSP respectively. It is found that CO is mainly emitted from domestic sector and transport sector. Main contributor of SO_2 and TSP is power plants. NO_2 is mainly emitted from transport, domestic sector and power plants and the contribution of transport sector has increased for NOx emissions from 2001-2008. Waste treatment is main contributor of CH_4 followed by enteric fermentation and then domestic sector. NH_3 emissions are mainly contributed by animal and waste sector. The sudden drop in CO emissions in the year 2006 could have been due to application of Euro –III emission norms in 2004-2005. The percentage increase in CO in past 8 years is 37%; the main contribution of CO is from Transport sector and as the numbers of vehicles are increasing every year the emission of CO is increasing. CH_4 has increased 21% in the past 8 years, mainly contributed from Waste Treatment Sector followed by Domestic Sector which is again a result of increasing population and the demands.

Figures 9, 10 and 11 also show observed annual average ambient concentrations of NOx, SO_2 and TSP along with emissions.

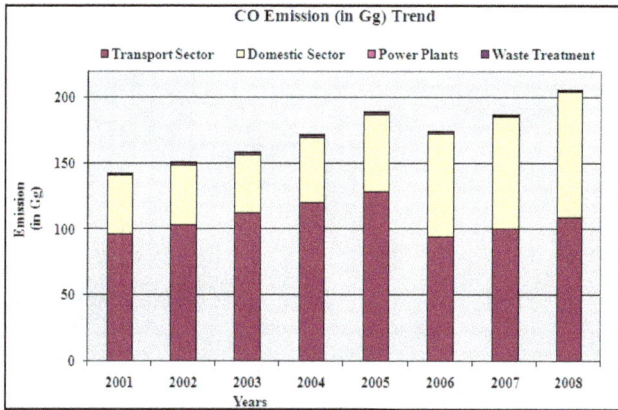

Figure 8. CO Emission Trends, 2001-2008

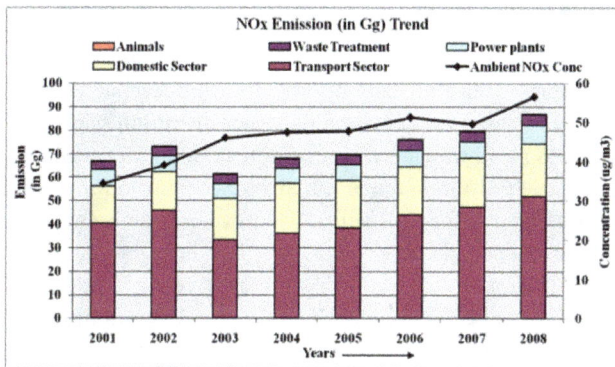

Figure 9. NOx Emission Trends, 2001-2008

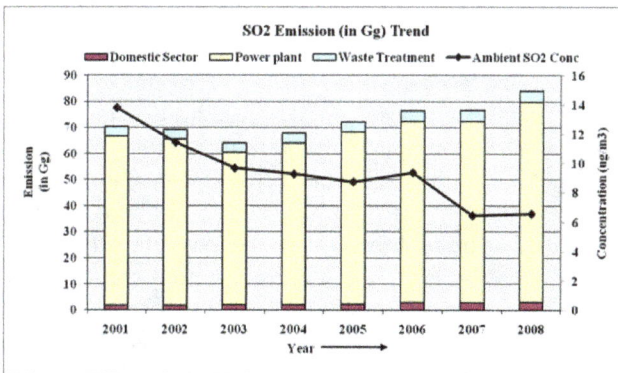

Figure 10. SO$_2$ Emission Trends, 2001-2008

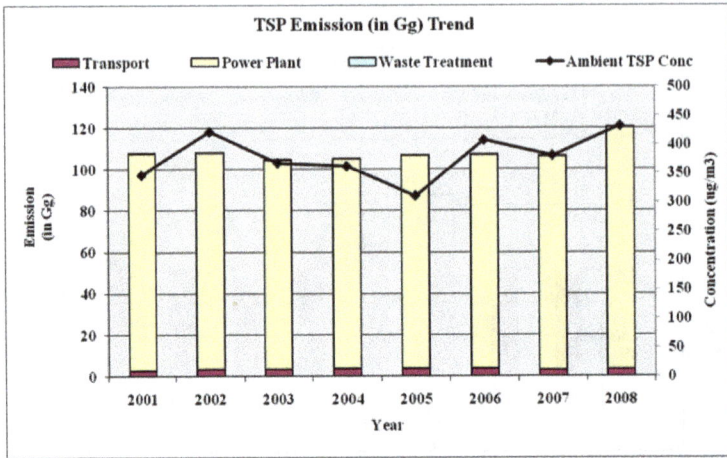

Figure 11. TSP Emission Trends, 2001-2008

5. Emission trends vis-à-vis control options

From time to time different regulatory actions have been initiated in Delhi for control of air pollution scenario. The major initiatives that have been undertaken are phasing out of old commercial vehicles, successive implementation of stringent emission norms, use of CNG and introduction of mass rapid transport means such as Delhi Metro and Bus Rapid Transit. In the period of year 2000-2001 many major initiatives were taken for control of air pollution such as introduction of Bharat Stage II (Euro-II equivalent) emission norms, reduction of Sulphur content in diesel and gasoline, replacement of pre 1990 three wheelers and taxis with new vehicles on CNG, phasing out of more than 8 year old buses and switch over to beneficiated coal in three coal based power plants. Table 2 shows timeline of interventions taken by government for abatement of air pollution [21]

Major problem due to dominance of transport sector

To some extent, the impact of control options is reflected in the emissions scenario. The period of 2001-2003 shows decrease in NOx, SO2 and TSP emissions. However, in the later years 2007-2008, an increasing trend emerges due to rise in number of vehicles on the road. Application of Euro –III emission norms on 2004-2005 has led to a marginal decrease in 2006-2007 emissions. Specially, in the transport sector, often introduction of a policy initiative like CNG, EURO-III norms are followed by decrease in emissions, but the ever increasing vehicles on road again leads to an increase. In terms of concentrations also, introduction of CNG fuel has been observed to be followed by a decrease in SO_2 and CO concentrations, while the NOx level was increased in comparison to those before the implementation of CNG. Further, total suspended particulate matter showed no significant change after the implementation of CNG [26]. Despite the rise of METRO ridership, the number of private and commercial vehicles are also increasing thereby negating the impact of control initiatives.

Year	Steps taken
2000	• Bhar at Stage II (Euro-II equivalent) emission norms introduced for all private vehicles • Sulphur content in diesel and gasoline reduced to 0.05 % • Replacement of all pre 1990 three wheelers and taxis with new vehicles on CNG • Switch over to beneficiated coal in three coal based power plants • Buses more than 8years old phased out or to ply on CNG
2001	• Bharat Stage II (Euro-II equivalent) emission norms introduced for all commercial vehicles • Replacement of all post 1990 three wheelers and taxis with new vehicles on CNG • Increase in number of CNG vehicles • Total 2210 industries closed/relocated between 1998-2001 • Piped natural gas to limited domestic and commercial establishment
2002	• All diesel buses phased out/converted to CNG • Increase in number of CNG vehicles • Increase in supply of Piped natural gas to more domestic and commercial establishment
2002-2003	• First route of METRO becomes operational
2003-2008	• Further increase in number of CNG vehicles
2005	• Euro-III emission norms introduced for all private vehicles, city public service vehicles and city commercial vehicles • Bharat Stage II emission norms introduced for 2/3 wheelers • Sulphur content in gasoline reduced to 0.035 % and 0.015% respectively • Phase I of metro rail project completed
2006	• Work of Bus Rapid Transit project started
2009	• 37 % completion of work on Phase II of metro rail started in 2006 • Construction of two new gas based power plants 330 MW and 1000 MW

Table 2. Regulatory actions taken by government for control of air pollution

CNG and NOx emissions

Ravindra et al [26] analysed impact of introduction of CNG on criteria pollutants by observing annual average ambient concentrations at Bahadur Shah Zafar Marg, one of the monitoring sites of CPCB. The introduction of CNG was followed by decrease in SO_2 and CO levels, however TSP and NOx levels were observed to have an increasing trend. The high CO/NOx ratio and low SO2/NOx ratio at BSZ indicated toward dominance of vehicular emission sources for CO and NOx while dominance of other sources for SO_2. Increasing NOx concentration were explained by [26] in relation to flash point of CNG (540°C) which is higher than that of diesel (232–282 °C). Thus due to higher temperature, combustion chamber of CNG vehicles allows more formation of nitrogen oxides.

Low levels of SO₂ in Delhi

A significant fall in SO₂ concentrations despite increase in emissions can be attributed to the use of refined (low S) coal in power stations, the reduction of the S content in diesel and the shifting/relocation of industries from residential to industrial sites [26].

Biswas et al [27] analysed CO concentrations at three monitoring sites in Delhi and observed that CO does not exhibit noticeable inter-annual variability at any of the three sites, signifying that there is not much variation in impact of emission sources from one year to the next. Table 3 discusses the trends of emissions of certain pollutants.

Pollutant	Years	Trend in emission	Remark	Corresponding Change in ambient concentration
SO₂	2001-2008	Increase	Main contributor is power plants. As the fuel consumption is increasing so the emission based on the fuel consumption in power plants is also increasing.	Decrease due to Improvement in coal quality combined with the fact that thermal power plants emit at a height thereby have lower contribution to surface level concentrations. Introduction of low sulfur diesel is also another cause.
NOx	2001-2002	Increase	The main cause is increasing number of vehicles.	Increasing emission have led to increase in concentrations also.
NOx	2002-2003	Decrease	This is outcome of outcome of decreasing number of taxies, three-wheelers, buses and goods vehicles based on diesel and gasoline which were phased out due to introduction of CNG	Even though fuel emissions have decreased , NOx concentration has increased due to facilitation of more formation of N oxides in high temperature CNG combustion chambers
NOx	2003-2008	Increase	As the number of all the types of vehicles are increasing (transport sector being the major contributor) also the fuel consumption in domestic sector, Power Plant fuel consumption and emission from Waste Treatment is also increasing.	Consistent increase in NOx emission as well as introduction of CNG is leading to increase in ambient NOx concentrations also.

Pollutant	Years	Trend in emission	Remark	Corresponding Change in ambient concentration
TSP	2001-2004	Decrease	Phasing out of diesel vehicles and introduction of CNG vehicles have led to an initial decrease in TSP emission.	Decrease followed by increase in 2001-2002. shows impact of CNG introduction in vehicles and beneficiated coal in power plants.
TSP	2005-2008	Increase	Consistent increase in number of vehicles as well coal consumption together led to an increase in overall TSP emissions.	No certain trend.
CO	2001-2004	Increase	The major source of CO is transport and Domestic sector. As the number of on road vehicles and fuels used in domestic sector is increasing, CO emission shows an increasing trend.	Decrease at traffic intersection [26]. Increase of CNG fuels usage and introduction of BS-II norms in 2001 leads to decrease in ambient levels.
CO	2004-2005	Decrease	The decrease in emission of CO is mainly from transport sector due to the different emission factors used in 2005 taking in view implementation of Euro IV standards.	No certain trend. Increase in number of vehicles counter the effect of EURO-IV norms and CNG introduction.
CO	2005-2008	Increase	The reason for increasing trend is the consistent increase in number of vehicles and domestic fuel consumption.	

Table 3. Emission Trends for year 2001 to 2008

6. Comparison with other inventories

6.1. Local city based data studies

Central Pollution Control Board along with National Environmental Engineering Research Institute prepared an emission inventory for Delhi based on data of collected between year 2006-2007 implying a bottom-up approach [5]. The emission inventory was based on revised USEPA AP-42 emission factors and those formulated by Automotive Research Association of India. Figure 12 displays emissions as estimated from present study with emission factors from various sources as described in section 3, and those using emission factors as per CPCB report [5]. It can be seen that emissions estimated from present study and those based on CPCB emissions are quite comparable. Moreover, the actual estimates of CPCB report emissions (from bottom up approach) are compared in Table 4 with those estimated in present study using a top down approach. Again the estimates from two approaches are of the same order with largest difference being in case of CO. While, bottom up approach incorporates detail sector information in an emission inventory, top down approach estimated are based on many assumptions. This can lead to differences in emission estimates. However, the present study shows the estimates from the two approaches are comparable.

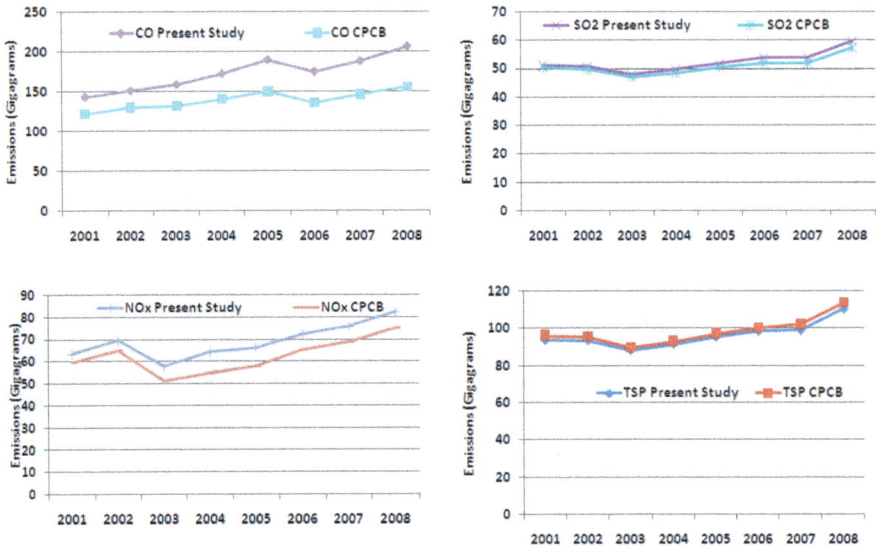

Figure 12. Comparison of emission estimates of some pollutants from present study and CPCB report.

Further, the differences between various emissions are largely attributed to emission factors. Emission factors developed for anthropogenic processes in one region at a certain time may not be applicable for other regions. Kansal et al [28] estimated emissions from power plants, vehicles and industries based on data of period 2005-2006. It was estimated that Vehicular emissions are the major sources of TSP concentrations (54%), followed by Thermal power plants (32%). For SO_2, the major contributors are TPPs (67%) and vehicles (33%). Further, vehicles and TPPs contribute 90% and 10% of NO_2 concentrations, respectively. However in this study transport emissions were considered only around certain monitoring sites of Delhi and emissions from power plant were based on stack monitoring and not on coal consumption which are likely to result into major differences between the two approaches for the power plant sector. The domestic and industry sectors emissions of Kansal et al 2010 are comparable with those of present study. Ramchandra et al, [29] developed a decentralized emission inventories for vehicular transport sector of India for different metropolitan cities based on various data of period 2003-2006. It was estimated that for the Delhi city annually 284.43×10^6, 87.74×10^6, 129.99×10^6, 9.13×10^6 and 42.38×10^6 g/km^2 for CO, HC, NOx, PM and SO_2 are emitted respectively. Table 4 lists some of these period specific studies for different pollutants and their comparison with emissions from present study. Results are comparable with other local studies.

Study	Pollutant-(Sector)	Based on data of Year	Emissions (Gg yr^{-1})	Corresponding emissions estimated in present study
CPCB, 2010 [5]	CO (Transport+Power Plants+Industries+Domestic)	2006-07	136	174
	NOx (Transport+Domestic)		41.7	64.8
	SO$_2$ (Transport+Power Plants+Industries+Domestic)		73.3	76.2
Chhabra et al [30]	CH$_4$ (Livestock)	2003	2.1	2.06
Jalihal et al [31]	TSP (Transport)	2002	4.6	3.3
	NOx (Transport)		40.3	45.9
	CO (Transport)		154	103
Kansal et al,[28]	SO$_2$ (Domestic+Industries)	2005-06	1.9	2.2
	NOx as NO$_2$ (Domestic+Industries		12.0	20.2
Ramchandra et al [29]	CO (Transport)	2003-06	122	101 (Year 2006)
	NOx (Transport)	2003-06	56	47 (Year 2006)
	TSP (Transport)	2003-06	4	3.7 (Year 2006)

Table 4. Comparison of emissions with other local studies

6.2. Global and regional inventories

Global inventories like EDGAR (Emission Database for Global Atmospheric Research) give emissions at 0.1 ° resolution for CO, SO$_2$ and NOx. EDGAR emissions calculated at country level for different sectors relevant to that country. These parameter data are based on evaluation of scientific literature, inventory guidance, inventory reports, industry reports, dataset documentation. Emissions by country and grid are allocated on a spatial grid to provide gridded emissions dataset for atmospheric modeling. the emissions are then spatially allocated using grid maps with 0.1°x0.1° resolution based on data such as location of energy and manufacturing facilities, road networks, shipping routes, human and animal population density and agricultural land use (EDGAR, 2008). Similarly REAS (Regional Emission Inventory in Asia) emissions use country region-specific emission factors for several emission species from subdivided source sectors to estimate emissions on state or and country levels. These emissions are divided into a 0.5 ×0.5 grid by using index databases, i.e. population data; information on the positions of large point sources; land cover data sets; and land area data sets [32]. GAINS-Asia emission inventories are calculated at country and sub-region level for China and South Asia (GAINS-Asia, 2011). Figure 13 compares emissions for CO, NOx and SO$_2$ for the city of Delhi as extracted from these databases for common sectors along with emissions estimated in present study.

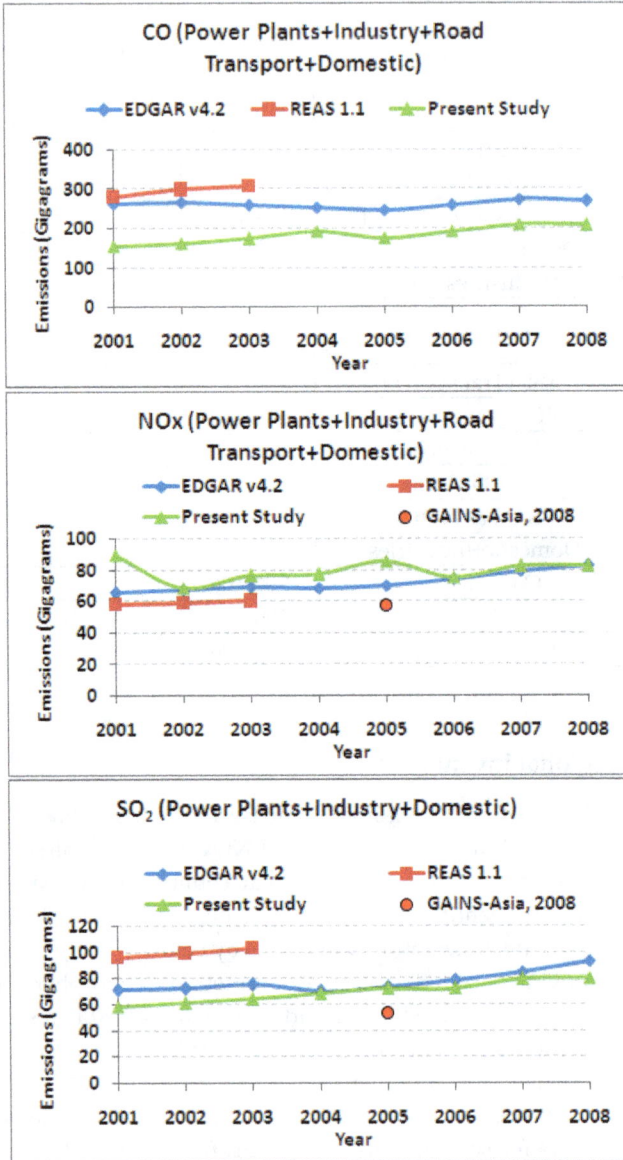

Figure 13. Emission estimates (Gigagrams yr[-1]) for Delhi extracted from different global/regional inventories

Estimates of global inventories range from a resolution of 0.1° to 1°. At these resolutions representation of emissions is usually average of a large region. Hence emission estimates of global inventories may be different when scaled to city level as compared to an emission inventory based on local data sources of a city. Granier et al,[33] compared global and regional inventories like EDGAR, REAS etc for different regions of the world and concluded that the identification of all the reasons for the differences between the inventories is difficult to establish quantitatively. One of the reasons could be that different inventories are updated at different intervals and therefore their respective reference activity data and emission factors could be significantly dissimilar. They observed that specially for India, there are large discrepancies between different inventories. For SO_2, CO and NOx emissions, the differences between global and regional scale inventories were largest for India. In India, the largest emissions were estimated from REAS database. Thus for the city of Delhi also, these features are being reflected in the present study.

7. Air quality analysis

Emission inventories are based on basic assumptions, available statistical data and formulations. Thus there are no standard methods for validation of emission inventories. In such cases, a qualitative validation can be attempted by making use of air quality models. Concentrations output of the models using estimated emissions as input can be compared with observed concentrations. Air quality models range from simple empirical relationships to sophisticated dispersion and chemical models. The latter ones are required for estimation of concentration levels on a fine temporal and spatial scale. However, in some cases, where relevant input data is not available or average concentrations are required, simple urban pollution models have also been found applicable [34]. The present study makes use of a simple fixed box model to estimate concentrations. The box model is applicable here as concentrations are being estimated for the city as whole on an annual average scale as a qualitative validation of emission trends in correspondence to total emissions per year for the entire city. In Delhi, frequent exceedence is observed in case of TSP and NOx. Thus in the present study, these pollutants have been considered for comparison with box model estimations. Figure 14 displays annual trend of observed concentrations of TSP and NOx with those estimated from box model.

Both TSP and NOx concentrations show reasonable trend with estimated concentration. The correlation coefficient for TSP observed and estimated concentration is 0.41 while that for NOx is 0.69. Both the trends and correlation coefficient are reasonable in case of NOx estimates.

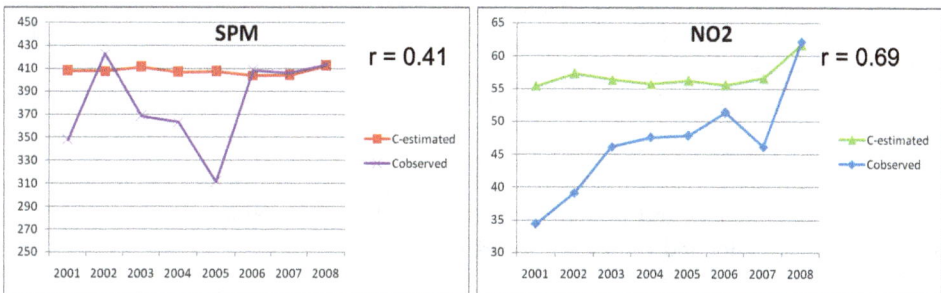

Figure 14. Comparison of TSP and NOx observed concentrations with estimated concentrations.

8. Conclusions and recommendations

An annual inventory for a period of 2001-2008 has been developed for major pollutants for the city of Delhi and compared with several global regional and national level local inventories including the one based on a bottom up approach. The results are comparable to these inventories barring one or two. Hence the methodology adopted here based on top down approach seems to have worked well and is promising in future for this purpose. The advantage here is that data requirements are less in comparison to bottom-up approach.

Emission estimates are for several years during the period when majority of the policy and control measures are implemented in Delhi and therefore the impact of these on emission trends is successfully studied. All criteria pollutants like CO, NOx , SO$_2$ and TSP have shown an overall increase in emissions. The major contributors remain the ever growing activities related to motor vehicle, and fuel consumption corresponding to population rise. High increase in number of vehicles often eliminates the influence of several other control options and requires a major policy intervention to circumvent this cause as also indicated by other studies [35]. This necessitates the augmentation of public transport and accessibility to modes of public transport as the need of the hour. Adoption of alternate fuel in vehicles and pollution free systems like electric vehicles also needs to be promoted. Studies have indicated that with adoption of electric vehicles cars in Delhi, an amount of Rs. 1225.25 crores (Approx. USD 28,16,663) can be reduced, which is now being annually spent on petrol [36]. Further, passage norms of vehicles from other states through Delhi also need to be made more restricted.

Coal based power plants are the major source of TSP and SO$_2$ emissions. However, SO$_2$ levels are under control in Delhi due to better fuel quality implemented through policy interventions. Methane levels from waste treatment in landfills are expected to increase as waste generation is increasing with rise in population. Shift of coal based power plants to natural gas will lead to further decrease in SO$_2$ and TSP emissions. The government is already in process of complete shift to natural gas in all power plants and new power plants are also coming up. Further, due to its high population density and a high percentage of people living in the slums, the extent of usage of cooking fuel in the form of biomass and kerosene is considerably high leading to high level of emissions from domestic sector also. Thus increase in supply of piped natural gas (PNG) and LPG to domestic sector is also recommended especially cleaner fuels are required in unorganized commercial sectors and slum areas.

Number of polluting industries in Delhi is decreasing due to their relocation. However, a large number of remaining industries still needs technology interventions like installation of electrostatic precipitator and venturi-scrubbers [5].

Delhi continues to be one of the most polluted megacities. However, the control measures are also being implemented in timely manner. Control strategies need to be in synchronization with other urban centers around Delhi so as to improve the efficacy of the regulatory measures.

Finally it is recommended that in future gridded emissions are estimated and air quality modeling be performed to study the impact of proposed policy interventions.

Author details

Manju Mohan*, Shweta Bhati and Preeti Gunwani
Centre for Atmospheric Sciences, Indian Institute of Technology, Delhi, India

Pallavi Marappu
Center for Global and Regional Environmental Research, The University of Iowa, Iowa City, USA

Acknowledgement

The authors would like to thank Ms Shraddha Sharma and Ms Medhavi Gupta for assistance in collecting relevant information and compilation for this study.

9. References

[1] Utah DEQ (2012) Emissions Inventory: Definitions. Available: http://www.airquality.utah.gov/Planning/Emission-Inventory/Define.htm.   Accessed 2012 Mar 31.

[2] Lents J, Walsh M, He K, Davis N, Osses M, Tolvett S, Liu H (2012) Estimating Emissions from Sources of Air Pollution. In: Handbook of Air Quality Management. Available: http://www.aqbook.org/read/?page=72. Accessed 2012 Mar 29.

[3] Majumdar D, Gajghate DG (2011) Sectoral CO_2, CH_4, N_2O and SO_2 emissions from fossil fuel consumption in Nagpur City of Central India. Atmos. environ. 45: 4170 – 4179.

[4] Bhanarkar AD, Goyal SK, Sivacoumar R, Rao CVC (2005) Assessment of contribution of SO_2 and NO_2 from different sources in Jamshedpur region, India. Atmos. environ. 39: 7745-7760.

[5] Central Pollution Control Board (2010) Air Quality Monitoring, Emission Inventory & Source Apportionment Studies for Delhi. Available: http://www.cpcb.nic.in/Delhi.pdf. Accessed 2011 Sep 16.

[6] Gurjar BR, van Aardenne JA, Lelieveld J and Mohan M (2004) Emission estimates and trends (1990–2000) for megacity Delhi and implications. Atmospheric Environment. 38: 5663-5681.

[7] Sharma C, Pundir R (2008) Inventory of green house gases and other pollutants from the transport sector: Delhi. Iranian j. environ. health sci. eng. 5: 17-124.

---

* Corresponding Author

[8] Mohan M, Dagar L and Gurjar B (2007) Preparation and Validation of Gridded Emission Inventory of Criteria Air Pollutants and Identification of Emission Hotspots for Megacity Delhi. Environ. Moni.t assesst. 130: 323-339.

[9] Mohan M, Bhati S, Sreenivas A, Marrapu P (2011) Performance Evaluation of AERMOD and ADMS-Urban for Total Suspended Particulate Matter Concentrations in Megacity Delhi. Aerosol and air quality research 11: 883-894.

[10] Mohan M, Bhati S and Rao A (2011) Application of air dispersion modelling for exposure assessment from particulate matter pollution in mega city Delhi. Asia-Pacific Journal of Chemical Engineering. 6: 85-94.
Ministry of Statistics and Programme Implementation (MOSPI) (2010) Number of motor vehicles registered in India. Available:
http://mospi.nic.in/Mospi_New/site/India_Statistics.aspx?status=1&menu_id=14.
Accessed 2011 Aug 19.

[11] Govt. of NCT of Delhi (2009) Delhi Statistical Handbook. Available:
http://www.delhi.gov.in/wps/wcm/connect/doit_des/DES/Our+Services/Statistical+Hand+Book. Accessed 2010 Oct 15.

[12] Talyan V, Dahiya RP, Anand S, Sreekrishnan TR (2007) Quantification of methane emission from municipal solid waste disposal in Delhi. Resour. conserve. recycle. 50: 240-259.

[13] EEA (European Environment Agency), 2001. Joint EMEP/ CORINAIR Atmospheric Emission Inventory Guidebook, third ed. European Environment Agency, Copenhagen.

[14] Foell W, Amann M, Carmichael C, Chadwick M, Hettelingh J, Hordijk L, Dianwu Z (1995) Rains-Asia: an assessment model for air pollution in Asia. Report on the World Bank sponsored project "Acid rain and emission reductions in Asia."

[15] Bouwman AF, Lee DS, Asman WAH, Dentener FJ, van der Hoek KW, Olivier JGJ (1997) A global high resolution emission inventory for ammonia Global Biogeochemical Cycles 11: 561–587.

[16] Reddy MS, Venkataraman C (2002) Inventory of aerosol and sulphur dioxide emissions from India Part II—biomass combustion. Atmos. Environ. 36: 699–712.

[17] Central Pollution Control Board (2001) Transport Fuel Quality for Year 2005. PROBES/78/2000-01.

[18] Sahu SK, Beig G and Parkhi NS (2011) Emissions inventory of anthropogenic PM2.5 and PM10 in Delhi during Commonwealth Games 2010. Atmos. environ. 45: 6180-6190.

[19] Central Electricity Authority (CEA) (2002-2009) Annual Thermal Performance Review. Available http://www.cea.nic.in/report.html. Accessed 2011 Sep 12.

[20] Govt of NCT of Delhi (2010) State of Environment Report for Delhi, 2010. Available http://www.delhi.gov.in. Accessed 2012 Feb 20.

[21] IPCC/OECD/IEA (1997). Revised 1996 IPCC Guidelines for National Greenhouse Inventories. OECD, Paris.

[22] Mohan M, Pathan SK, Narendrareddy K, Kandya A, Pandey S (2011) Dynamics of Urbanization and its Impact on Land-Use/Land-Cover: A Case Study, of Megacity Delhi. J. environ. prot. doi:10.4236/jep.2011.29147.

[23] Chakraborty M, Sharma C, Pandey J, Singh N and Gupta PK (2011) Methane emission estimation from landfills in Delhi: A comparative assessment of different methodologies. Atmos. environ. 45: 7135-7142.

[24] Mor S, Ravindra K, De Visscher A, Dahiya RP and Chandra A (2006) Municipal solid waste characterization and its assessment for potential methane generation: A case study. Sci. total environ. 371: 1-10.

[25] Ravindra K, WautersE, Tyagi SK, Mor S (2006) Assessment of air quality after the implementation of compressed natural gas (CNG) as fuel in public transport in Delhi, India. Environ. monit. assess. 115: 405–417.

[26] Biswas J, Upadhyay E, Nayak M, Yadav AK (2011) An Analysis of Ambient Air Quality Conditions over Delhi, India from 2004 to 2009. Atmos. climate sci. 1: 214-224.

[27] Kansal A, Khare M, Sharma CS (2011) Air quality modelling study to analyse the impact of the World Bank emission guidelines for thermal power plants in Delhi . Atmos. pol. res. 2: 99-105.

[28] Ramachandra TV and Shwetmala (2009) Emissions from India's transport sector: Statewise synthesis. Atmos. environ. 43: 5510-5517

[29] Chhabra A, Manjunath KR, Panigrahy S, Parihar JS (2009) Spatial pattern of methane emissions from Indian livestock. Current Science. 96: 683-689.

[30] Jalihal SA and Reddy TS (2006) Assessment of the Impact of Improvement Measures on Air Quality: Case Study of Delhi. Journal of Transportation Engineering. 132: 482-488. Ahmad I and Dewan KK (2007) Electric vehicle: a futuristic approach to reduce pollution (A case study of Delhi). World Review of Intermodal Transportation Research. 1: 300-312

[31] Ohara T, Akimoto H, Kurokawa J, Horii N, Yamaji K, Yan X, et al (2007) An Asian emission inventory of anthropogenic emission sources for the period 1980-2020. Atmos. Chem. Phys. 7: 4419-4444.

[32] Granier C, Bessagnet B, Bond T, D'Angiola A, Denier van der Gon H, Frost G, et al (2011) Evolution of anthropogenic and biomass burning emissions of air pollutants at global and regional scales during the 1980–2010 period. Climatic Change. 109: 163-190.

[33] Venegas LE, Mazzeo N.A., Rojas A.L.P. (2011) Evaluation of an Emission Inventory and Air Pollution in the Metropolitan Area of Buenos Aires. Air Quality-Models and Applications, Nicolas Mazzeo (Ed.), ISBN: 978-953-307-307-1, InTech.

[34] Mohan M, Kandya A (2007) An Analysis of the Annual and Seasonal Trends of Air Quality Index of Delhi. Environ. monit. assess. 131: 267-277.

[35] Ahmad I and Dewan KK (2007) Electric vehicle: a futuristic approach to reduce pollution (A case study of Delhi). World Review of Intermodal Transportation Research. 1: 300-312.

Chapter 8

Microscopy and Spectroscopy Analysis of Mems Corrosion Used in the Electronics Industry of the Baja California Region, Mexico

Gustavo Lopez Badilla, Benjamin Valdez Salas,
Michael Schorr Wiener and Carlos Raúl Navarro González

Additional information is available at the end of the chapter

1. Introduction

MEMS are very important devices used in the electronic industry of the northwest of Mexico where are located Mexicali as arid zone and Tijuana as marine region. In boths enviroments the humidty and temperature are factors which have influence in the operation of these microcomponents. MEMS are utilized principally in electronic systems as acoustic, audio and video, communication, industrial and medical, as a control and operation activities. In the industrial plants are primordial in the manufacturing process to improve the production yielding, with automatized systems (B.G. Lopez et al, 2007). These microdevices have advantages as a lot operation and, need small spaces in the electronic boards, but microcorrosion appears very easy and quiclky in short spaces between electrical connections. At RH levels higher than 80%, an electrochemical process ocurr in the metallic surfaces of connectors of MEMS, and its arevery difficult detect it, to the naked eye. For this reason MEMS presents deterioration in the electrical connections, and originates electrical failures, and it is not detected until are checked in the last steps of the manufacturing process or when buyers are used, causing economic losses. In base of this, an analysis was made to know the behaviour of corrosion phenomena in the MEMS used in the development of new technologies in the electronics, mechanical and electromechanical systems (Tanner et al, 2000).

2. History of MEMS

MEMS technologies have evolved to develop miniaturize complex systems integrating multiple functions in small or simple packages. In Europe these microsystems are known as MST or "Micro Technology Systems", however the term of MEMS has become increasingly prevalent. Helvajian (Helvajian, 1999) and Vittorio (Vittorio, 2001) are agree that MEMS manufacturing, emerged with the development of novel semiconductors in the late of XXI century. The first was manufactured semiconductor transistor at Bell Laboratories by Shockley, Bardeen and Brattain in 1947 (Lopez B. G. et al, 2011). This led to an unparalleled development in the technology semiconductors, which subsequently creation of electronic systems increasingly faster, smaller and less expensive manufacturing. However, this same manufacturing process created a vacuum so fast in knowledge of their operations (Hindrichsen t al, 2010). This was because more and smaller circuits were manufactured with process information faster, and the interfaces of these circuits, sensors and actuators, could not easily attached to their functions in their applications. Moreover, in a principle of the manufacture of MEMS not take into account their protection against corrosion. The efficiency of MEMS is a major motivationafter development, because these micro devices proved to be more rapid, inexpensive and efficient their macroscopic devices. However the development of such solutions has been envelope by the technological limitations (Plass etal, 2003). The first research in the area of MEMS was oriented to the micro sensors obtained as a result of these investigations: the discovery of piezoelectricity. This feature present in the silicon and germanium, allows the development of micro silicon pressure sensor. According to Vittorio, the sensor silicon pressure was the first of the sensors and micro more successful. Additionally the availability of silicon as the material raw encouraged the development of micro techniques semiconductor whose range has grown to include measurements physical, chemical and biomedical. Table 1 shows a list of some of the most significant findings in the evolution of the MEMS. Among them is in the development of solid state transducers and micro sensors with micro machining techniques, producing micro actuators that can lead to the appearance of the first mechanisms and engines microscopic level (Zawada et al, 2010, Jin, 2002).

Historial development of MEMS	Period, Year
Anisotropic silicon device with attack	Season of 1940
Piezoresistive silicon component	1953
Semiconductor extensometers	1957
Silicon pressure sensors	Season of 1960
Solid state transducers	Season of 1970
Microactuators	Season of 1980
Micro mechanisms and micro motors	From 1987 to 1989
Micro Electromechanical Systems	After 1988

Table 1. History of manufacturing process of MEMS

The term MEMS emerged in 1987 as part of a series of workshops held in Salt Lake City, Utah, Hyannis, Massachusetts and in 1988 in Princeton, New Jersey (Pedersen et al, 2007). These workshops were the forerunners in the development and adoption of this technology. MEMS are currently considered as an interdisciplinary field of knowledge that makes use of many areas of science and engineering to solve problems. Helvajian, considers some advantages of MEMS technology on macroscopic aspects that determine its development today: definition of small geometries, precise dimensional control, design flexibility and low cost and interface The term MEMS emerged in 1987 as part of a series of workshops held in Salt Lake City, Utah, Hyannis,Massachusetts and in 1988 in Princeton, New Jersey (Olson III et al, 2007). These workshops were the forerunners in the development and adoption of this technology. MEMS are currently considered as an interdisciplinary field of knowledge that makes use of many areas of science and engineering to solve problems. Helvajian (1999) considers some advantages of MEMS technology on macroscopic aspects that determine its development today: definition of small geometries, precise dimensional control, design flexibility and low cost, interface with control electronicsand easy manufacturing procceses (Christian C et al, 2010, Kihira, 2005).

3. MEMS in the industry in the northwest of Mexico

Actually the industries need more efficiency in their manufacturing processes using MEMS for the diversity of activities and the small spaces in the electronic boards. In the northwest of Mexico, the majorly of the indusrial plants manufacture electronic equipments and have a lot electronic systems with MEMS in their production operations (Lopez-Badilla Gustavo et al, 2011). This study has relevance today because it is a developing topic that improves the efficiency of the products. It is important in the societyto be applied in various daily activities to ensure the reliability of the products manufactured.

4. Fabrication of MEMS

The evolution of MEMS is closely related to advances of semiconductor industry, in many process steps are similar, or identical, and adapted to the requirements (Roger, 2010). The principal features of the MEMS manufacturing processes are:

- *Miniaturization.* The reduction in the size to be smaller and lighter with shorter response times.
- *Multiplicity.* The capacity to produce tens, hundreds or even thousands of products in parallel, being inherited characteristic of the semiconductor production processes.
- *Microelectronics.* The intelligence of MEMS which allows control implemented as closed systems with integrated microsensors and microactuators.

Theses advantages shows the great influence of the manufacturing processes of the electronics integrated circuits. However, it is important to note that not all microdevices have a benefit of miniaturization (Baldwin, 2009). One of the major limitations of the techniques inherit from semiconductor processes, is the works in planar scale, difficult the design of devices in three dimensions (Figure 1).

5. MEMS in automatized control systems

An automatized control system (ACS) is equipment which operates with a lot functions to control industrial machines. An ACS is an interconection of elements relationed to manipulate activities and control the behaviour of other devices which not has control as devices of high power to work for a purpose to control the inputs nd outputs of manufacturing processes (Lopez B. G., 2008, Cai, 2005). In ACS are used MEMS to a lot functios to check, adjust and operate with the industrial regulations of each products, with a feedback equipments to regulate the variations of the output which are signal of the inputs of the activities. An ACS may be open or closed depending on if it makes a feedback system, using the same system output as input to make decisions based on past states of the system. For example, an open system may be one that receives the amount of light of the environment and this intensity can be controlled in any objet, but uncontrolled of the source (Bateson, 1999). By other way, the closed systems can control which works regulating the effect in the objet and the source. It makes the automation as group of technologies that are used in different areas of knowledge as electronic, mechanical and electromechanical mahines, controled by computer systems, for a more autonomous or independent functions. The benefits of this are flexibility, lower cost, higher capacity and quality. An ACS with MEMS is an elemental sensor that performs the function of measuring the manufacturing process, while active an actuator which implements the control action on the industrial operations, and change the behavior of the system accodring the adequate regulations (Beeby, 2004, Chongchen, 2003). An ACS performs the actions of process control, manipulates the actuators and bases its decision on information received by the sensors. In addition MEMS are used in real life control systems to perform tasks that require powerful output, meaning that it is easier for robust systems. A diagram of control system, below showed that exprese the importance of sensors and actuators in a control system (Figure 2).

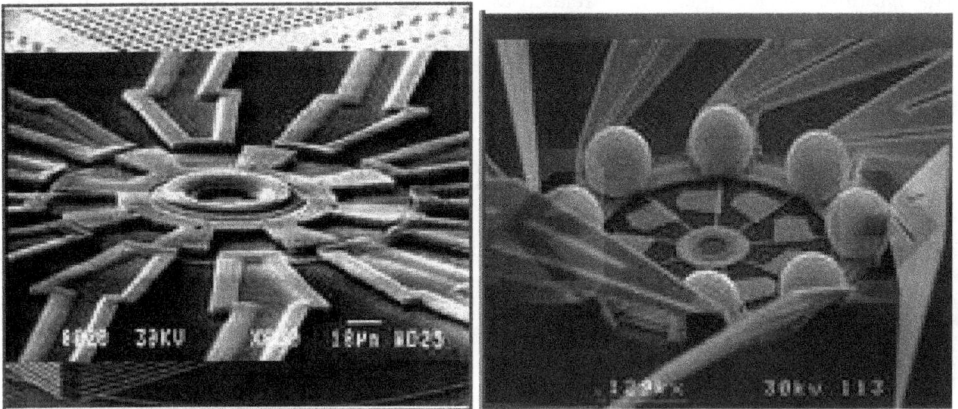

Figure 1. Microdiagram of MEMS utilized in industrial operations

The minimized ACS with MEMS in the last twenty years revolutionated the technologies, by the diminute spaces and the great quantity of funtions in the microelectronic boards, but its necessary a specific and adequate design to avoid the acummulation of visible and invisible water by the condensation of humidity and the variations of temperature. It originates the corrosion phenomena and deteriorates rapidly the metallic surfaces principally of copper. This generates electrical failures in the indusrial equipments and products manufactured and reduce the productive yielding (Yao, 2000). MEMS for their mechanical properties can receive physical and chemical agents from the environment, which deteriorate thier electrical pins and are complicated, analyze it.Based on this, a study was made to determine the principal physicochemical agents which react with the metals, determine the types of corrosion and propose methods to reduce or avoid the corrosion process (Tuite, 2009).

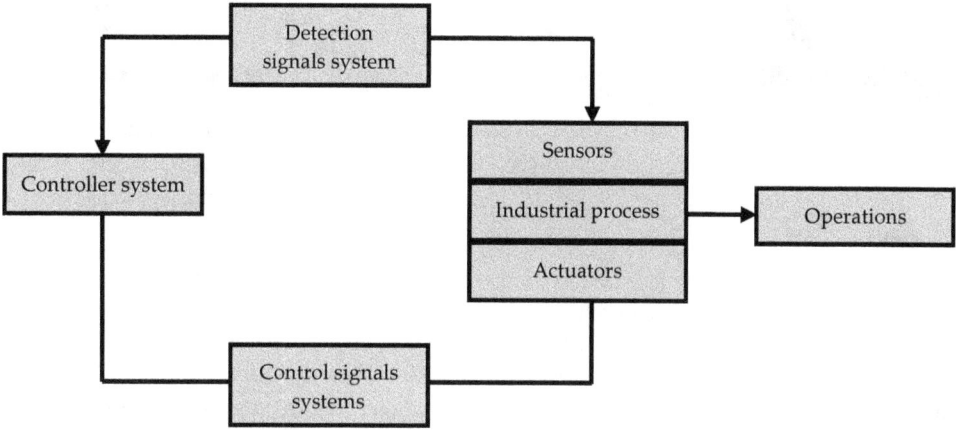

Figure 2. Diagram of a control system used in industrial plants in the Baja California, Mexico

6. Innovative applications

Many people dont know about MEMS technology that occupe a place in our lives, such as the use of cellular phones, which has a microsystem with MEMS. Other examples is the airbag of cars that have a MEMS equipment, which determine the exact time that a collision occurs and triggers the air bag states (Nise et al, 2004, Stanley, 2006). MEMS are micromachines that will change the world, and are currently used in different activities or areas such as industrial, medical, automotive and related technologies including in the electronics systems. Below is explained a brief description of MEMS in each of the areas being developed (Wojciechowski, 2007, Virmani, 2006).

6.1. Medicine

In the medical field, MEMS are implanted in body parts such as the heart, brain and other parts that are difficult to diagnose or study (Roger, 2010), and their implementation are used to prevent diseases like cancer,cardiovascular diseases, lupus and others. Today many companies are developing devices with this technology, for monitoring patients with heart and cancerigen conditions, and also utilized as a prototype chip to test for the presence of substances such as viruses or diseases like flu. Actually are used as microdevices to screen blood, inject drugs or doses of medicine to improve the heath of people (Figure 3).

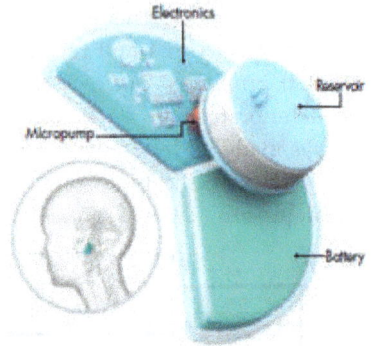

Figure 3. MEMS used in the medical applications: (a) microelectronic circuit and (b) micromedical equipment.

6.2. Automotive

In the automotive area, the electronic systems with MEMS are an important key to the smart cars with microsensors, with different functions for security in the operations of cars (Figure 4). In the automotive applications, MEMS are used to analyze and respond to a variety of mechanical and electrical actuators (Smart et al, 2007). Successful applications of MEMS are presented in the automotive market as accelerometers bags air and micro-lenses to projectors light of cars and the future applications in tihs area will reduce costs and improve the performance of the devices.

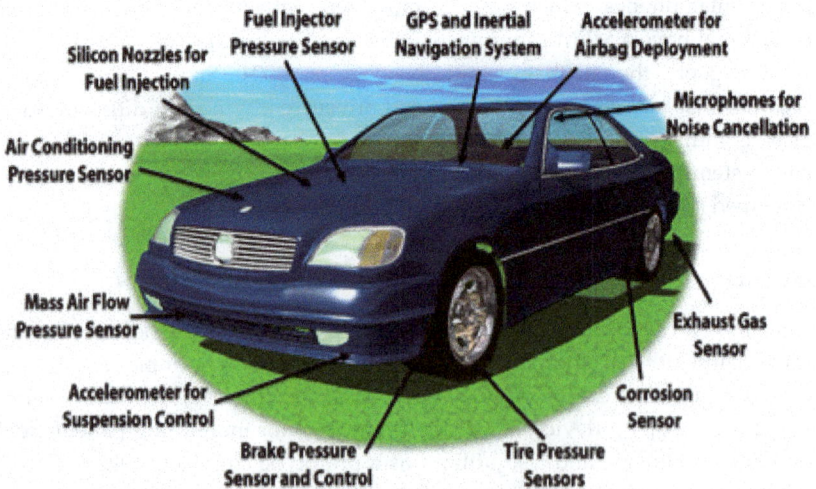

Figure 4. MEMS used in the automotive applications with lot applications

Microscopy and Spectroscopy Analysis of Mems Corrosion Used in the Electronics Industry of the Baja California
Region, Mexico

149

6.3. Industrial

In the industrial area, MEMS generates more efficient heating systems, air conditioners, refrigerators, freezers, principally, to integrate these equipments, being an intelligent system, to monitoring their functions of climatic parameters as humidity, temperature levels, and pollution agents that react with metallic surfaces of systems in the robust industrial machines (Figure 5). The MEMS are used to a great variety of industries, as military (principally to embed in the body to prevent diseases) and aerospace industries (principally to monitore air pressure and vibration process in the fligths (Laurent, 2008).

Figure 5. MEMS used in the industrial operations.

6.4. Acoustic

MEMS promises revolutionize products allowing complete systems on a chip, and be an integration of mechanical elements, sensors, actuators, and electronics on a common silicon substrate through microfabrication technology. Using MEMS cell phones have less power consumption, greater flexibility, lower costs and better features. Other application of MEMS is the replacement of electromagnetic condenser microphone for silicon microphones. Motion capture is a vital application for accelerometers (utilized as acceleration sensors for machine interface / human) and gyroscopes (used for image stabilization by the image sensor), applied in air bagas of cars (Tuite, 2009). A more recent use of these systems are applied in the Iphone and Ipod Touch from Apple and game consoles like the Wii (Figure 6).

7. MEMS in micro and nano electronic systems

MEMS is the branch of micro electronic systems that operates at low level of electrical signals in any type of activities, where scientifics of this area are working every day to improve thier functions. At this level of scale analyzes with MEMS are very important for the amount of information that can be obtained, as the eye can not observe. Due to the large surface to volume ratio of the MEMS, surface effects such as electrostatic and viscosity effects dominate such as volume or mass thermal inertia (Van Spengen, 2003, Cole, 2004). The finite element analysis is an important part of the MEMS design. Sensor technology has made significant progress due to MEMS. The complexity and advanced performance of MEMS sensors has evolved with the different generations of MEMS sensors. The potential of very small machines was appreciated long before there was technology that could develop with experiments of scientifcs in the electronics area, which were presented as electronic devices with a very little space. MEMS became practical once they could be manufactured using modification of semiconductor manufacturing technologies, normally used in electronics (Maluf et al, 2004). These include molding and plating, wet and dry etching, electro discharge machining and other technologies capable of manufacturing very small devices (Gerhardus, 2006). There are different sizes of companies with strong MEMS programs. Larger companies specialized in the manufacture of components or high volume at low cost, manufacture MEMS to principally of automotive, biomedical, aerospatial, communications,and indusrial operations. The success of small businesses is to provide value in innovative solutions and absorb the cost of manufacture with high margins of sales.One of biggest problems is the absence of autonomous MEMS with micro power sources of high current density, power and electricity.Advances in the field of semiconductors are resulting three-dimensional integrated circuits with features and even moving parts (Ohring, 2003) (Ohring, 2003) (Ohring, 2003).

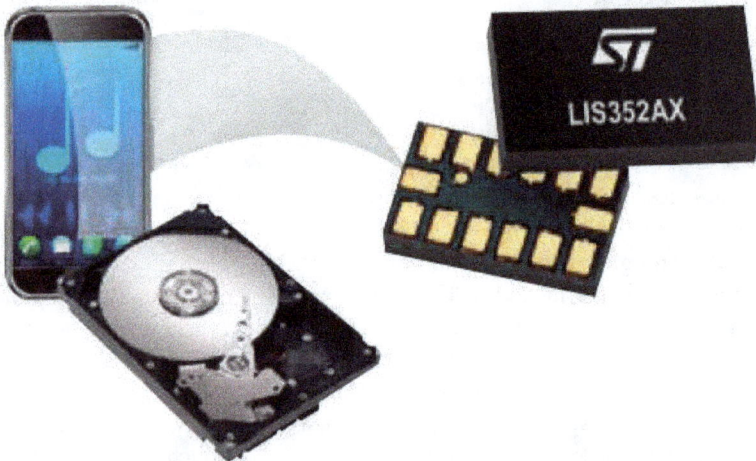

Figure 6. MEMS used in acoustic systems as cellular phones.

MEMS can solve many problems that a microprocessor with software or configuration as ASIC (Application Specific Integrated Chip) can't develop. MEMS technology can be applied using a number of different materials and manufacturing techniques, depending of the type of device that is being created and the commercial sector which has to operate (Zawada, 2010). Researchers in MEMS use various software engineering tools to their design

Microscopy and Spectroscopy Analysis of Mems Corrosion Used in the Electronics Industry of the Baja California Region, Mexico

151

with simulation, prototyping and testing operations. The finite element analysis is an important part in the design of MEMS. Dynamic simulation of heat and electricity, mechanical properties can be performed by specialized softwares as ANSYS and COMSOL. Other software, such as MEMS-PRO, is used to produce a composition suitable design for be easy processes of manufacturing of electronic devices with MEMS (Figure 7) used in indusrial plants. After the process of software activities, the prototypes are ready to be tested by the researchers using various tools, as scanning laser Doppler vibrometers, microscopes, and strobes (Tuite, 2009).

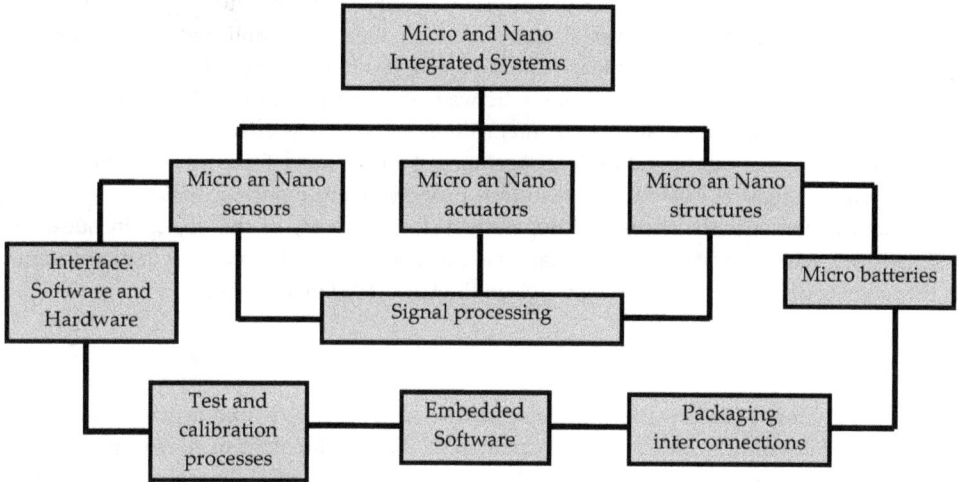

Figure 7. MEMS used in acoustic systems as cellular phones.

8. Operations of MEMS

The miniaturization of electromechanical machines or MEMS is a reality today. Indeed, these microdevices are already used for the performance of accelerometers, found in airbags in cars to determine the right time when a collision occurs and shoot like the inflation mechanism of the bags. This same type of MEMS is used as navigation elements, particularly in the aerospace industry, but also provides applications such as pressure, temperature and humidity (Lope-Badilla Gustavo et al, 2011). They have been incorporated into pacemaker to sense the patient's physical activity and change your heart rhythm. To prevent forgery of signatures has been designed to incorporate these accelerometers in pens. Thus, not only be registered owners of the firm stroke but also the velocities and accelerations imparted to the hand to the pen while signing, which would make it much more difficult to counterfeit. Also MEMS are used in heads of ink jet printers, producing the controlled evaporation of the ink at the right moment, thanks to the localized delivery of heat. Besides the advantage of size of these devices is the fact that they can be manufactured by the thousands lowering manufacturing cost significantly (Madou, 2002). MEMS, like any new technology, have had a major impact when it comes to promoting access to new scientific knowledge. This is the case called adaptive optics. The light from astronomical objects coming to ground-based telescopes necessarily passes through the atmosphere, changing its optical path by changes in air density and temperature. The result is a blurred image with poor angular resolution. To avoid this problem, an expensive solution is to place telescopes in space (such as Hubble). Another interesting and least expensive solution for its ability to use large telescopes, not limited by the dimensions that can be handled in space transport, is what brought the development of mirrors whose surface is deformed by MEMS, correcting distortions produced by the atmosphere land (Tanner et al, 2000). Another scientific application of MEMS was the realization of instruments for measuring forces between two objects whose surfaces are at submicron distances (<1um). One objective was to highlight possible deviations from the law of universal gravitation on the law established by Newton, as predicted by some theoretical models. According to these models, these deviations could be made more evident the smaller the distance between the objects (Vittorio, 2001). The problem is that at short distances are also other interactions, such as that arising from the so-called Casimir effect. This effect, linked to the appearance of an attractive force between conducting objects, whose origin is related to a quantum property (zero-point oscillations), is mainly manifested nanometric distances and depends on the geometry of the objects in question. MEMS have provided the tools to evaluate these forces and corroborate laws and deviations in the range of distances explored so far (Olson III et al, 2007, Lewis, 2004).

9. Corrosion in MEMS

From its beginnings MEMS was an emerging technology and is very used actually. The use of MEMS in the electronics inustry has contributed to the development of new tecnologies to improve the microelectronic devices (MED) (Schoedrer A. et al, 2003). In the last decade, has increased its use in MED in any operations. Exposure of MEMS to aggressive environments, decrease its operating yielding (Lopez-Badilla Gustavo et al, 201) for the presence of corrosion, for the chemical and electrochemical processes. For this reason, scientifics were analyzed new designs and materials to avoid the generation of corrosion. Is necessary the use of specialized packaging in the manufacturing processes for mantain optimal functions (Yao, 2000, L. Veleva, 2008). To technical and manager people, corrosion in MEMS is very concerned factor because its working in harsh or corrosive enviroments in the packaging processes of MEMS and its functionability in a lot operations (Ashrae, 1999). There are some parameters as variations of humidity and temperature, bad design with small and curve spaces and at the exposition of air pollutants in indoor of industrial plants acelerates the kinetics of the chemical reactions of corrosion proccess. Sometimes hermetecally sealed packages contain drops of watter vapour or any aggresive acids, which increase the corrosion rate (CR) and deteriorate very fast the electrical connections of MEMS. The most common packages of MEMS are plastic, and when these are not well sealed, water penetrates very slowly to the internal parts of these microdevices (Wojciechowski, 2007, Abdulaziz, 2003).

10. Methods and materials

In the northwest of the Mexico ountry, in the state of Baja California, are located around 400 electronics industries in Mexicali (AMM, 2011) as arid zone where predominate sulfurs

(ASTM, 2004) as air pollutants and Tijuana (AMT , 2011) and Ensenada (AME, 2011) as marine regions with sulfurs and chlorides (ASTM G140, 2008) corrosive agents. This research makes an evaluation of the operatig of MEMS which are of importance for the scientific techniques used for analysis, manufacturing and electrical tests. The industrial plants intestered in this study required specialized techniques and methods analysis of MEMS (Robertson et al, 2005). This study is of great importance for the electronics industry of the northwest of Mexico and other environments similar of this region, where the influence of climatic factors as relative humidity (RH) and air pollutants as sulfurs and chlorides are present for long periods of time and at concentrations higher than the air quality standars (ASTM G 4, 2008). This is important because MEMS are very efficient systems that generate a lower cost in production and more reliability (Lopez, 2008). The application of MEMS is an important issue in the electronics industry of the orhwest of Mexico, which is available for use in the improvement and optimization of systems (Lopez et al, 2010).The sudy focuses on analyzing the factors that generate corrosion in MEMS with the Scanning Electron Microscopy (SEM) method and the Auger Electron Scanning (AES) technique. These techniques were applied to determine the agents which react with the copper surfaces of the electrical conexions and connectors and propose the design of new methods of manufacture of micro devices with MEMS and the use of new materials, to avoid or reduce the generation of corrosion (Lopez et al, 2010, Leidecker, 2006). This study was made in three steps:

1. Evaluation of electrical failures of MEMS in industrial plants of video games in Mexicali, cellular phones in Tijuana and portatile computers in Ensenada.
2. Micro and nano analysis of surfaces ith the SEM and AES techniques.
3. Developed of an electronic system to control humidity and temperature and make periodic monitorings to detect very fast principally the high concentration levels of sulfurs and chlorides.

11. Evaluation of electrical failures

The measuring of electrical signals is important in the evaluation of productive yielding of MEMS, and its process consists in the analysis of the period of oscillation, whcih is the time for one complete oscillation and its unit is second. The frequency is the number of cycles, oscillations or vibrations in the unit time and its unit is the Hertz or cycles / sec. The amplitude of oscillation is the maximum elongation or detachment of the particles with respect to its middle position and due to the oscillation introduced. The wavelength is space for the propagation speed v for a time corresponding to the period T, using measurements of length. The propagation velocity is that which reaches the oscillatory state at various points on the source of the disturbance. The amplitude of the oscillations depends on the power used to produce the initial disturbance (Figure 8).

12. Micro and nano analysis of surfaces

SEM analysis represents the evaluation of surfaces with microphotographies at surface level indicating the grade of deterioration of the internal and external electrical conexiones and connectors of MEMS. This shows the pathways damaged and the agents that react with the metallic surfaces added and blocked the eletrical conductiviy (ISO 9223, 1992, ISO 11844-1, 2006, ISO 1844-2, 2005). AES technique analyze at surface level and in of this, representing the chemical agents that reacted with the copper surfaces and the dimension of thin layers formed in different areas of the electrical conexions of connectors of MEMS (Figure 9).

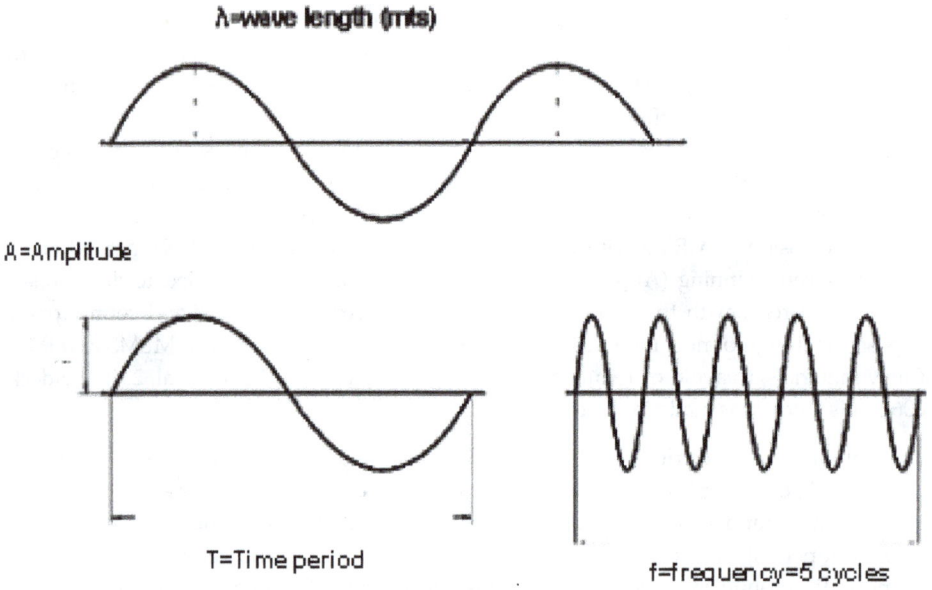

Figure 8. Electrical signal of MEMS with wave length and different periods of time.

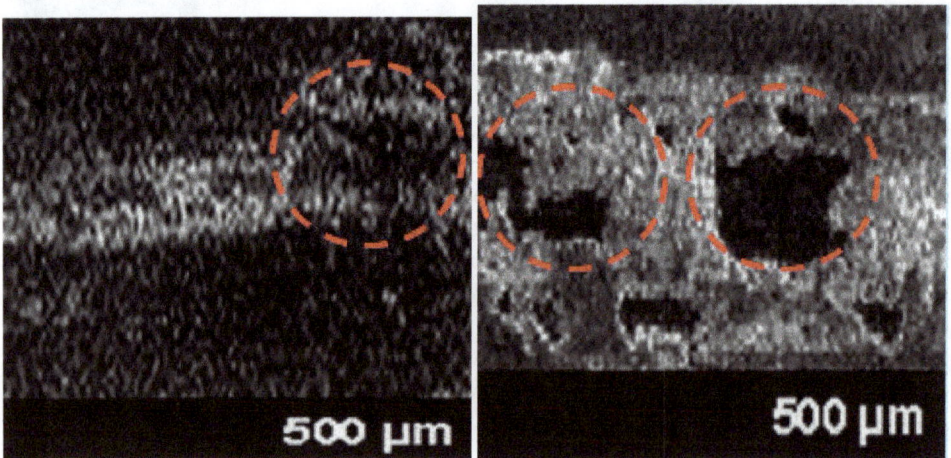

Figure 9. Electrical conexions and connectors deteriored in the electronics industry of the northwest of Mexico.

13. Electronic system to detect and control the atmospheric corrosion

The design, fabrication and use of an electronic system to mantain the humidity and temperature at 70% and 35ºC , which are the standard levels to reduce the corrosion process, was made. This equipment is very cheap and easy to apply in any place of the industrial plants evaluated. Its system detects small variations of values of RH and temperature and when any aggresive agent reacts with the metallic surfaces of conexions and connectos of MEMS (Figure 10). The fisrt step is the power supply that gets electrical energy to the system (Lopez B. Gustavo, 2011). Next are the electronic devices which detect the variations of humidity, temperaure and the electrical current to determine if any chemical agent was reacting with the copper surfaces. Also is the indication step to shows the good or bad function of the control system and in the last step is the operation that active a power electrical actuator as electrical fan or other mechanism to maintain the values of climati and electrical currect factors (Lopez B. Gustavo, 2010).

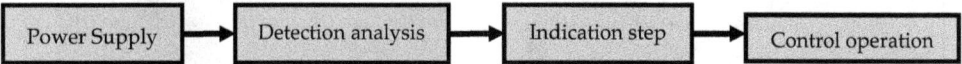

| Power Supply | → | Detection analysis | → | Indication step | → | Control operation |

Figure 10. Steps of electronic and automatic control system.

14. Results

The deterioration of MEMS used in the electronics industry was affeting to their operatibility and increasing the electrical failures at around 35%, representing the major factor in the download of manufacturing areas of the electronic industry. The electrochemical process was accelerated in winter very fast than in summer by the condesantion of humidity and low temperatures. There were presented differen types and procceses of deterioration: uniform, piiting and crevice corrosion. A small and big tarishings were presented in the electrical conexions and conectors of the industrial electronics equipments that were represented in the figures 11 to 13 at 500 X of the SEM evaluation and figures 14 to 16 at 5 X in the same technique (Van Ingelgem, 2007). The Auger analysis showed in combination with the SEM technique (G. L. Badilla et al, 2011), the interested zones of the metallic surfarces to determine the chemical agents which reacted with the copper surfaces (figures 17 to 19). As the theory considerations mention, the sulfurs in Mexicali as arid zone was the higher air pollutant that affected increase the corrosion rate and the chlorides in Tijuana and Ensenada as marine regions. Also the traffice vehicles and chemical agents to verted in water of cities were other pollutants to contribuite to the air pollution and deterioration of metals. Figures 11 to 13 in both pictures showed tarnishings that are an obstruction to the electrical pathways of MEMS. In Mexicali was represented the bigger tarnishing, in Tijuana some pathwys affected and in Ensenada less intensity.

At 5X level of microscopy the same analysis as figures 11 to 13, were presented the figures 14 to 16 with better view as its is showed and with a better descroption of the deterioration. In figure 14 that shows the ocurred in Mexicali, an electrical pathway was deteriored and braked. The same deterioration process was in Tijuana and Ensenada but without the brake of the pathays. With this microanalysis the description of the negative effect of the air pollutants mentioned above and the drastic variations of humidity and temperature were the principal parameters involved in the corrosion processes in each type of corrosion to determine the methods of control to avoid and reduce this phenomenon.

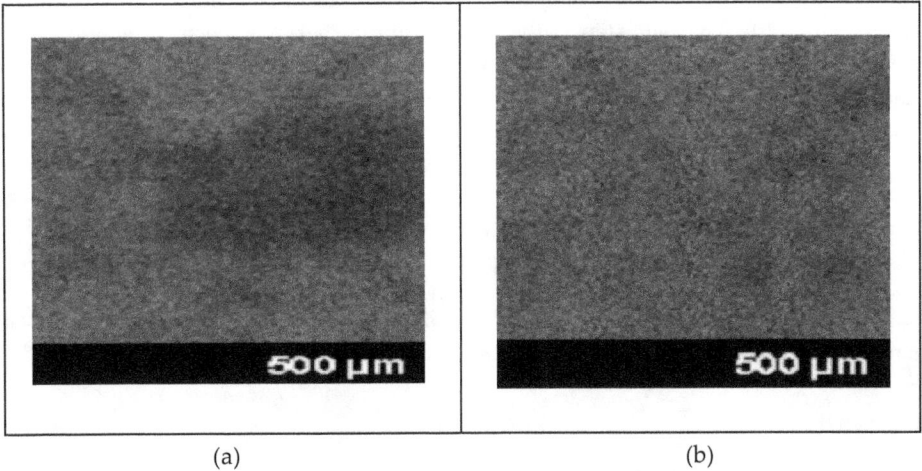

(a)   (b)

Figure 11. SEM 500X microphotographies of pathways damaged by the atmospheric corrosion in (a) summer.and (b) winter in Mexicali (2010).

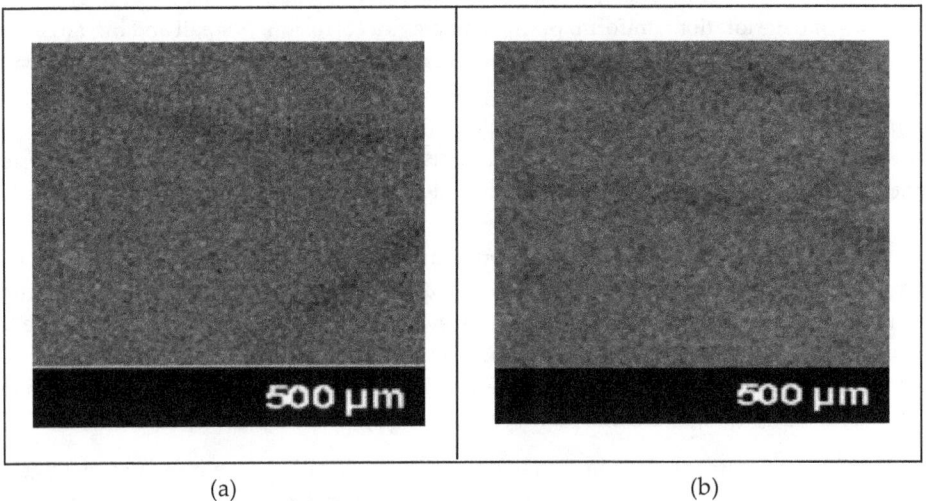

(a)   (b)

Figure 12. SEM 500X microphotographies of pathways damaged by the atmospheric corrosion in (a) summer.and (b) winter in Tijuana (2010).

Microscopy and Spectroscopy Analysis of Mems Corrosion Used in the Electronics Industry of the Baja California Region, Mexico

157

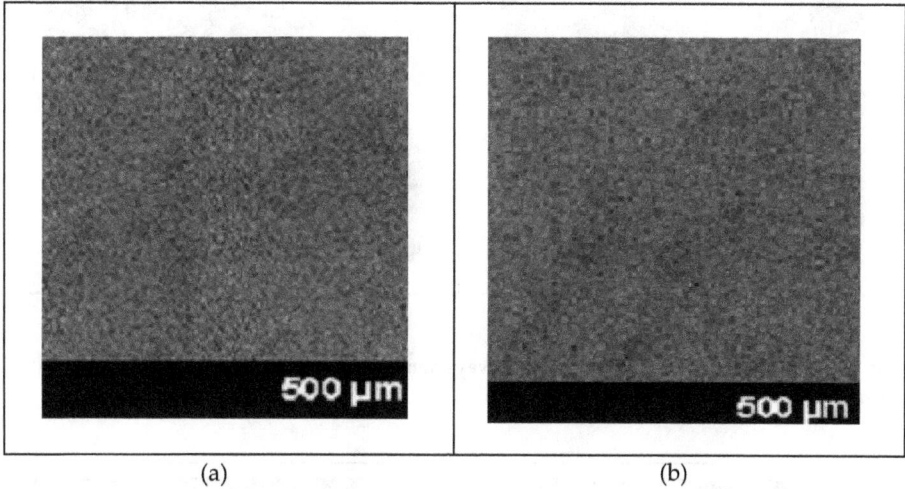

(a)   (b)

Figure 13. SEM 500X microphotographies of pathways damaged by the atmospheric corrosion in (a) summer.and (b) winter in Ensenada (2010).

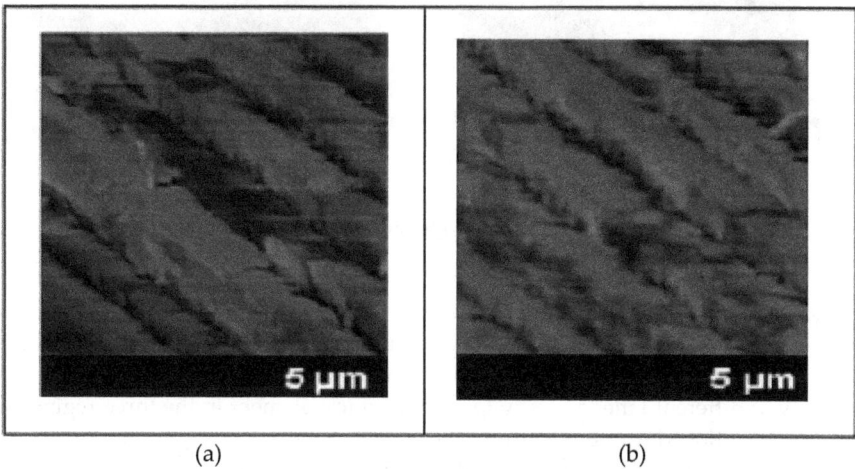

(a)   (b)

Figure 14. SEM 5X microphotographies of pathways damaged by the atmospheric corrosion in (a) summer.and (b) winter in Mexicali (2010).

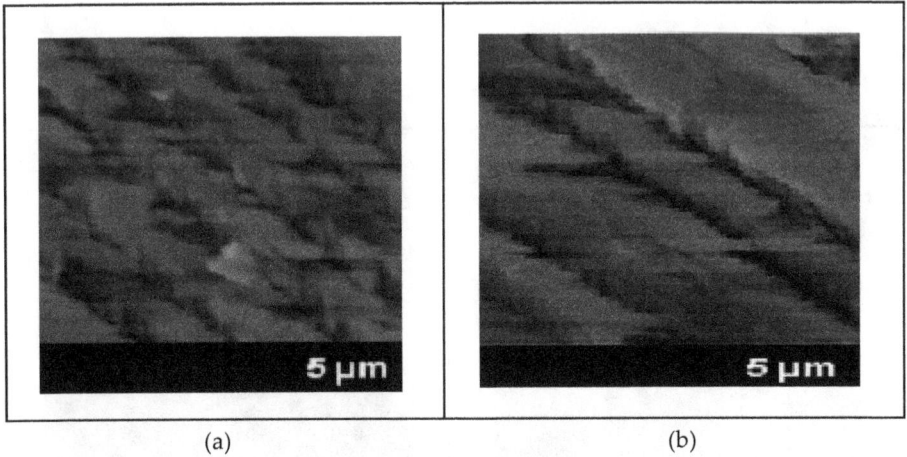

Figure 15. SEM 5X microphotographies of pathways damaged by the atmospheric corrosion in (a) summer.and (b) winter in Tijuana (2010).

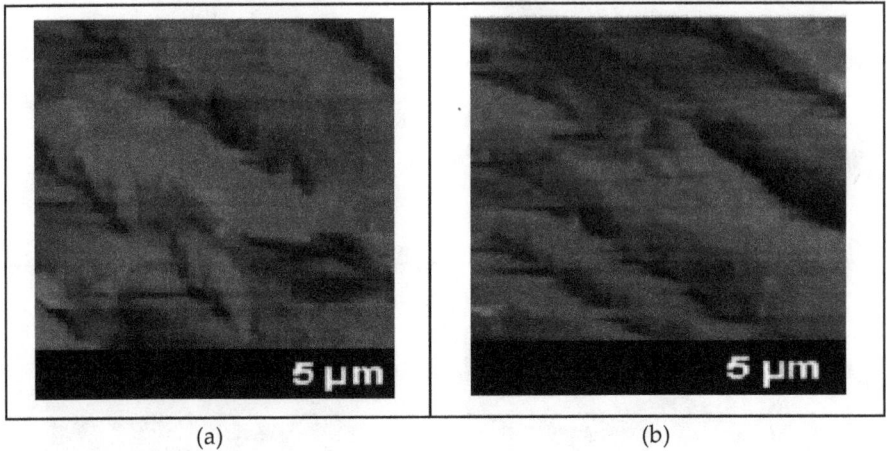

Figure 16. SEM 5X microphotographies of pathways damaged by the atmospheric corrosion in (a) summer.and (b) winter in Ensenada (2010).

The Auger analysis determined the principal chemical agents that reacted with the copper surfaces and the percentages of these air pollutants combined with the carbon dioxide of the environment and the description of the level of copper. The evaluation shows with spectra the intensity at different kinetic energy to each chemical element in the three regions of the interest. As mentioned carbon is from the atmosphere and oxygen can be from the enviroment or of the oxidation process. SEM evaluation represents a microscopy analysis and Auger technique at nanoscale penetrating the surface and has an operation to clean individual surfaces to a specfic analysis.

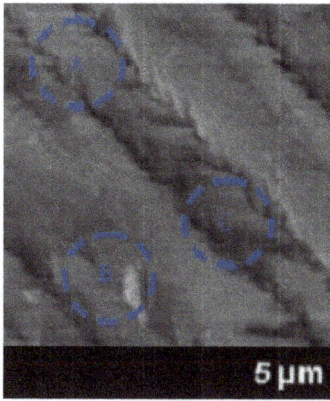
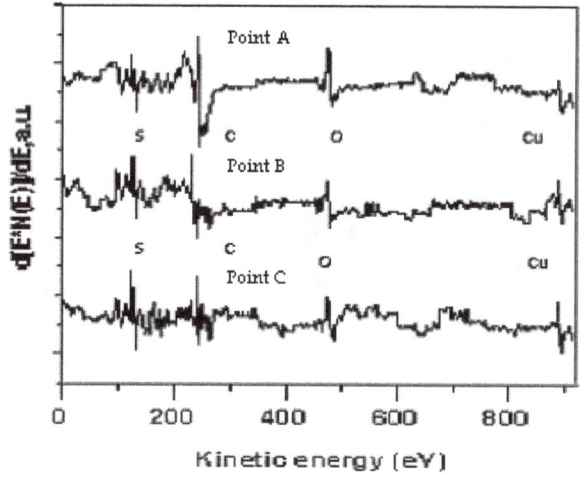

(a)   (b)

Figure 17. AES analysis: (a) Auger map and (b) spectrum evaluation of pathway damaged by the atmospheric corrosion in Mexicali.

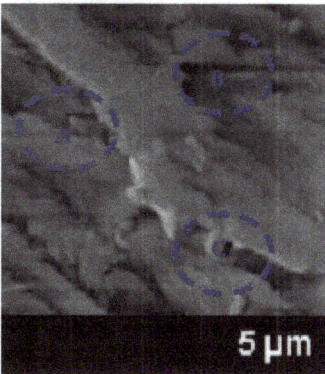
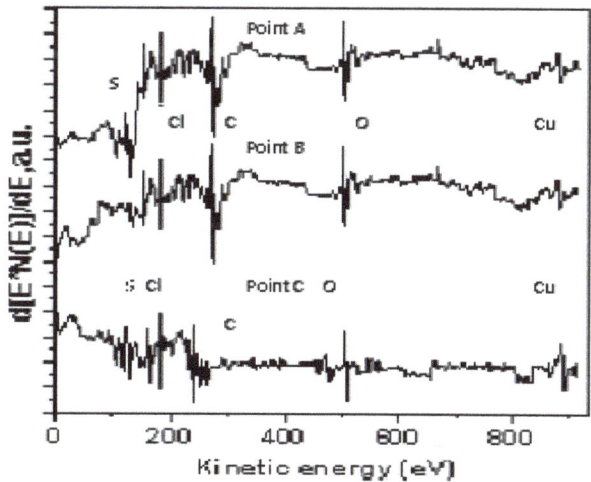

(a)   (b)

Figure 18. AES analysis: (a) Auger map and (b) spectrum evaluation of pathway damaged by the atmospheric corrosion in Tijuana.

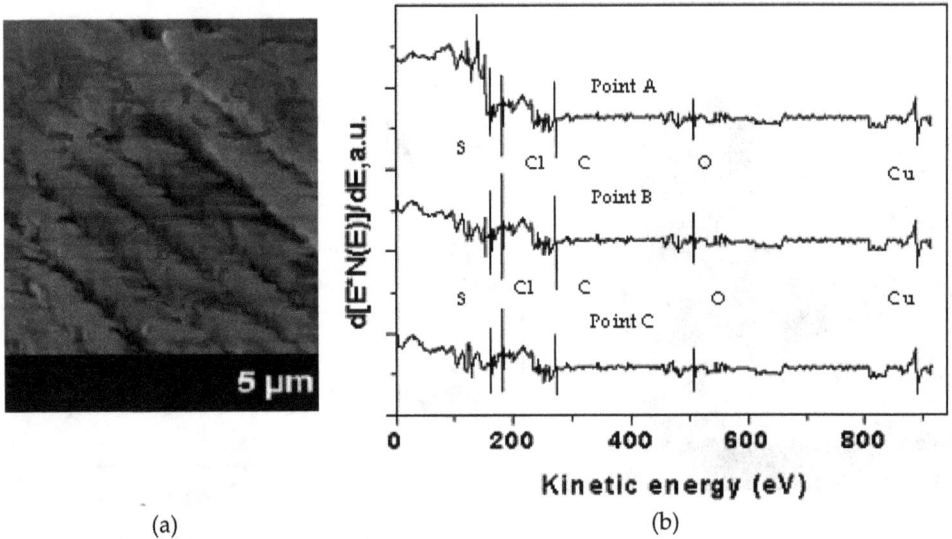

Figure 19. AES analysis: (a) Auger map and (b) spectrum evaluation of pathway damaged by the atmospheric corrosion in Tijuana.

15. Conclusions

The use of MEMS in the electronics industry is an important technology applied to improve the manufacturing processes and their commercial products. This microdevices support to diverse operations in the industrial plants increasing the productive yielding. With climatic factors and pollution parameters controlled is possible avoid the presence of aggresive environments and the generation of corrosion in the electrical conexions and conectors of MEMS used in indoor of the electronics industry. This contrbuite to mantain in good conditions the indusrial equipments and machines of companies. In Mexical considered as arid zone, the principal air pollutants that react with the metallic surfces are sulphurs and in Tijuana and Ensenada, that are marine regions, the chlorides are the principal chemical agents that deteriorate the copper surfaces of electrical conexions and connectors. This study represents an analysis of the presence of corrosion in MEMS of industrial electronic systems, evaluated by the SEM and Auger techniques, and describe the percentage of each chemical agent that react to estimate the type of corrosion ocurred and know the grade of deterioration that origiante the electrical failures. Also was designed, fabricated and tested an electronic system to detect drastic variations of humidity and temperature and low concentrations of air pollutants with a filter that determine a bad operation of electronic equipment, detecting and indicating the begin of the corrosion process that can ocurr. At levels 80% to 90% of RH and 30 °C to 35 °C of temperature, the corrosion was uniform and was presented in winter and some periods of summer especially in some days of July and August. At levels around 60% to 80% of RH and 35 °C to 45 °C in Mexicali principally; the deterioration was as a pitting corrosion. In the ranges of RH and temperature mentioned above, the concentration levels of sulphurs in Mexicali and chlorides and Tijuana and Ensenada overpass the air quality standards and the corrosion rate was higher. The thin films formed in the metallic surfaces of electrical conexions and conectors acelerate the electrochemical process and increase the CR very fast. Mexicali was the city where the CR was higher followe of Tijuana and at the final case was Ensenada. A factor important is the traffic vehicle that emmit sulfurs, nitrogen oxides and CO, and Mexicali being the city where the corrosion process affect with a major intensity, is smaller in population around 3 times less than Tijuana and have less automoviles, but the emission of sulfhidric acid of the geothemoelectrical plant that generate electricty to Mexicali and other small cities in this region, is an important factor, which cause the deterioration of metallic surfaces as copper used in MEMS. With electronic system controller (ESC) of climatic and enviromental parameters in indoor of the electronics industry, the corrosion decrease and the electrical failures was reducing, obtained a good productive yielding, and this manitained less concerned to the manager and technical people. The indusrial plants of this region principally in Mexicali make new rules and proccedings to avoid the corrosion process, as control of microclima in indoor of the electronics industry, install filtes with better quality to detect finite particles and the use of this ECS to detect gases with a special filter of air pollution. Also were developed courses to managers and technical people to avoid the corrosion pehnomena and if this appears, was suggesting some methods to control. With these resources the corrosion decreasing in the last six months and the manufacturing processes were in good conditons with efficient productive yielding.

Author details

Gustavo Lopez Badilla
Universidad Politecnica de Baja California, Calle de la Claridad SN,Col. Plutarco Elias Calles, Mexicali, B.C., Mexico

Benjamin Valdez Salas, Michael Schorr Wiener and Carlos Raúl Navarro González
Instituto de Ingenieria, Departamento de Materiales, Minerales y Corrosion, Universidad Autonoma de Baja California, Mexicali, Baja California, Mexico

16. References

Abdulaziz A. and Maher A.; "Atmospheric corrosion investigations of iron using quartz crystal microbalance"; The Electrochemical Society; consultado en: http://www.electrochem.org/dl/ma/204/pdfs/0483.PDF; 2003.

Ashrae; Handbook; Heating, Ventilating, and Air-Conditioning; applications; American Society of Heating, Refrigerating and Air-Conditioning Engineers, Inc.; 1999.

Asociación de Maquiladoras de Ensenada-AMAE; Departamento de Estadística; Reporte Anual de la Industria en Ensenada; Gobierno Municipal; 2011.

Asociación de Maquiladoras de Mexicali-AMAM; Departamento de Estadística; Reporte Anual de la Industria en Mexicali; Gobierno Municipal; 2011.

Asociación de Maquiladoras de Tijuana-AMAT; Departamento de Estadística; Reporte Anual de la Industria en Tijuana; Gobierno Municipal; 2011.

ASTM G140–02, Standard Test Method for Determining Atmospheric Chloride Deposition Rate by Wet Candle Method, 2008.

ASTM G4–01, Standard Guide for Conducting Corrosion Tests in Field Applications, 2008.

ASTM G91–97, Standard Practice for Monitoring Atmospheric SO_2 Using the Sulfation Plate Technique, 2004.

B.G. Lopez, S.B. Valdez, K. R. Zlatev, P.J, Flores, B.M. Carrillo and W. M. Schorr, Corrosion of metals at indoor conditions in the electronics manufacturing industry; AntiCorrosion Methods and Materials, 2007.

Baldwin Howard; "New Generation MEMS Promise Energy Efficiency";Physcak and Chemistry J.; 2009.

Bateson, R.; "Introduction to control system technology"; Upper Saddle River, New Jersey; Prentice Hall. 5ta Edición; 1999.

Beeby S., Ensell G., Kraft M.,White N.; "MEMS Mechanical Sensors". USA: Artech House Incorporated; 2004.

Cai J.-P. and Lyon S.B.; "A mechanistic study of initial atmospheric corrosion kinetics using electrical resistance sensors with ASTM standards"; Corrosion Science; Vol. Vol. 47; pages 2957-2973; 2005.

Chongchen Xu; Corrosion in Microelectronics; Partial Filfillment of MatE 234; 2003.

Christian C Hindrichsen, Ninia S Almind and Simon H Brodersen; " Triaxial MEMS accelerometer with screen printed PZT thick film, 108-115; *Journal of Electroceramics* 25 (2-4); 2010.

Clark A. E., Pantan C. G, Hench L. L.; "Auger Spectroscopic Analysis of Bioglass Corrosion Films"; Journal of the American Ceramic Society Vol. 59 Issue 1-2 pp 37–39; 2006.

Cole S. and Paterson D. A.; Relation of atmospheric pollution and the generation of corrosion in metals of copper, steel and nickel; Corrosion Engineering; 2004.

G. L. Badilla, B. V. Salas, M. S. Wiener, "Micro and nano corrosion in steel cans used in the seafood industry", INTECH Chapter, Book "Food industry"2011.

G. Lopez, A. Vega, D. Millan, J. Gonzalez, and G. Contreras; "Effect of Corrosion on Control Systems in the Electronis Industry of Mexico", Materials Performance, 2012.

G. Lopez, B. Valdez, M. Schorr; "Analysis of ", Materials Performance, 2012.

Gerhardus H. Koch, Michiel P.H. Brongers, and Neil G. Thompson (NACE); Historic Congressional Study: Corrosion Costs and Preventive Strategies in United States; Materials Performance, Corrosion Prevention and Control Worldwide; 2006.

Helvajian, H, Mehregany M. and Roy, S.: Chapter 1: Introduction to MEMS, Microengineering Aerospace systems. USA: Aerospace Press, 1999.

Hindrichsen C. G., Lou-Møller R., Hansen K.l.; "Advantages of PZT thick film for MEMS sensors"; 9-14. In Sensors and Actuators A: Physical *163 (1); 2010.*

ISO 11844-1:2006. Corrosion of metals and alloys - Classification of low corrosivity of indoor atmospheres- Determination and estimation of indoor corrosivity. ISO, Geneva, 2006.

ISO 11844-2:2005. Corrosion of metals and alloys - Classification of low corrosivity of indoor atmospheres - Determination and estimation attack in indoor atmospheres. ISO, Geneva, 2005.

ISO 9223, Corrosion of Metals and Alloys. Corrosivity of Atmospheres. Classification, International Organization for Standardization, Geneve, Switzerland, 1992.

Jin S., Van Dover R. B., Zhu W., Gasparyan A., and Shea H., "Co-Fe-O conductive oxide layer for stabilization of actuation voltage in optical MEMS," unpublished.[19] J. Ehmke, C. Goldsmith, Z. Yao, and S. Eshelman, "Method and apparatusfor switching high frequency signals," U.S. Patent 6,391,675, May 21, 2002.

Kihira H., Senuma T., Tanaka M., Nishioka K., Fujii Y. and Sakata Y.; A corrosion prediction method of weathering steels; Corrosion Science; Vol. 47; pag 2377-2390, 2005.

L. Veleva, B. Valdez, G. Lopez, L. Vargas and J. Flores; "Atmospheric corrosion of electro-electronics metals in urban desert simulated indoor environment"; Corrosion Engineering Science and Technology; 2008.

Laurent, F. ;"Design of a MEMS sensor for medical products containers"; Applied Physics J.; 2008.

Leidecker H., Brusse J; Tin Whiskers: A History of Document Electrical System Failures; Technical Report to Space Shuttle Program Office; 2006.

Lewis B.J., Houston P.N. & Bladwin D.F.; Flip Chip Processing for SIP Applications;2004.

Lopez B. Gustavo, Valdez S. Benjamin, Schorr W. Miguel, Rosas G. Navor, Tiznado V. Hugo, Soto H. Gerardo; Influence of climate factors on copper corrosion in electronic equipment and devices, AntiCorrosion Methods and Materials; 2010.

Lopez B. Gustavo, Valdez S. Benjamin, Schorr W. Miguel, Zlatev R., Tiznado V. Hugo, Soto H. Gerardo, De la Cruz W.; AES in corrosion of electronic devices in arid in marine environments; AntiCorrosion Methods and Materials; 2011.

López B.G.; Tesis de Doctorado; Caracterización de la corrosión en materiales metálicos de la industria electrónica en Mexicali, B.C., 2008.

Lopez B.G.; Valdez S. B.; Schorr M. W.; "Spectroscopy analysis of corrosion in the electronic industry influenced by Santa Ana winds in marine environments of Mexico"; INTECH Ed. INDOOR AND OUTDOOR POLLUTON, 4; Edited by Jose A. Orosa, 33, Book, 2011.

López-Badilla, Gustavo; González-Hernández, Catalina; Valdez-Ceballos, Antonio; Análisis de corrosión en MEM de la industria electrónica en ambientes árido y marino del noroeste de México; Revista Científica, Vol. 15, núm. 3, julio-septiembre, pp. 145-150; Instituto Politécnico Nacional, Distrito Federal, México; 2011.

Madou M. J.; "Fundamentals of Microfabrication: The Science of Miniaturization", 2nd ed. Cleveland, OH: CRC Press; pp. 59; 2002.

Maluf N. and Williams K.; "An Introduction to Microelectromechanical Systems Engineering", Second Edition.Prentic Hall; 2004.

Nise H., Norman S.; Control Systems Engineering. Wiley International Edition. 4ta edición; 2004.

Ohring M., "Reliability and Failure of Electronic Materials and Devices"; .New York: Academic; Physics and Electronics J.; pp. 310–325; 2003.

Olsson III R.H., Fleming J. G., El-Kady I. F., Tuck M.R and McCormick F.B.; "Micromachined Bulk Wave Acoustic Bandgap Devices," International Conf. on Solid-State Sensors, Actuators, and Microsystems, pp. 317-321,2007.

Pedersen T., Hindrichsen C. C., Thomsen E. V.; "Investigation of Top/Bottom Electrode and Diffusion Barrier Layer for PZT Thick Film MEMS Sensors", 201-213; IEEE Sensors 367; 2007.

Plass R. A., Walraven J. A., Tanner D. M., and Sexton F. W.; "Anodic oxidation-induced delamination of the SUMMiT poly 0 to silicon nitride interface," Testing, and Characterization of MEMS; Tanner Ed.,Vol. 49; pp. 81–86; 2003.

Roger Allan; "MEMS and Nano Push for Higher System Integration Levels; Electronic Design": Vol. 58 Issue 1; pp 777; 2010.

Smart K. J., Olsson III R. H., Ho D., Heine D. R., and Fleming J. G.; "Frequency Agile Radios Using MEMS Resonators," Govt. Microcircuit App. and Critical Tech. Conf.; pp. 409-412; 2007.

Stanley s. Wong, Ngai T. Lau, Chak K. Chan, Ming Fanng, Lap-In Chan; "Effect of atmospheric particles on corrosion"; 2006.

Tanner D. M., Smith N. F., Irwin L.W., Eaton W. P., Helgesen K.S., Clement J. J., Miller W. M, Walraven J. A., Peterson K.A.,Tangyunyong P., Dugger M.T., and Miller S.L., "MEMS reliability: Infrastructure,test structures, experiments, and failure modes," Sandia National Laboratories,Albuquerque, NM, Sandia Rep. SAND2000-0091. http://mems.sandia.gov/search/micromachine/docs/000091o.pdf, 2000.

Tuite, Don; Motion-Sensing MEMS Gyros And Accelerometers Are Everywhere; Mc Graw Hill Ed.; 2009.

Van Ingelgem Y., Vandendael I.,Vereecken J., Hubin A.; "Study of copper corrosion products formed during localized corrosion using field emission Auger electron spectroscopy"; Surface and Interface Analysis Volume 40 Issue 3-4 pp 273–276, 2007.

Van Spengen W. M/; "MEMS reliability from a failure mechanisms perspective" Microelectron. Reliail., vol. 43, no. 7; pp. 1049–1060; 2003.

Virmani Y. P.; "Corrosion Costs and Preventive Strategies in the U.S."; Publication No. FHWA-RD-01-156; HRDI, (202) 493-3052; 2006.

Vittorio, S.; Microelectromechanical Systems, Ed. Oceanica, 2001.

Wojciechowski K. E., Olsson III R. H., Baker M. S., Wittwer J. W., Smart K. and Fleming J. G.;, "Low Vibration Sensitivity MEMS Resonators," *IEEE Frequency Control Symposium*, In-Press, 2007.

Yao J. J., "RF MEMS from a device perspective," J. Microelectromechan. Syst., vol. 10, pp. R9–R38, 2000.

Zawada Tomas, Lou-Moeller Rasmus, Hansen Karsten; "High coupling piezoelectric thick film materials for MEMS-based energy harvesting devices"; In Energy Harvesting Storage Conference Munich; 2010.

Chapter 9

Short-Term Effect of Changes in Fine Particulate Matter Concentrations in Ambient Air to Daily Cardio-Respiratory Mortality in Inhabitants of Urban-Industrial Agglomeration (Katowice Agglomeration), Poland

Małgorzata Kowalska

Additional information is available at the end of the chapter

1. Introduction

The impact of ambient air pollution on human health and the role of particulate matter in this relationship have been discussed for many years, and have been documented by numerous published studies (Pope & Dockery, 2006; Schwartz, 2004; CAFE, 2005; Brunekreef & Holgate, 2002; Katsouyanni et al., 2001; WHO, 2004). However, we still don't have enough evidence to clear explain the possible mechanisms that induce air pollutants related events (mortality or other health effects). The main reasons of these disturbances are: difficulties in understanding the mechanism of damaging impact on human body, but also variety of particulate pollutant sources, their various chemical composition and physical properties and their variability in time (Pope & Dockery, 2006; Schwartz, 2006; Wichmann et al., 2000). Moreover, the majority of evidence comes from studies conducted in North American or West European countries, only few concern the region of middle-east European (Katsouyanni et al., 2001; Jędrychowski, 1999; Krzyżanowski & Wojtyniak, 1991).

In proven health problems assessed with particulate air pollution are short-term effects such us increase of daily mortality. The presented work has been applied exactly to short-term health effects. The potential health risk in population living in urban-industry agglomeration exposed to $PM_{2.5}$ in Poland has not been studied yet. The work is complement of the lack in public health in range of risk communication to improvement of health status of population.

2. Results of Polish study – Purpose and accepted methods

The aim of presented study was assessment of an impact of short-term ultrafine particle concentration changes in ambient air to daily number of deaths due to cardio-respiratory diseases in population living in Katowice Agglomeration. In specific goals of the study was: assessment of daily risk of death due to cardio- and respiratory diseases related to an increase of PM_{10} and $PM_{2.5}$ concentration. Furthermore the most precise scenario of acute exposure to particulate air pollution in relationship with specific (cardio- respiratory) mortality in Katowice Agglomeration, expressed as concentration in day of death and as moving averages concentrations, was determined. Finally the quality of ambient air was rated according to Air Quality Index useful in environmental risk communication process. The established goals allowed to verify thesis very often described in many publications: deterioration of ambient air quality leads to increase health risk in exposed population. Simultaneously, the obtained results allowed to complement data in Poland in range of relative risk of daily death and related to $PM_{2.5}$ exposure.

To realize established goals of the work a time-series analyses was used, covering six calendar years in the period 2000-2005. Data concerning daily number of deaths and daily averages of ambient air quality applied Katowice Agglomeration in Silesian voivodeship and separately large city located in described region (Zabrze).

Mortality data was obtained from the registry at the Central Statistical Office in Warsaw. The records were analyzed according to the classification scheme of the International Classification of Diseases – 10th edition (ICD-10). The category of deaths due to cardiovascular diseases included deaths with codes I00-I99, and to respiratory diseases included deaths with codes J00-J99, database contained the records in two age categories: less than 65 years and 65 or more years.

Data on ambient air pollution was provided by the State Environmental Agency in Katowice. For each day the 24-hour spatial average concentrations of PM_{10}, SO_2 and NO_2 were calculated as the average of all site-specific measurements, monitored in the Katowice Agglomeration. Additionally daily average concentrations of $PM_{2.5}$ measured in Zabrze were obtained. Moreover, for each study day the Institute of Meteorology provided additional data, such as daily mean temperature, daily mean relative humidity and daily mean atmospheric pressure. The data set included also a variable describing a climatic season and influenza epidemic.

Variability of the daily number of deaths and daily concentrations of pollutants was presented in two ways. First was used methods of descriptive statistics, where presentation of the pollutant concentration was based on tertile value of concentration. A low level concentration was established at a value below 33% percentile; a medium concentration between the range of 33% and 66% percentile, and high concentration above 66% percentile. Next, was assessed the relationship between daily mortality and daily concentrations of air pollutants using GLM procedure. Variables describing ambient air pollutant were expressed as average 24-hours concentration on the death day, day preceding death, and 3-day or longer moving average concentration of each pollutant (PM_{10}, $PM_{2.5}$, SO_2, NO_2) (Kowalska et al., 2010). The formula (1) used in multivariate Poisson regression model was:

$$\text{Deaths} = b_0 + b_{1*}\text{season} + b_{2*}\text{air temperature} + b_{3*}\text{air humidity} +$$
$$+ b_{4*}\text{atmospheric pressure} + b_{5*}\text{pollutant}$$

(1)

Model tested the effect of the increase of air pollution concentration (single pollutant) by 10 $\mu g/m^3$ on a daily number of deaths in different specific scenarios: impact of concentrations of pollutants recorded on the day of death, on the day preceding a death and pollutions expressed as the moving average in the period preceding deaths and longer (3, 5, 7, 14, 30, 40, 50 and 60 days).

The relative risk (RR) estimates of the total and circulatory deaths in relation to a 10 $\mu g/m^3$ increase in each pollutant was calculated using the formula (2):

$$RR = e^{b\,\text{delta}}$$

(2)

where b is the regression coefficient of the pollutant in question and delta is its increase by 10 $\mu g/m^3$. Additionally, a percentage change in mortality was calculated as the (RR-1) times 100%.

To assess the quality of ambient air pollution in Katowice Agglomeration according to Air Quality Index (AQI) different methods of indexation (American, French, British and German) were used. Daily PM_{10} and sulphur dioxide (SO_2) concentrations were transformed to qualitative variable expressed by: good, medium and unhealthy air quality. Finally, mean value of daily number of deaths characteristic for days with particular categories quality of air was calculated. All analyses were performed using procedures available in the Statistica 7.0 or SAS statistical packages.

2.1. Quality of ambient air pollution in study agglomeration (Katowice Agglomeration)

Available published data suggests that in Poland, after the political and economical change in 1990 the quality of environment became one of the national target priorities, and the effort towards environmental clean-up has resulted in a substantial improvement of ambient air quality (National Health Program, 1996). The largest effect was seen in the urban area of Katowice, known for high levels of industry-related air pollution. A time-series analysis performed in years 1994-5 revealed the most significant effect of exposure to sulphur dioxide (SO_2), followed by particulate matter (PM_{10}) (Zejda, 2000). Since then the quality of ambient air has significantly improved. Between 1994 and 2005 ambient air daily average concentrations of pollutants have decreased by 38% in the case of SO_2 and by 28% in the case of PM_{10}. The apparent decline in ambient air pollution created an opportunity to find out if the pattern of acute mortality has responded to the decreased exposure, under 'natural experiment' scenario.

Database was constructed so that it was possible to analyze the number of deaths from the regard of the season of year: winter – months from 01 January to 31 March, the spring –

months from 01 April to 30 June, summer – months from 01 July to 30 September, and the autumn – months from 01 October to 31 December. The highest concentrations of ambient air pollutants were in cold season (winter and autumn), while the lowest related hot season (summer). Figure 1 shows daily concentrations of pollution in particular seasons in Katowice Agglomeration.

Moreover, the implementation of $PM_{2.5}$ concentration measurements in large city of region (Zabrze) makes it possible to estimate the risk of deaths related to fine particulate exposure, not unexplored in Poland yet. Table 1 presents data on concentrations of pollutants in this city. It was documented that in the study period the total number of days with exceedance of limit value (Polish Ministry of Environment, 2002) for $PM_{2.5}$ concentration (25 $\mu g/m^3$), PM_{10} (50 $\mu g/m^3$) and SO_2 (125 $\mu g/m^3$) in Zabrze city were 295, 448 and 15 days respectively. These exceedances were always associated with the cold season (autumn-winter).

The major source of particulate matter, SO_2 and NO_2 in the environment remained combustion of vehicle fuels (petrol, oil), burning of wood in fireplaces, as well as stoves and gasoline burning (Knol et al., 2009). Current data confirmed that fine particulates were the predominant fraction of suspended particles in air measured in the Katowice Agglomeration (Klejnowski et al., 2007). Simultaneously, other authors indicated that elemental and organic carbon are significant proportion (60 to 80%) of fine particles, which indirectly points to the advantage of municipal pollution sources (Pastuszka et al., 2003; Rogula et al., 2007). Finally, it should be noted, that extensive use of coal for heating, industrial purposes and mass production of energy in the Baltic countries of Eastern Europe makes another profile of air pollution in this region with dominant role of PM and sulphur dioxide (Jędrychowski, 1999; Medina et al., 2002).

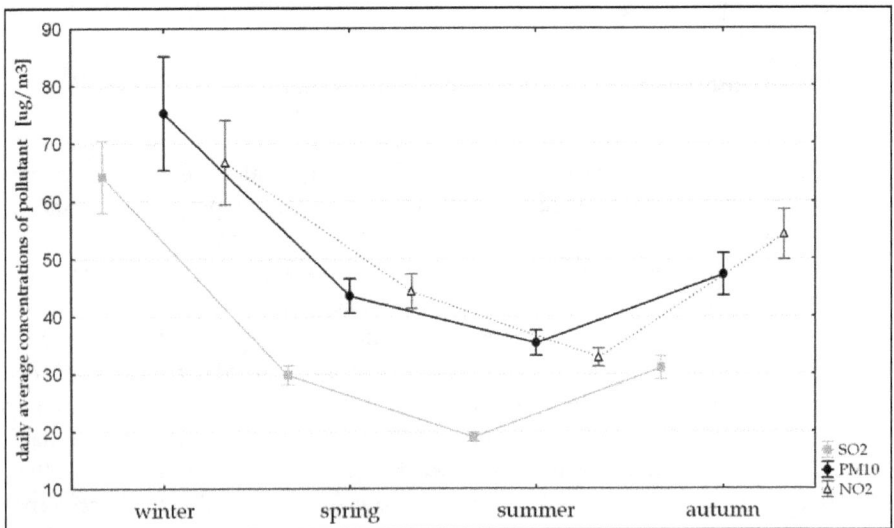

Figure 1. Daily concentrations of PM_{10}, SO_2 and NO_x in ambient air, Katowice Agglomeration in the study period (01.01.2000-31.12.2005).

Pollution	Descriptive statistics of air pollutants concentrations [$\mu g/m^3$]					
	Mean	Min	Max	Perc 33%	Perc 66%	SD
SO₂ (daily limit value 125 $\mu g/m^3$)	35.7	2.6	266.2	21.6	36.8	25.2
PM₂.₅ (annual limit value 25 $\mu g/m^3$)	40.5	4.9	323.7	21.8	40.0	34.5
PM₁₀ (daily limit value 50 $\mu g/m^3$)	40.8	5.6	407.5	26.0	42.4	28.1
NO₂ (annual limit value 40 $\mu g/m^3$)	26.9	6.0	87.0	21.3	29.2	11.2

Table 1. Descriptive data of pollutants concentrations in ambient air, Zabrze in the study period 2000-2005

2.2. Daily specific mortality

Among documented short-term health problems associated with fine particulate air pollution are mentioned increase the daily number of deaths, especially deaths due to cardio-respiratory diseases (WHO, 2004). Systematic epidemiological studies indicated the occurrence of this relationship even when PM concentrations are at the levels safe for human health (Krzyżanowski & Wojtyniak, 1991; Anderson et al., 2001; EPA 2004). Some studies suggest the importance of two pollutants ie particulate matter and sulphur dioxide for a daily increase in mortality (Dockery et al., 1992; Ostro, 1993). Since the composition of air pollution in Polish agglomerations differs from profile in other countries (dominant role of SO₂) and measured levels are usually higher, it is necessary to estimate the relative risk of daily mortality related to ambient air pollutants in selected industrial region (Katowice Agglomeration).

During the study period (from 2000 to 2005 years), 146 592 people died due to cardio-respiratory diseases in Katowice Agglomeration, on average 67 people per day. Analysis showed that out of this cases, 115 635 (78.9%) was deaths in older population (people aged 65+ years). It was noted that only 8.3% of cardio-respiratory deaths (n=12 179) was related with respiratory diseases. Further analysis of available register data of infectious diseases incidence in Poland confirmed the occurrence of two influenza episodes, in the years 2003 and 2005. In both cases, the influenza episode occurred between 16th February to 22nd March (National Institute of Public Health, 2010) with no observable increase in the rate of mortality due to influenza. Seasonal diversity of average value of specific mortality shows figure 2. The highest level of deaths due to cardio-respiratory diseases was related to cold season (winter) and the lowest was characteristic for summer.

Figure 3 shows changes in daily number of deaths due to cardio-respiratory diseases in inhabitants of Zabrze according to the fine particulate concentrations. The obtained results suggest that the highest mortality were associated with the highest level of PM₂.₅ concentration, the results of ANOVA test confirmed statistically significant relationship.

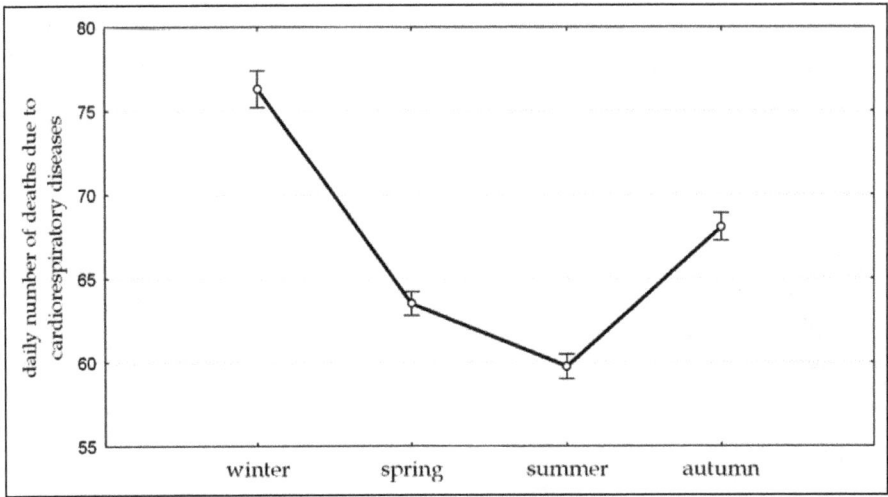

Figure 2. Daily average number of deaths due to cardio-respiratory diseases in Katowice Agglomeration according to season of year, study period 2000-2005.

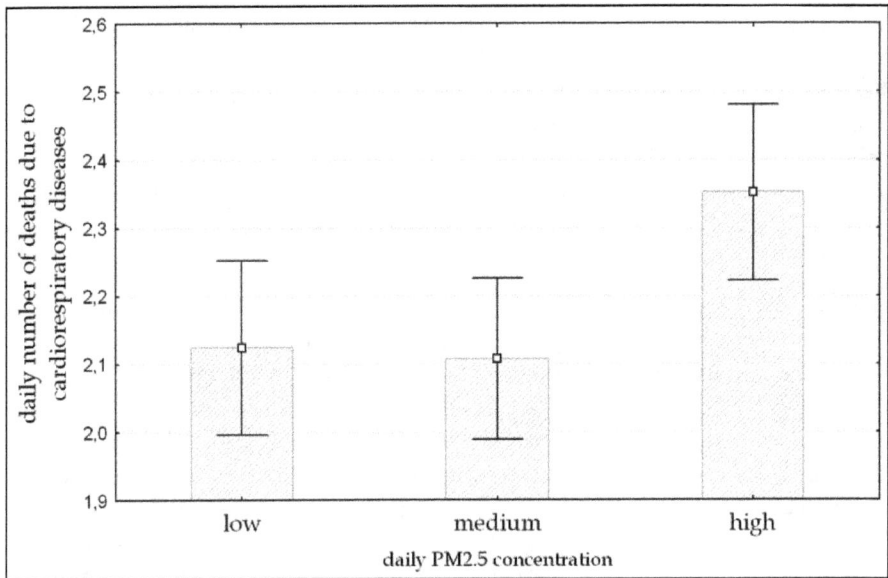

Figure 3. Daily average number of cardio-respiratory mortality in Zabrze city according to quality of daily concentration of $PM_{2.5}$, study period 2000-2005.

The results of multivariate analysis confirmed increase of cardio-respiratory death in both, total and older population of inhabitants living in Katowice Agglomeration in relation to increases in the daily average pollutant concentration by 10 $\mu g/m^3$. The values of relative risk depend on the time of exposure with a comparably higher risk for a longer time of exposure. Simply, as the time of exposure gets longer, the risk gets higher. Figure 4 shows changes in relative risk of cardio-respiratory mortality in older population of Katowice Agglomeration related to changes of moving average concentrations of pollutants (PM10, SO2 and NO2).

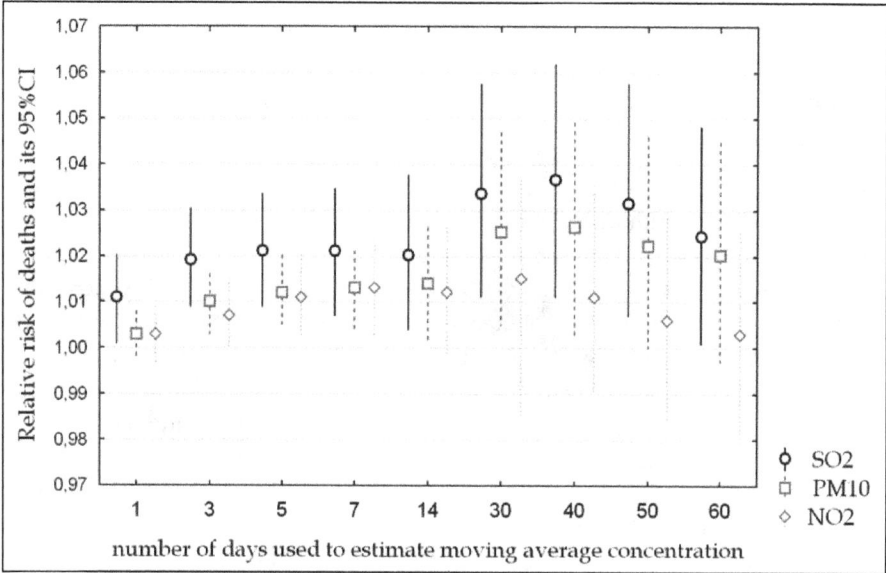

Figure 4. Relative risk of deaths due to cardio-respiratory diseases in older (65+) inhabitants of Katowice Agglomeration , study period 2000-2005.

Pollution	Moving average concentration	Relative risk of death	Author
PM10	24-hours	0.2 – 0.8%	Pope & Dockery, 2006
PM10	24-hours	0.3%	Kowalska et al., 2010
PM10	5-40 days	1.3 -1.8%	Pope & Dockery, 2006
PM10	5-40 days	0.7-1.3%	Kowalska et al., 2010
PM2.5	24-hours	0.6-1.2%	Pope & Dockery, 2006
PM2.5	3-day moving average, older population in Zabrze	0.5-2.0%	Kowalska, 2011

Table 2. The increase in relative risk of total mortality in relation to the increase in moving average concentration by 10 μg/m³, comparative data.

A similar effect has been documented for cardio-respiratory mortality in the population of Zabrze, however a small number of inhabitants significantly reduces the precision of the risk estimation. The obtained results suggest that the most precise scenario useful for assessment of short term health effects (e.g. specific daily mortality) related to particulate ambient air pollution in Katowice Agglomeration is presentation of longer exposure expressed by moving average concentration, from 3-day to 14-days. Similar observation is applicable to gaseous air pollutants: sulphur and nitrogen dioxide, relative risk of death was larger for longer time of exposure. The results of own research suggest that the relative risk of daily mortality related to short term of particulate matter (PM_{10} and $PM_{2.5}$) concentration in ambient air is similar to those given by others authors (Table 2).

2.3. Air Quality Index and its significance in environmental health risk communication, Polish study

Air Quality Index (AQI) is a standardized summary measure of ambient air quality used to express the level of health risk related to particulate and gaseous air pollution. The construction of AQI allows distinction between "good" and "dangerous" air quality. The index, first introduced by US EPA in 1998 classified ambient air quality according to concentrations of such principal air pollutants as PM_{10}, $PM_{2.5}$, ozone, SO_2, NO_2 and CO (EPA, 2003). Subsequently similar, index-based approach to express health risk, was developed in France, Great Britain and in Germany. Table 3 shows the cut-of PM_{10} values for the specific air quality zones used in European and US standards. No such environmental warning system exists in Poland, although some test-trials took place in Katowice area and in the city of Gdańsk. However, the operational value of AQI under environmental circumstances in Poland remains unknown. The aim of study was to examine current air pollution levels in Katowice Agglomeration and to confront AQI categories with local air quality, also in terms of health impact on the population as expressed by daily cardio-respiratory mortality.

Category of air quality	Cut-of values for daily PM_{10} concentration [$\mu g/m^3$]			
	USA	France	Great Britain	Germany
Good	0-54	0-39	0-49	0-34
Moderate	55-154	40-79	50-74	35-99
Dangerous	155 and more	80 and more	75 and more	100 and more

Table 3. Categorization of ambient air quality according to diferent methods of AQI indexation based on PM10 daily concentration.

The obtained results suggest significant discrepancy in range of air quality categories depending on applied system of classification. Percent of days with 'unhealthy' air quality (in the period 2001-2002) was running from 0.1% (American method of indexation) to 11.2% (British method) and usually applied winter season. The frequency of days with dangerous air quality for health (PM_{10} concentrations) calculated by French and German AQI were similar and amounted near 6% (Kowalska et al., 2009). Statistically significant Spearman correlation coefficients was obtained for relationship between air quality and total number of deaths, as well as number of deaths due to cardio-respiratory diseases in total and older population (aged 65+). However, the observed values of correlation coefficients are very low and don't exceed value 0.2 for each chosen method of indexation (Table 4).

Daily mortality	Population aged	British AQI	French AQI	American AQI	German AQI
Cardio-respiratory diseases	0-64	0.00 (NS)	0.01 (NS)	0.04 (NS)	0.00 (NS)
	65 +	0.15 (p<0.05)	0.15 (p<0.05)	0.16 (p<0.05)	0.14 (p<0.05)
	Total	0.13 (p<0.05)	0.13 (p<0.05)	0.16 (p<0.05)	0.12 (p<0.05)
Total number of deaths	0-64	0.03 (NS)	0.03 (NS)	0.04 (NS)	0.04 (NS)
	65 +	0.17 (p<0.05)	0.17 (p<0.05)	0.13 (p<0.05)	0.16 (p<0.05)
	Total	0.15 (p<0.05)	0.15 (p<0.05)	0.11 (p<0.05)	0.14 (p<0.05)

Table 4. Spearman correlation coefficients for relationship between daily number of deaths and air quality index by different method of indexation, p value in the bracket; NS- not statistically significant.

Moreover it was calculated mean value of daily cardio-respiratory mortality characteristic for days with particular AQI defined as: good, moderate and dangerous category of air quality, determine by particular methods of indexation. It was observed that the highest mortality concerned days with dangerous quality of air. Figure 5 shows results of this analysis. The association between mortality and quality of air was similar for German, British and French method of indexation, but finally the obtained results confirm that the highest mortality concerned days with dangerous quality of air and the lowest concerned days with good quality of air. The observed variability was statistically significant in each AQI categories.

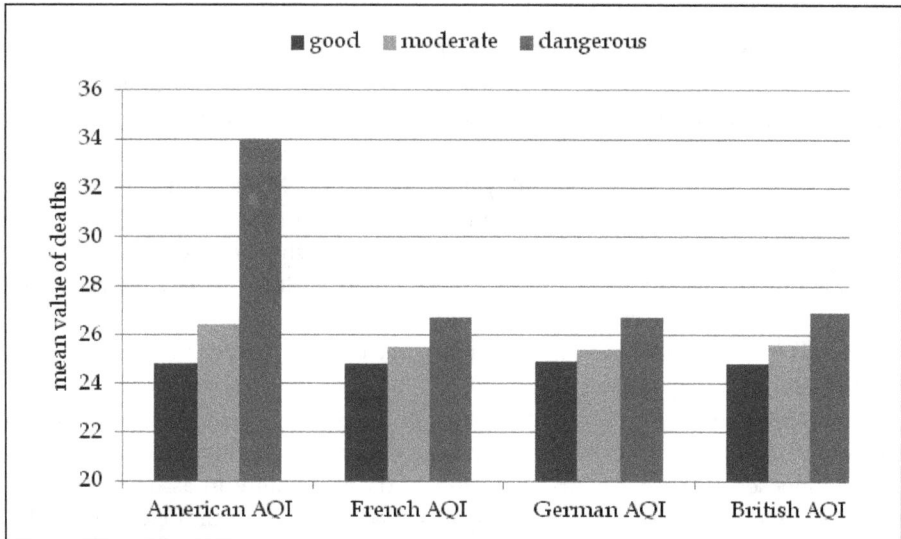

Figure 5. Number of daily deaths due to cardio-respiratory diseases in Katowice Agglomeration during days with different air quality, study period 2000-2005.

According to poor ambient air quality in Katowice Agglomeration, especially during the winter time, it is essential to inform inhabitants about environmental health hazard.

Confrontation data of local air quality with AQI categories and with daily total and specific (cardio-respiratory) mortality confirm, that British and French method of AQI indexation are the best way to risk communication in Poland. Probably, similar climate conditions and specific of air pollution are comparable in all described countries, so the association between air quality index and health effect is similar too. It is necessary to disclose the knowledge about air quality index and their association with health effect. Very important source of this information are medical doctors, especially general practitioner. Moreover well known websites or regional televisions are very useful sources to transmit important information about environmental health risk. Current position of public health experts suggest the need for a debate on the communication of real health risk associated with ambient air pollution (Schwartz, 2006).

3. Summary and conclusions

The impact of ambient air pollution, including fine particulate matter, to the health of the population in Poland was not often an undertaken research study. However, slow increases in the number of measurement data for fine particulate matter concentrations and its chemical composition allow for the estimation of environmental health risks. The lack of direct access to data describing the health status of population expressed by cardio-respiratory mortality, especially those which deals with lack of relevant rules of public health (e.g. the right of the Act), makes it difficult to obtain reliable information relating the impact of air quality to the real health risk of population living in a particular region of Poland.

This paper explains in details the health risks (expressed by daily mortality due to cardio-respiratory diseases) in response to increased concentrations of particulate pollutants, including $PM_{2.5}$ in Katowice Agglomeration, the most polluted region in Poland. The results obtained in own study confirm a slight decrease in the relative risk of death from cardio-respiratory causes in response to air quality improvement in PM_{10} concentrations during the last decade. The simultaneous observed improvement in air quality in the range of SO_2 concentrations in the study area did not change the relationship between air pollution and daily mortality. The higher profile of air pollution and health risk may be attributed to higher gas and other alternative sources of heat and electricity costs. Not without significance is the relatively low degree of environmental risk hazards in Polish population and the resulting consequences in the form of a number of behaviors such as grass and garbage burning, a high percentage of smokers in the country and lack of solutions for a rapidly growing flow of traffic. Even though the source of air pollution may be different in Poland as opposed to other countries, the risk of daily mortality in association with the exposure to fine dust is similar to those observed in other countries. Our findings, explaining the effect of seasonal influence on the size of the daily deaths due to cardio-respiratory diseases are consistent with those documented in literature. As a rule, greater number of adverse health effects in populations (mostly marked for deaths after the age of 65) varies with the different seasons; winter associated with the highest adverse effect more than the summer. It can be concluded that the relationship between concentrations of air pollution and daily deaths probably reflect exacerbation of existing disease in people suffering from respiratory or circulatory system diseases (e.g. coronary artery disease and chronic obstructive pulmonary disease, but also arrhythmia, atherosclerosis or diabetes) and instead of new diseases.

Another issue was to determine the best possible presentation of short-term exposure of air pollution in the population. The obtained results suggest that the most precise scenario useful for assessment of short term health effects (e.g. specific daily mortality) related to particulate ambient air pollution in Katowice Agglomeration is the presentation of longer exposure expressed by moving average concentration, from 3-day to 14-days. A similar observation is applicable to gaseous air pollutants such as sulphur and nitrogen dioxide showing the relative risk of death being larger when the duration of exposure is longer. Furthermore, the results of our research suggests that the relative risk of daily mortality related to short term of particulate matter (PM_{10} and $PM_{2.5}$) concentration in ambient air is similar to those given by others authors.

Finally, due to poor ambient air quality in Katowice Agglomeration, especially during the cold season, it is essential to inform inhabitants about real environmental health risk. An important role in risk communication can play the Air Quality Index. Comparison of air quality data with data on daily cardio-respiratory mortality confirms that the British and French method of AQI indexation are the best way to communicate risk in Poland. Due to similar climate and cause of air pollution in these described countries, it may be possible to make a similar association between air quality index and health effect. It is necessary to integrate ideas and opinions from researchers in environmental epidemiology and public health as well as journalist and politicians in the evaluation of air quality in order to develop an effective and proper health policy.

Author details

Małgorzata Kowalska
Medical University of Silesia, Poland

Acknowledgement

The study was supported by the Medical University of Silesia, Katowice, Poland (Grant: KNW-1-034/P/1/0)

4. References

Anderson H.R.; Bremner S.A.; Atkinson R.W.; Harrison R.M.; Walters S. (2001). Particulate matter and daily mortality and hospital admission in the west midlands conurbation of the United Kingdom: associations with fine and coarse particles, black smoke and sulphate. *Occupational and Environmental Medicine*, Vol.58, pp. 504-510

Brunekreef B. & Holgate S.T. (2002). Air pollution and health. *Lancet* , Vol.360, pp. 1233-1242

CAFE, (February 2005). CAFE Programme: Baseline scenarios for Clean Air for Europe. Final report. Luxemburg, Austria , 07.02.2012, Available from http:// ec.europa.eu/ environment/archives/cafe/activities/pdf/cafe_scenario_report_1.pdf

Dockery D.W.; Schwartz J.; Spengler J.D. (1992). Air pollution and daily mortality: Associations with Particulates and Acid Aerosols. *Environmental Research*, Vol.9, No.2, pp.362-373

Jędrychowski W. (1999). Ambient air pollution and respiratory health in the east Baltic region. *Scandinavian Journal of Work, Environment & Health* , Vol.25, Suppl 3, pp. 5-16

Katsouyanni K.; Touloumi G.; Samoli E.; Gryparis A.; Le Tertre A.; Monopolis Y.; Rossi G.; Zmirou D.; Ballester F.; Boumghar A.; Anderson H.R.; Wojtyniak B.; Paldy A.; Braunstein R.; Pekkanen J.; Schindler Ch.; Schwartz J. (2001). Confounding and effect modification in the short-term effects of ambient particles on total mortality: results from 29 European cities within the APHEA2 project. *Epidemiology*, Vol.12, No.5, pp. 521-531

Klejnowski K.; Krasa A.; Rogula W. (2007). Seasonal variabitity of concentrations of total suspended particles (TSP) as well as PM10, PM2,5 and PM1 modes in Zabrze, Poland. *Archives of Environmental Protection*, Vol.33, No.3, pp. 15-29

Knol A.B.; de Hartog J.J; Boogaard H.; Slottje P.; van der Sluijs J.P.; Lebret E.; Cassee F.R.; Wardekker J.A.; Ayres J.G.; Borm P.J.; Brunekreef B.; Donaldson K.; Forastiere F.; Holgate S.T.; Kreyling W.G.; Nemery B.; Pekkanen J.; Stone V.; Wichmann H.E.; Hoek G. (2009). Expert elicitation on ultrafine particles: likelihood of health effects and causal pathways. *Particle and Fibre Toxicology*, Vol.6, pp. 19, 13.09.2010, Available from http://www.particleandfibretoxicology.com/ content/6/1/19

Kowalska M.; Zejda J.E.; Skrzypek M. (2010). Short-term effects of ambient air pollution on daily mortality. *Polish Journal of Environmental Studies*, Vol.19, No. 1, pp. 101-105

Kowalska M.; Ośródka L.; Klejnowski K.; Zejda J.E.; Krajny E.; Wojtylak M. (2009). Air quality index and its significance in environmental health risk communication. *Archives of Environmental Protection*, Vol.35, No.1, pp. 13-21

Kowalska M. (2011). Short-term effects of changes in the levels of fine particulates in ambient air on daily mortality and hospitalization due to cardio-respiratory diseases in urban-industrial population (Katowice Agglomeration). Medical University of Silesia, Katowice, Poland, ISNN 1689 6262 [In Polish]

Krzyżanowski M. & Wojtyniak B. (1991). Air pollution and daily mortality in Cracow. *Public Health Review*, Vol.19, pp. 73-81

Medina S.; Plasencia A.; Artazcoz L.; Quenel P.; Katsouyanni K.; Mucke H.G.; DeSaeger E.; Krzyżanowski M.; Schwartz J. and the contributing members of the APHEIS group. (2002). APHEIS Health Impact Assessment of Air Pollution in 26 European Cities. Second Year Report 2000-2001. Institut de Veille Sanitaire, Saint-Maurice, 06.02.2012, Available from http://www.apheis.org/Pdf/ Apheis_1_60.pdf

National Health Program. Polish Ministry of Health. (1996), 02.02.2012, Available from
http://www.mz.gov.pl/wwwfiles/ma_struktura/docs/zal_urm_npz_90_15052007p.pdf
[In Polish]

National Institute of Public Health in Warsaw. (2010). Influenza and new cases, 23.07.2010,
Available from http://www.pzh.gov.pl/oldpage/epimeld/ grypa/index.htm [In Polish]

Ostro B.D. (1993). The association of air pollution and mortality. Examining the case for
inference. *Archives of Environmental Health: An International Journal*, Vol.48, No.5, pp.336-
342

Pastuszka J.S.; Wawroś A.; Talik E.; Paw K.T. (2003). Optical and chemical characteristics of
the atmospheric aerosol in four towns in southern Poland. *Science of the Total
Environment*, Vol.309, pp. 237-251

Polish Ministry of Environment. (June 2002). The Minister of Environment dated 6 June
2002 on the assessment of levels of substances in the air. Official Set 2002 year No 87,
item   798   and   796,   07.02.2012,   Available   from   http://isap.sejm.gov.pl/
VolumeServlet?type=wdu&rok=2002&numer=087 [In Polish]

Pope, C.A. & Dockery D.W. (2006). Health effects of fine particulate air pollution: lines that
connect. *Journal of the Air & Waste Management Association*, Vol.56, pp. 709-742, ISSN
1047-3289

Rogula W.; Pastuszka J.S.; Talik E. (2007). Concentration level and surface chemical
composition of urban airborne particles near crossroads in Zabrze, Poland. *Archives of
Environmental Protection*, Vol.33, No.2, pp.23-34

Schwartz, J. (2004). Is the association of airborne particles with daily deaths confounded by
gaseous air pollutants? An approach to control by matching. *Environmental Health
Perspectives*, Vol.112, No.5, pp. 557-561, PMC1241921

Schwartz J., (2006). Air pollution and health: do popular portrayals reflect the scientific
evidence? *American Enterprise Institute for Public Policy Research*, Vol. 2, 13.09.2010,
Available from http://www.aei.org/article/energy-and-the-environment/ contaminants/
air/air-pollution-and-health/

US EPA. (2003). Air Quality Index. A Guide to Air Quality and Your Health. EPA-454/K-03-
002, 15.06.2008, Available from http://www.njaqinow.net/ App_AQI/AQI.en-US.pdf

US EPA. (October 2004). Final Report: Air quality criteria for particulate matter. Vol. I.
EPA/600/P-99/002aF. US Environmental Protection Agency, 02.02.2012, Available from
http://cfpub.epa.gov/ncea/cfm/recordisplay.cfm?deid= 87903#Download

WHO, (June 2004) WHO Report: Health aspects of air pollution. Results from the WHO
project Systematic review of health aspects of air pollution in Europe. WHO Regional
Office for Europe, document E83080, Copenhagen 2004a, Denmark, 07.02.2012,
Available from http://www.euro.who.int/__data/assets/ pdf_file/0003/74730/E83080.pdf

Wichmann H.E.; Spix C.; Tuch T.; Wolke G.; Peters A.; Heinrich J.; Kreiling W.G.; Heyder J.
(2000). Daily mortality and fine and ultrafine particles in Erfurt, Germany. Part I: role of
particle number and particle mass. *Research Report Health Effects Institute* Vol.98, pp. 5-86

Zejda J.E. (2000). Health effects of ambient air pollution – the magnitude of risk and current hazard in Poland. In: *Environment and Health*, K. Janicki; W. Klimza; J. Szewczyk, (Eds.), Cmyk-Art, pp.221-235, Częstochowa, Poland [In Polish]

Chapter 10

Air Quality in Urban Areas in Brazilian Midwest

Edmilson de Souza and Josmar Davilson Pagliuso

Additional information is available at the end of the chapter

1. Introduction

The terrestrial atmosphere includes a vast region, mainly dominated by gas that surrounds the planet. One of the most relevant features of the Earth's atmosphere is its selectivity to energy, while remaining transparent to the significant portion of wavelengths sent from the sun, and opaque to others arising from the de-excitation, as heat from the surface. The climate system and the thermal equilibrium is maintained by the transfer of solar energy for the Earth's surface, and from one region to another surface of the planet, in addition to referral to the outer space. This dynamic is intrinsically related to selective qualities present in the architecture and physical chemistry of Earth's Atmosphere.[1-3]

According [1] due to the atmosphere to be in a fluid system, it is capable of supporting a wide range of movements, ranging from a few meters turbulent eddies to the movements of air masses having a size of the planet. This mobility of the fluid system makes the description of complex behavior, she being responsible for the redistribution of mass, energy and other constituents in infinite configurations.

Current knowledge about the characteristics of the atmosphere is derived from several experimental observations. The composition of gases as to its constitution and concentration is essential for sustaining life on the planet, so the oscillation percentage, even of trace gases, can cause damage to populations of living beings or phenomena closed imbalance in the atmosphere.

The variability of meteorological parameters as temperature, wind speed and direction and relative humidity, among others, are closely associated with fluctuations in the concentration of trace constituents. However, the composition of the gaseous atmosphere is substantially constant between the major gas as nitrogen (78%), oxygen (21%) and argon (0.93%). [3,4,5]

A very small extent the composition of atmospheric gases, about 0.07%, is so-called trace gas. Some of these gases due to the action of human means of production, mainly in urban areas, concentrations higher than those found in remote areas of anthropogenic influence.

A finding in recent decades, due to restrictions imposed by law, is to reduce the rates of pollutant emissions in developed countries. In Germany, pollutants as carbon monoxide, sulfur dioxide and particulate matter, fell sharply, while others, such as nitrogen oxides and volatile organic compounds have undergone modest reductions. According to [6] between 1975 and 1996 emissions of carbon monoxide decreased from more than 16 Mton to negligible values after 1994.

In this manner, there is a close relationship between the level of development of a nation and control pollutant concentrations monitored in regions with large population.

In Brazil, the time series of Air Quality are restricted temporally, that is, there are few cities in the country that maintain monitoring network of air pollutants. The Metropolitan Region of Sao Paulo (MRSP) is one of those few cities which, through the State Company of Environmental Technology and Sanitation (CETESB) since the 1980s has received continuous monitoring of air quality and increasing control and inspection.

2. Air quality in Brazil: Some features

In Air Quality Report of the State of Sao Paulo [7], the decrease in concentration of pollutants in the Greater Sao Paulo, as a result of improvement in the control and supervision, are not evident for all pollutants monitored. Figures 01 and 02 illustrate, respectively, the maximum annual average concentrations of over 10 years of Carbon Monoxide and Ozone in different monitoring stations in the MRSP. In both figures the scales are below the safety limit determined by the National Standards for Air Quality.

In Figure 01 the annual average concentrations of carbon monoxide related to maximum of 8 hours in some regions such as Central and Congonhas, decreased by over 50% in 10 years. The same occurs with less intensity, to the Center of Campinas that, although with a smaller series, the values fluctuate around 2.0 ppmv. However, after 2003, all stations show stabilization of concentrations in a range between 1.0 and 3.0 ppmv.

This, possibly, part of his explanation is related to the implementation of emission control technologies in automotive vehicles in the MRSP. In the early 1980s the estimated emission factor for carbon monoxide, light fleet was 54 g/km in the last year this index is less than 0.5 g/km. [8]

The annual average concentrations of 1-hour maximum ozone, Figure 02, show small displacement amplitude, for most areas, between the years 2000-2003 to 2005-2006. In the first period, the mean concentration varied between 90 and 110 μg/m3 subsequently between 70 and 90 μg/m3. But this does not mean that there is a tendency to decrease. Often, the limits for ozone standards are exceeded in the MRSP, and although between 2002 and 2006 this rate has fallen nearly 50%, caution is recommended in the analysis of this information since this pollutant is strongly influenced by meteorological parameters. [7]

On cities in the interior, networks monitoring air quality are rare. In the State of Sao Paulo, the CETESB maintains a network covering 12 cities outside the Metropolitan Area, but, in general, stations are limited in two or three pollutants. The State Foundation of Environmental Engineering of the State of Rio de Janeiro (FEEMA), also maintains monitoring stations within the state, but only in three cities, and generally monitors inhalable particles and/or ozone. [7,9]

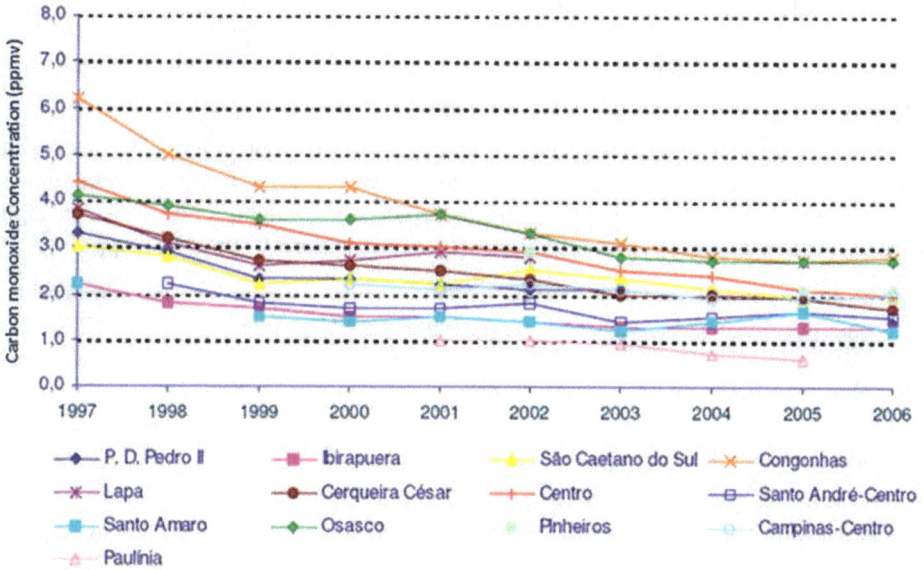

Figure 1. Annual Average Concentration of CO in the MRSP. [7]

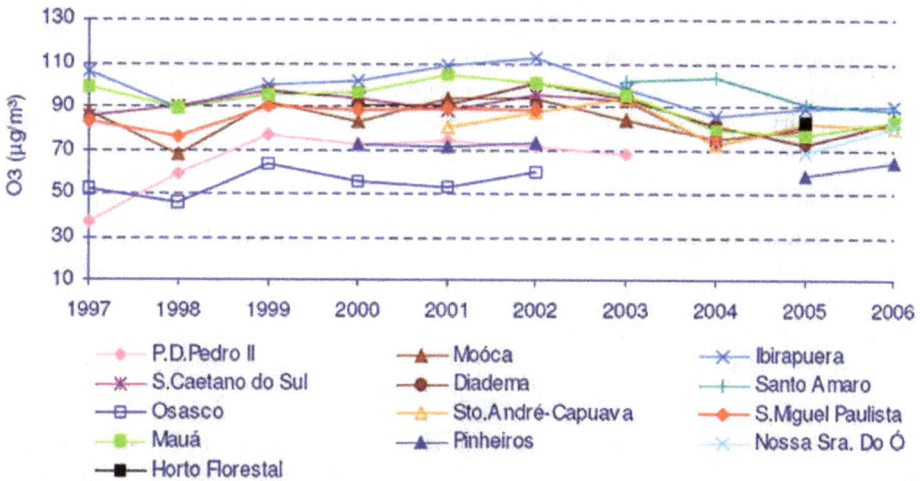

Figure 2. Annual Average Concentration Ozone in the RMSP.[7]

3. Midwest region: A case of Mato Grosso do Sul state

In Mato Grosso do Sul (MS), research in urban air quality is scarce. There is not continuous air quality monitoring stations on MS maintained by the Government. Some isolated initiatives take advantage of the existence of some databases of companies, which in accordance of their activities has monitoring stations in cities including Campo Grande and Tres Lagoas.

[10] developed inventory of pollutant emissions for the fleet of public transportation buses and minibuses in Campo Grande. Depending on the reference adopted for calculation, such as emissions of carbon monoxide are estimated at up to 46 tons/month, while hydrocarbons and nitrogen oxides, 12.7 and 80.6 tons/month, respectively.

[11] investigated the daily and seasonal variability of surface ozone in a region near the city of Campo Grande. The hourly variability in September 2005 indicates more intense concentration in the afternoons due to the action of sunlight, and lower at night. Concentrations were recorded, the minimum and maximum of 5.6 ppbv to 71.9 ppbv.

In the Midwest region of Brazil studies on emissions of air pollutants are sometimes related to seasonal fires in states as Amazonas (AM), Rondonia (RO), Acre (AC), Goias (GO), Mato Grosso (MT) and Mato Grosso do Sul (MS) very significant part of these fires reach a global scale . However, for the regional context, there is evidence that, particularly at the time of the dry season in the Midwest, the presence of ozone at low altitude may be related to the transport of biomass burning regions. [12]

[13,14] present studies on the high concentration of ozone in the dry regions, respectively, Goiania (GO) and Maringa (PR). In Goiania, balloons were launched to monitor weather conditions and the concentration of ozone. The average altitude of 2.0 km were recorded peaks of concentration of the gas up to 112 ppbv.

In Maringa, located in northern Parana, therefore, more distant regions where fires occur, surface ozone measurements seem to confirm the influence of long-range transport during the dry season. When confronted measures in January (rainy season) and August (dry season) during the afternoon, the variation can be greater than 80 ppbv between stations.

In this chapter we present some results of an extensive research done on one of the Brazilian regions,the Midwest region, which in the last ten years has increased its participation in the industrial economy. Tres Lagoas (20.75 S, 51.68 W) represents the largest urban area in the eastern state of Mato Grosso do Sul, the gateway to two major biomes the Cerrado and the Pantanal, that have rich biodiversity and plant and animal species found only in this part of the world. The acelerated economic growing of this region is due to installation of factories of various industries, especially pulp , thermoelectric power generation and steel mill.

Some regions of Brazil are passing by economic changes, which easily imply the production of various effluents as solids, liquids, or, as is the focus of this chapter, gas, and additionally, many of these regions outside the Rio-Sao Paulo, where studies in air pollution are not consolidated, legislation and supervision are less intensive and restrictive and there is not relevant networks of air quality monitor.

As these are regions with no history of industrial impacts and that are at the beginning of its activity that the research results are intended as an instrument parameter for new scientific research in Brazil and elsewhere in the world. Figure 03 illustrates a clipping from the eastern to the urban area of Tres Lagoas featured, the Parana River and the areas of Sao Paulo State border, and with eucalyptus plantation for the pulp industry.

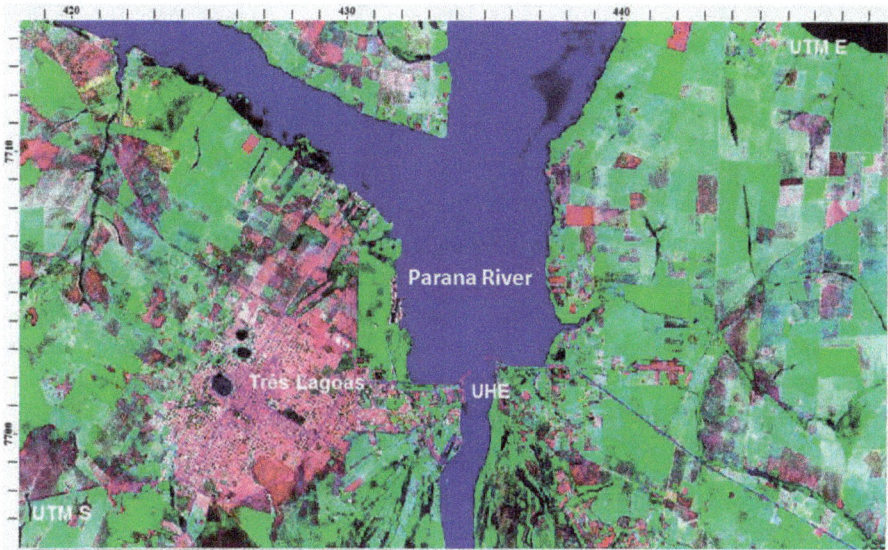

Figure 3. Image of Tres Lagoas City and Parana River.

4. Developing of the research

The sources of emissions of pollutants in the atmosphere, three groups were selected for investigation: Vehicular Sources, Biogenic Sources and Industrial Sources. The source vehicle through urban light fleet, is recognized as the greatest contribution in large urban centers to the outside world. Also, industrial sources, depending on the productive sector, effluent types and sizes, have a relevant role in the total emissions of a city. The biogenic sources in the present study are motivated primarily by its contribution to emissions of hydrocarbons, which, given their high reactivity, has a potential to contribute to the processes of formation and/or removal of secondary pollutants.

The research used different methodological resources to lead to results. In regions such as those examined did not have a monitoring network maintained by the public power, only stations, which by law, and due to the nature of the activity, have such monitoring stations in urban area of Tres Lagoas. Gas parameters monitored are carbon monoxide, nitrogen monoxide, nitrogen dioxide and ozone, meteorological parameters are wind speed, wind direction, relative humidity, temperature, solar radiation and rainfall. In addition to monitoring were visited several companies and government agencies with the purpose of obtaining information to produce an inventory to assist in estimating emissions.

4.1. Inventory of gaseous emissions

In 1971, the U.S.Environmental Protection Agency (EPA) established the U.S. American National Emissions Data (NEDS), a kind of database on the sources of emissions. This measure directed the organization of an annual report, an inventory estimates of emissions from sources potentially polluting, helping directly subsequent measures to control air quality.

The Air Emissions Inventory comprises, strictly speaking, a list containing the information gathering that characterize the pollution potential of a geographic region or sector-specific human activity or natural.

Among the main information that derive from an inventory of emissions are the pollutants selected for evaluation, the emission sources that best represent the scenario being studied, the activity level of each emission source selected, the most appropriate emission factors for the productive sector or source in focus, the range of uncertainty associated with estimates of each source and, finally, estimates of emissions of each pollutant within their respective categories of sources. [15-18]

Some definitions of Atmospheric Emissions Inventory in the literature addressing its broader nature, this document is a response to the relationship between emissions and concentration of pollutants in a region. Among these definitions:

"It is a compilation of pollutant emission estimates classified acoording to the different sources of emissions" [17]

"Approach is based on an understanding of the quantitative relationship between atmospheric emissions and ambient air quality" [19]

Other definitions address the relationship of the information generated in the inventory with the need to manage air quality, control of emissions from sources, adequate supervision regulations. Two of these definitions are:

"An air emissions inventory can range from a simple summary of estimated emissions compiled from previously-published emissions data to a comprehensive inventory of a facility using specific source test data that will be used to support compliance activities." [18]

"An emissions inventory is the foundation for essentially all air quality management programs. Emissions inventories are used by air quality managers in assessments of the contributions of and interactions among air pollution sources in a region, as input data for air quality models, and in the development, implementation, and tracking of control strategies." [16]

The emissions inventory can be classified as to the procedures adopted for the preparation in two approaches: top-down and bottom-up. The top-down approach, more general, used as input data for calculating the total quantity of a given fuel consumption and associated with an emission factor that is representative of the type of fuel measured. This approach is useful in situations where information is scarce or not available, or the collection costs are high or the end use of the inventory does not justify the investment. In general, this approach examines the sources and areas, and in this case, the use of average emission factors and activity level obtained from national or regional reports reduce the accuracy of estimates. [20,21]

The bottom-up approach considers a large number of details of the source evaluated in the input data for calculation. Typically it is used for point sources. However it can be used for the source area when data are relative to the source in question by means of prior research. Requires more resources to collecting information for the specific location. In general, are more accurate than top-down approach, because it does not use global information reporting national and regional agencies, but information more directly related to the source evaluated. [20,21]

Different authors have prepared inventories of emissions of pollutants in urban areas such as Bogota, Buenos Aires, Mexico City, etc. and recommended the use of both approaches. The organization of information may involve the use of different tools to meet the approaches, such as using simulation programs, remote sensing images, emission factors from the literature, data from continuous monitoring of pollutants. [21-24].

The results of this research used both approaches to the emission inventories of vehicles, emissions from the power plant and gas phase biogenic hydrocarbons in the forest of eucalyptus and other plants. Details of the methodology can be found in documents of the Intergovernmental Panel on Climate Change (IPCC), and table of emission factors in AP42 U.S. Environmental Protection Agency.

4.2. Monitoring of ambient air - gaseous and meteorological parameters

Concentration measurements of gaseous pollutants were conducted on three different monitors of the mark Horiba, APOA models 360 (Ozone), APNA 360 (NOX) and APMA 360 (CO), installed in a Fixed Station Monitoring of Air Quality and Meteorological Parameters in the outskirts of the city, northeast to the central area, coordinated UTM: 427,944 mE, 7,702,272 mS, altitude 325 meters. These instruments are company-owned Petroleo Brasileiro S.A. and their data were kindly ceded, in its raw form, to the Engineering School of USP and Center for Environmental Research of UEMS for the realization of this work.

The data of air quality monitoring and meteorological were submitted to calculations using basic statistic and multivariate statistical analysis of principal components. The basic statistics was applied to all series of the parameters measured and accepted as validated for the study, after coherence analyze. A multivariate analysis was applied to estimate the correlations between meteorological parameters and the average concentrations of ozone, carbon monoxide and nitrogen oxides.

5. Results and discussion

The results presented, in sequence, are part of investigation developed in eastern of MS, and it was gathered a group of information about the monitoring performed in primary pollutants for ozone formation and the meteorological behavior, and, also, the use a consolidated inventory of mobile and fixed sources.

Seasonal variations of NOx and CO for the daily cycle indicate similarities in the behavior of both pollutants. Their concentrations are higher at certain times of day (between 7 and 8 a.m. and 18 and 21 hours) in any month of the year, however, in the months between April and September peak periods are increased. Probably, the reduction of the average height of the mixed layer and/or temperature inversions of this period of months may explain, at least in part, increased concentration of CO and NOx. Figures 04 and 05 illustrate the seasonal behavior of NOx and CO for 2006 to 2005, years when there is less loss of data for both pollutants.

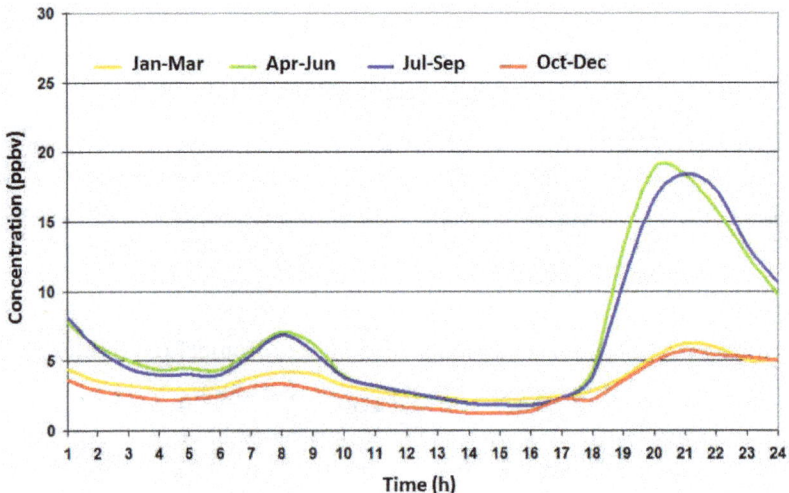

Figure 4. Seasonal behavior daily cycle of NOx at 2006.

The peak periods in both figures, are directly associated to the time of entry and exit of vehicular fleet of urban movement. This behavior indicates that the NOx and CO are available from the local atmosphere, probably mostly from vehicle emissions, given that other sources, such as ceramics and thermoelectric operate under almost continuous. The working of the ceramics kiln, on average during the firing process, 50 hours and, interchangeably, which ensures the discharge of pollutants in a nearly constant.

The Thermoelectric Unit of Tres Lagoas in a 24-hour cycle suffers almost no fluctuation in the levels of NOx that are emitted in their chimneys. On 11 October 2005, with ozone level recorded in 80 ppbv (threshold standard of air quality) to the average concentration of 16.9 ppmv NOx, the standard deviation did not exceed 3.6 ppmv.

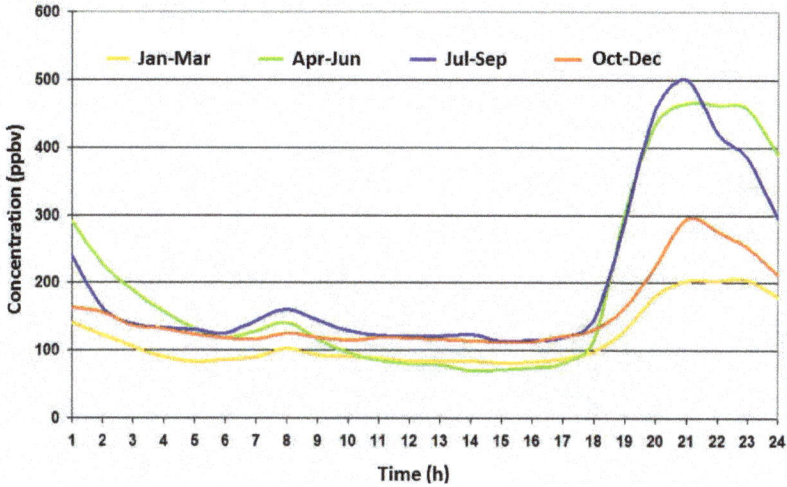

Figure 5. Seasonal behavior daily cycle of CO at 2005.

The behavior of NOx and CO, ozone precursors, are observed for some periods of the year for which data are available for both pollutants. The relationship kept between both NOx and CO, can be an indicator of a common source of emission. In 2006, the highest monthly average ozone occurred in September, and the week between 3 and 9 showed a strong linear relationship (R^2 = 0.85) between NOx and CO, Figure 06 illustrates this result.

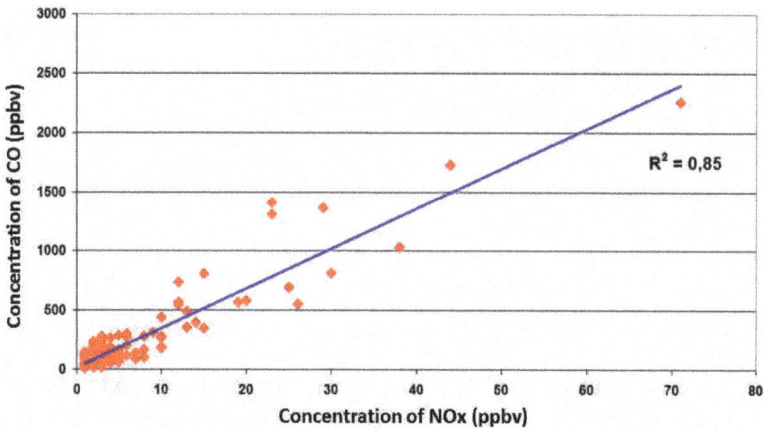

Figure 6. Relationship between NOx and CO for the period of 3 to 9 September 2006.

Moreover, in 2005, the month in which occurred the largest monthly average, in October, this behavior is not observed with the same intensity. The linear relationship between NOx and CO cannot be considered strong for the series of data from weeks 9-15 and 16-22 October 2005. In Figure 07 the ratio is more representative of a common source of both pollutants than in Figure 08, where $R^2 = 0.12$ shows weak linear relationship. In a test is disregarded one of the points of Figure 08, far away, with a concentration of NOx = 33 ppbv and CO =194 ppbv. This procedure resulted in an increase of R^2 (0.33) however did not alter the condition of weak linear relationship between NOx and CO for this period.

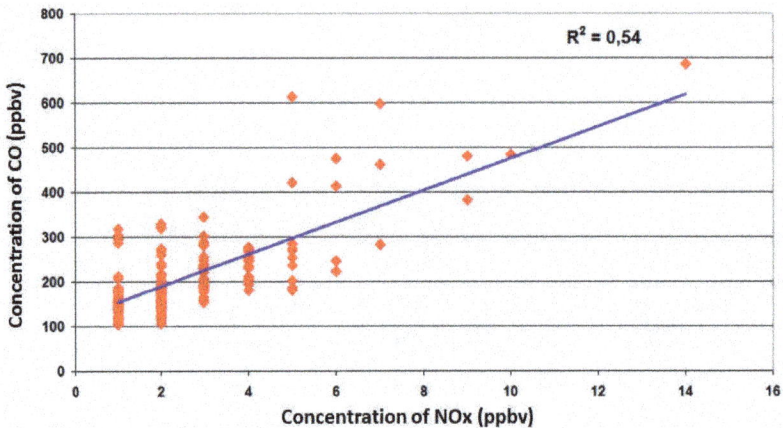

Figure 7. Relationship between NOx e CO for the period of 9 to 15 in October 2005.

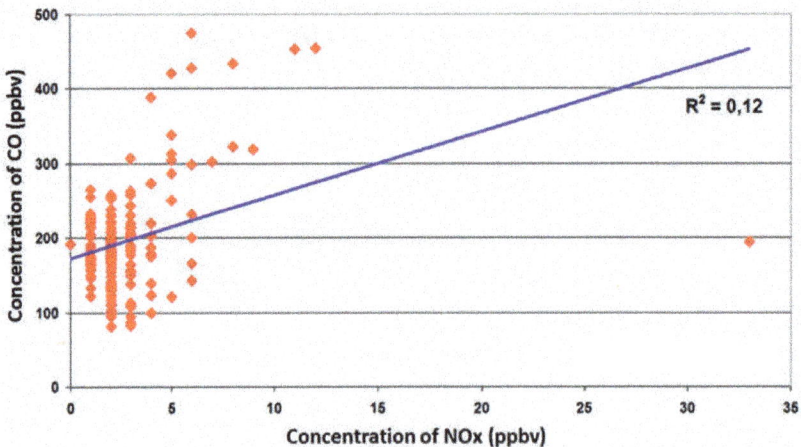

Figure 8. Relationship between NOx e CO for the period of 16 to 22 in October 2005.

Figure 09 illustrates the behavior of the average annual daily cycle of NOx and O3 for the year 2006, and Figure 10 for the series of CO and O3 in 2005. The choices of years in correspondence with the pollution are due to the larger number of valid data.

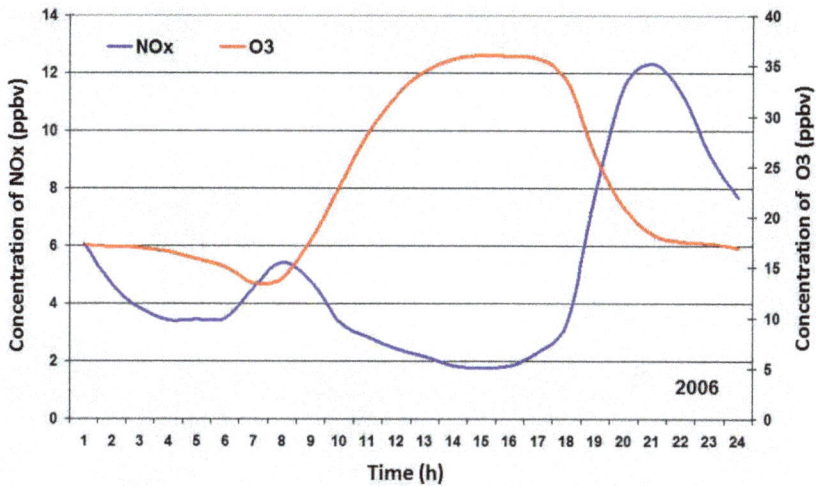

Figure 9. Daily Cycle of NOx and Ozone at 2006.

The NOx emitted initially in the form of NO from exhaust of the vehicle fleet later in the atmosphere, chemical reactions with peroxyl radical results in the formation of NO_2. As the intensity of solar radiation increases, photons with wavelengths below 420 nm break NO_2 molecules that allow, in the presence of oxygen (O_2) the formation of ozone.

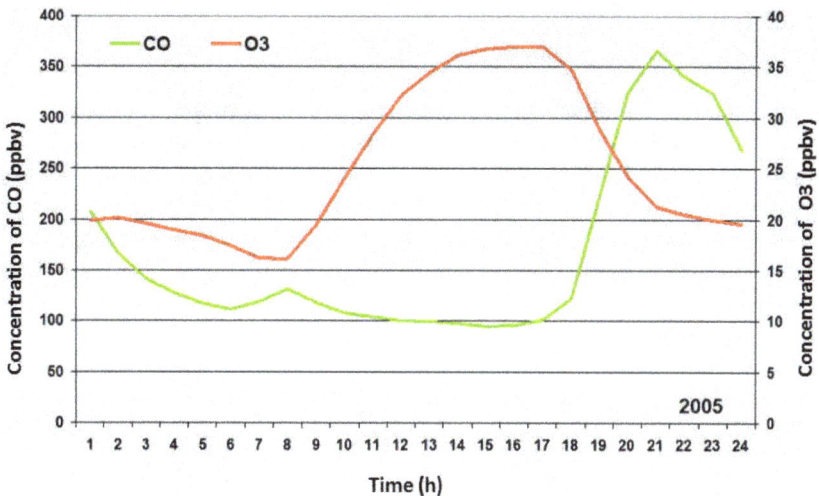

Figure 10. Daily Cycle of CO and Ozone at 2005.

NOx levels are directly related to the concentration of ozone, is to intensify the formation of O3 or even limit its formation. In the case of Tres Lagoas NOx levels are considered low, almost equivalent to concentrations found in rural areas, according to [2], which may partly explain the low ozone concentration.

According to [2] the concentration of carbon monoxide in the environment, for the Southern Hemisphere, should range between 60 and 200 ppbv. Mean levels of Tres Lagoas are close to this age, presenting, in a few months, comparable to peak concentrations of large cities such as Ribeirao Preto, in Sao Paulo State, which according [25] may have concentrations up to 3 ppmv.

Carbon monoxide while relatively stable (residence time 30 to 90 days) may react with the hydroxyl radical, which, in the presence of oxygen (O_2), carbon dioxide and form hydroperoxide radical which therefore reacts with NO to NO_2 in converting , triggering the process of formation of ozone molecules.

5.1. Relationship of the NOx, CO, ozone and wind conditions

Two periods were selected to analyze the relationships between pollutants, wind direction and speed, temperature and relative humidity. They are 6 to 9 January and 1 to September 4, 2005. Tables 01, 02, 03 and 04 illustrate results of the first period.

Wind Direction	Average Concentration of Pollutants (ppbv)			
	%	CO	NOx	O₃
≤ 90	60.4	108.6	0.4	16.6
90-180	10.4	207.8	2.4	12.6
180-270	7.3	264.3	4.0	7.3
> 270	21.9	143.7	0.9	16.8

Table 1. Pollutants Concentration and Wind Direction. (6-9 /January/2005).

The wind direction, predominant in the period 6-9 January, occurs toward Lake-town (NE-SW) by almost 70% of the time, being higher towards the city, 60.4%. In this direction the average concentrations of CO and NOx are lower and higher ozone when the wind blows from the lake to the city (Direction ≤ 90), and the process is reversed when the winds blow from the city to the lake, when the CO (264.3 ppbv) and NOx (4.0 ppbv) have their highest average concentrations and the minimum ozone (7.3 ppbv). The monitoring station is virtually between the lake and the city, thus the data suggest that higher concentrations are caused by sources, whether mobile or fixed, present in the city.

Wind Speed	Average Concentration of Pollutants (ppbv)			
	%	CO	NOx	O₃
≤ 1,0	31.3	206.1	2.4	10.9
1,1-2,0	55.2	108.1	0.3	17.4
2,1-4,0	13.5	102.2	0.2	18.8
> 4,0	-	-	-	-

Table 2. Pollutants Concentration and Wind Speed. (6-9/January/2005).

The occurrences of wind speeds below 2.0 m/s represent 86.5% of cases, 55.2% between 1.1 and 2.0 m/s that are typical of the daytime period. For this last category of wind speed average concentrations of CO, NOx and O3 are similar to those for the wind that blows from the lake to the city. The reduction of the concentrations of CO and NOx as the wind speed increases indicate better dispersion of pollutants, so conversely, lower speeds, require greater stability of the atmosphere, the smaller the dispersion. Speeds below 1.0 m/s occur, mostly at night. At this speed category the average concentrations of NOx and CO reach the highest values, respectively, 2.4 and 206.1 ppbv, while ozone is reduced to the lowest level, of course, in case night due to the absence of the photochemical process.

5.2. Relationship between NOx, CO, ozone and relative humidity

The relative humidity can influence the flow and hence the actin photochemical processes. The average concentration of ozone decreases with increasing relative humidity, however for percentage of moisture above 80%, ozone, and also other pollutants may suffer reductions due to removal by rainfall.

The Table 03 illustrates that, indeed, the ozone is reduced due increasing the percentage of moisture, but should not be, in every day, directly because the action of precipitation, even as the measures rainfall occurrences indicate only the days 7,8,9, or before 8 am or after 19 hours. In six days the precipitation occurred in the period (afternoon), resulting in lower values of O3, CO and NOx. Table 04 illustrates pollutants for days 6-9 January.

% Relative humidity	Average Concentration of Pollutants (ppbv)			
	%	CO	NOx	O_3
≤ 40	0	-	-	-
40-60	14.6	93.6	-	21.1
60-80	32.3	127.4	0.5	18.1
> 80	53.1	156.5	1.4	12.4

Table 3. Pollutants Concentration and Relative Humidity(6-9/January/ 2005).

Days	Daily Average Concentration of Pollutants		
	CO	NOx	O_3[1]
6	114.7	0.6	17.0
7	148.7	1.1	19.1
8	180.3	2.0	20.4
9	108.0	0.1	21.4

Table 4. Average Concentration of Pollutants relating to day 6 to 9 January 2005.
[1] Mean values for the period between 11 and 17 hours.

It is possible verify that the mean values of CO and NOx are smaller, also on day 9 in which no precipitation occurred during the day. However, between 20 hours (day 8) and 8 o'clock in the next morning, the 9th, it rained almost 11.2 mm, assisting in the removal of Local air pollutants

5.3. Relationship between NOx, CO, ozone - conditions of temperature

Although the temperatures for the period in focus, are all greater than 20 ºC, it is possible to note differences between the average concentration of pollutants in the range of 20-30 ºC and above 30 °C. The ozone precursors undergo a reduction in its concentration with increasing temperature, while the level of ozone rises. There are several parameters which influence the concentration of pollutants, but the higher temperature can also assist the formation of ozone, and therefore the consumption of CO and NOx due to the role of the temperature in photochemical processes. Table 05 illustrates the influence of temperature on the concentration of pollutants.

		Average Concentration of Pollutants (ppbv)		
Temperature	%	CO	NOx	O₃
≤ 10		-	-	-
10-20		-	-	-
20-30	76.0	151.3	1.2	13.9
> 30	24.0	95.4	0.04	20.8

Table 5. Pollutant Concentration and Temperature. (6 9/January/2005).

5.4. Relationship between NOx, CO, ozone - conditions of the wind

The results of the wind for the period September, Table 06, indicate again the predominant direction of Lake-town, about 80.2% of the time analyzed. However, the direction of the wind flow was reversed to Town-lake, 53.1% of the time. The highest average concentrations of ozone, 40.5 and 32.9 ppbv, still, respectively, occur in directions ≤ 90 and> 270, as from January. By comparison, in January the values of the average concentrations of ozone in these directions were similar, but in September the average concentrations of ozone, winds coming from the lake were higher, with differences of up to 23%.[26]

Carbon monoxide has higher average concentrations for January for directions> 270° (417.0 ppbv) and between 90° and 180° (666.8 ppbv). However, these points wind directions in 1/5 of the evaluation period, as represented in January 1/3. The NOx exhibits the same behavior pattern in January, but with higher average concentrations up to 14 times.

Winds off the lake from the city to carry higher concentrations of NOx (55.6 ppbv), while in the opposite direction, the average is not more than 8.0 ppbv. The ratio of inversion between O3 and NOX, regardless of direction, seems to remain, while carbon monoxide undergoes variations between the direction > 270 and 180-270 °.

The directions in which the CO has the highest concentrations occur at night, in which, according to Figure 05, even for seasonal variations in the period of early evening CO concentrations rise, and especially in the month on September they reach levels above 500 ppbv. The reduction of the mixed layer in the evening may result in the transport of pollutants from higher regions to the surface. And in the second half, in Brazil, there are high numbers of fires in the Amazon and the Midwest regions that can affect distant areas, as indicated by [15]. Table 06 illustrates the information talked about above the average concentration of pollutants as a function of wind direction.

Wind Direction	Average Concentration of Pollutants (ppbv)			
	%	CO	NOx	O₃
≤ 90	27.1	235.7	8.0	40.5
90-180	10.4	666.8	28.5	14.2
180-270	53.1	147.4	55.6	16.6
> 270	9.4	417.0	16.8	32.9

Table 6. Pollutant Concentration and Wind Direction.(1-4/September/2005).

Table 07 illustrates the average concentrations of pollutants and the average speed of the winds. Carbon monoxide and nitrogen oxide reducing their concentrations with increasing wind speed; this is due to better dispersion of pollutants. However, the same does not occur with ozone, which for speeds greater than 4.0 m/s concentration rises to 30.7 ppbv. According [26] at high speeds can reduce the stability of the mixed layer allowing the entry of additional upper ozone layer, however higher speeds should alleviate this scenario.

High concentrations of CO and NOx emissions associated with velocities below 1.0 m/s must occur, in most cases, during the night, in view of low media concentrations of ozone, 8.8 ppbv.

Wind Speed	Average Concentration of Pollutants (ppbv)			
	%	CO	NOx	O₃
≤ 1,0	20.8	639.2	31.5	8.8
1,1-2,0	24.0	232.5	4.7	33.4
2,1-4,0	36.5	114.9	7.4	24.0
> 4,0	18.7	106.2	3.1	30.7

Table 7. Pollutant Concentration and Wind Speed.(1-4/September/2005).

5.5. Relationship NOx, CO and ozone - relative humidity

Table 08 concerns the relationship of pollutants to the air humidity reflects similar behavior to that found for the period 6-9 January. The increase in humidity implies the reduction of ozone and NOx and CO increased. As mentioned earlier, the presence of water in the air flow affects the actin and thus the photochemical processes and the formation of ozone.

% Umidade	Average Concentration of Pollutants (ppbv)			
	%	CO	NOx	O₃
≤ 40	10.4	90.3	2.3	51.4
40-60	29.2	341.7	15.2	29.4
60-80	46.9	282.2	12.3	18.2
> 80	13.5	69.0	2.2	13.9

Table 8. Pollutant Concentration and Relative Humidity.(1-4/September/2005).

The intensity of solar radiation for the two periods is similar, with January showing slightly higher values. In percentages above 80%, the rapid reduction of NOx and CO, and O3 less intense, are related, with higher chances of occurrence of rainfall. But, during the evaluation period precipitation occurred within 20 hours of the day 1 and 14 hours of the day 2, it represents about 6% of the period. This situation favors the removal of pollutants from the atmosphere, thereby reducing its concentration.

5.6. Correlation between a group of pollutants and meteorological parameters

Another statistical analysis was the principal component analysis, considering the time series of meteorological parameters of wind direction and speed, temperature, relative humidity and radiation pollutants CO, NOx and O3.[26] The periods, above, 6-9 January and 1-4 September in 2005 were used to compose the correlation matrices. Tables 09 and 10 illustrate the matrices.

	CO	NOx	O₃	D	V	T	U	R
CO	1							
NOx	0.90	1						
O₃	-0.65	-0.70	1					
D	0.26	0.26	-0.06	1				
V	-0.49	-0.54	0.59	-0.15	1			
T	-0.08	-0.11	0.50	0.19	0.32	1		
U	0.25	0.28	-0.65	-0.15	-0.43	-0.96	1	
R	-0.26	-0.16	0.39	0.17	0.28	0.54	-0.55	1

Table 9. Principal Component Analysis for the period from 6 to 9 January 2005

In both tables, 09 and 10, the strongest correlations between pollutants, occur with CO and NOx.Correlations with values of 0.89 and 0.90, as occurred with CO and NOx, indicate, in fact, compared strong linear between the two pollutants, and that possibly arise from the same source.

	CO	NOx	O₃	D	V	T	U	R
CO	1							
NOx	0.89	1						
O₃	-0.38	-0.54	1					
D	-0.07	0.02	-0.40	1				
V	-0.50	-0.66	0.28	-0.11	1			
T	0.19	-0.03	0.76	-0.35	0.02	1		
U	-0.09	-0.05	-0.67	0.28	0.31	-0.77	1	
R	-0.31	-0.38	0.58	-0.38	0.14	0.43	-0.56	1

Table 10. Principal Component Analysis for the period 1 to 4,September, 2005.

Both the CO and NOx average correlation remains strong with two other parameters in both periods, the ozone concentration and wind speed, and inverse correlations were negative. This fact, combined with the strong linearity for each other (CO and NOx) and ozone concentrations remain at average levels up to 3 times lower than the established by legislation (close to levels from rural areas) seem to confirm the greater weight of the dynamics of pollutants are related to sources and local meteorological conditions.

Ozone-average correlation remains strong in both periods, with the temperature, relative humidity and solar radiation, which were expected. However, regarding the direction and wind speed suffers considerable variations. In January, the correlation of ozone with the wind, virtually nonexistent points was the statistic (-0.06), while for the period September to correlate with the wind direction is close to the average (-0.40). Table 06 showed that the highest average concentration of ozone for the period 1 to September 4 were from the region of Jupia Lake (between 0° and 90° and 270° and 360°), no such sources directions.

5.7. Analysis of the results of simulation training ozone

Additionally, a computer program, for simulation, provided by EPA was able to estimate the concentrations of ozone in urban areas. The input data for simulation were taken from the Atmospheric Emission Inventory produced on this investigation and the database of monitoring the local atmospheric concentration of CO and NOx. As no data are available concentration of volatile hydrocarbons were adopted in Tres Lagoas three different concentration levels: 800 ppbC (polluted urban region), 150 ppbC (typical urban region), 50 ppbC (rural), for the performed simulation.

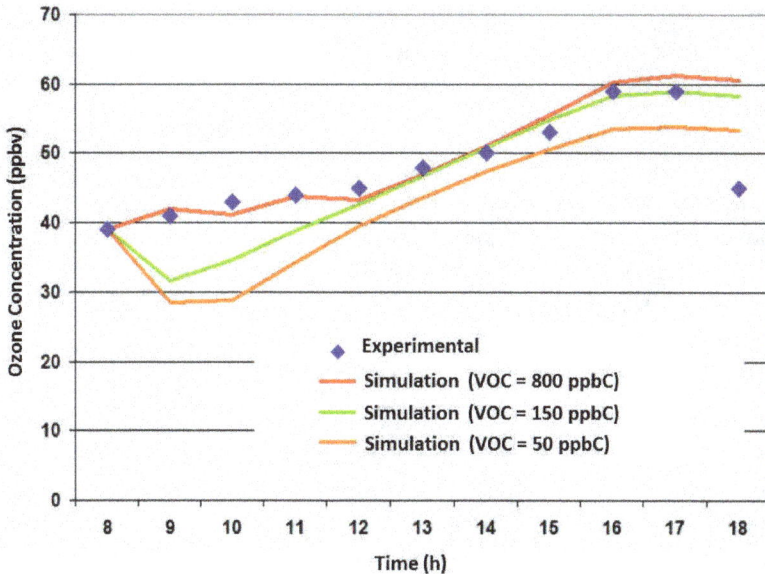

Figure 11. Simulation of the formation of ozone at different concentrations of Volatile Organic Compounds.

The results obtained, and illustrated in Figure 11, indicate that variations of ozone with the different levels of hydrocarbons do not undergo significant changes, indicating a possible limitation of the formation of ozone due to the NOx. The concentration curve that best fits the experimental data was 800 ppbC, which it can be consider high value, and maybe no real, additional investigations about hydrocarbon gas, with monitor devices must be performed to elucidate this question.

In considering much lower concentrations, as 50 ppbC, the difference, near the midday, in which solar radiation is more intense, is not significant. But for the initial hours of the day the difference increases. In the morning, in general, the mixing height layer is at its lower levels and may cause the entry of the ozone layer to the upper surface.

6. Conclusion

The highest emissions were carbon monoxide, with the majority coming from the vehicle fleet. Estimates indicate that in 2005 and 2006 emissions of the vehicle fleet accounted for 97% of emissions of carbon monoxide, and about 53% of nitrogen monoxide from the sources analyzed, which generally agrees with results other urban areas where emissions from the fleet are dominant. The total emissions of both pollutants CO and NOx emissions in 2006 were, respectively, 290.00 and 2,614.40 tons, whose magnitude is low, of course, compared to other cities that have air emissions inventory, which in general are of size greater the Tres Lagoas, or are attached to larger ones.

The statistical results show that the concentrations of pollutants, CO and NOx are well correlated with meteorological parameters, demonstrated that their sources must be local. The statistical correlation between CO and NOx reaches 0.90, confirming the linear dependence between them.

The computer simulation in order to evaluate the role of hydrocarbons showed that the local atmosphere, although with potential sources of NOx, this source is an agent for limiting the formation of local ozone. Compared with simple equations under steady state, it shows similar results, with a slight discrepancy in the morning. The biggest difference for the model that uses the steady state condition is in September, dry season. In January, the correlation between the curves suggests low concentrations of hydrocarbons in the atmosphere, but also, the rainfall can cause wet deposition, or the reduction of the action of actin flow. This issue must be better investigated.

The assessment of emissions and behavior of the concentration of primary and secondary pollutants arising out of or caused by existing sources so far are not sufficient to cause degradation scenarios of air quality. However weather conditions may lead to intensification of the action of pollutants in urban areas. The information available to indicate that the air quality in Tres Lagoas, the scope for analysis, is within the standards acceptable by current brazilian law, and some pollutants at concentrations equivalent to the average levels in rural areas indicated in the literature.

The original contribution of this work is to be a reference for future assessments of air quality in this region, where industrial growth is accelerated and the government has not yet manifested in monitoring policies and actions for monitoring the impacts on local and regional atmosphere, or its possible consequences for the environment.

Author details

Edmilson de Souza
Mato Grosso do Sul State University, UEMS, Brazil

Josmar Davilson Pagliuso
University of Sao Paulo, USP, Brazil

Acknowledgement

Our sincere gratefully to FUNDECT for funding and supporting research and PETROBRAS S.A. for their collaboration.

7. References

[1] Salby, M. L. Fundamentals of Atmospheric Physics. Academic Press. 1995

[2] Seinfeld, J. H. and Pandis, S. N. Atmospheric Chemistry And Physics: From Air Pollution to Climate Change. John Wiley & Sons. 1997

[3] Wallace, J. M. and Hobbs, P. V. Atmospheric Science: An Introductory Survey. Academic Press. 2006.

[4] Warneck, P. Chemistry of the Natural Atmosphere. Academic Press. 1988

[5] Finlayson-Pitts, B.J. & Pitts, J.N. Chemistry of the Upper and Lower Atmosphere. Acamedic Press. 2000

[6] Mayer, H. Air Pollution in Cities. Atmospheric Environment. Vol. 33.p.4.029-4.037. 1999

[7] CETESB. Relatório de Qualidade do Ar do Estado de Sao Paulo. 2006

[8] IBAMA. http://www.ibama.gov.br/proconve/login.php (accessed from to february 2006)

[9] FEEMA. Inventário de Fontes Emissoras de Poluentes Atmosféricos da Região Metropolitana do Rio de Janeiro. 2006

[10] Kozerski, G. R. & Hess, S. C. Estimativa dos Poluentes Emitidos Pelos Ônibus e Microônibus do Campo Grande/MS, Empregando Como Combustível Diesel, Biodiesel ou Gás Natural. Engenharia Sanitária Ambiental. Vol. 11. p.113-117. 2006

[11] Pavao et. al. O ozônio de superfície: variações diárias e sazonais para Campo Grande, MS. Simpósio de Geotecnologias no Pantanal. 2006.

[12] Freitas, S. R. et. al. Emissões de Queimadas em Ecossistemas da América do Sul. Estudos Avançados. Vol.19. p.167-185. 2005

[13] Kirchhoff, V.W.J.H. and Marinho,V.A. Layer enhancements of tropospheric ozone in regions of biomass burning. Atmospheric Environment. Vol. 28. p.69-74. 1994

[14] Boian, C. and Kirchhoff, V.W.J.H. Surface ozone enhancements in the south of Brazil owing to large-scale air mass transport. Atmospheric Environment. Vol. 39. p.6140-6146. 2005

[15] IPCC. Revised Guidelines for National Greenhouse Gas Inventories: Reference Manual. 1996

[16] EEA – European Environment Agency. Good Pratice for Emission Inventories. 2001

[17] Power, H. & Baldasa, J. M. Air Pollution Emissions Inventory. Witpress. 1998

[18] EPA – U. S. Environmental Protection Agency. Introduction to the Emission Inventory Improvement Program. Energy. 1997

[19] NRC-NATIONAL RESEARCH COUNCIL. Rethinking the Ozone Problem in Urban and Regional Air Pollution. National Academy Press. 1991

[20] EPA – U. S. Environmental Protection Agency. Handbook for Criteria Pollutant Inventory Development: A Beginner's Guide for Point and Area Sources. Office of Air Quality. 1999

[21] Puliafito, S. E. Emissions andA ir Concentrations of Pollutant for Urban Área Sources. Mecânica Computacional. Vol. 24. p. 1389-1408. 2005

[22] Zarate, E. et. al. Air Quality Modeling over Bogota, Colombia: Combined Techniques to Estimate and Evaluate Emission Inventories. Vol. 41. p. 6302-6318. 2007

[23] Arriaga-Colina, J. L. et. al. Measurements of VOCs in Mexico City (1992-2001) and Evaluation of VOCs and CO in the Emissions Inventory. Vol. 38. p.2523-2533. 2004

[24] Sawyer, R. F. et. al. Mobile Sources Critical Review: 1998 NARSTO Assessment. Vol. 34. 2161-2181. 2000

[25] Freitas, M. K. & Pagliuso, J. D. Investigação da Produção e Dispersão de Poluentes do AR no Ambiente Urbano: Determinação Empírica e Modelagem em Rede Neural da Concentração de CO. EESC/USP. 2003

[26] Jun Tu et. al. Temporal Variations in Surface Ozone and its Precursors and Meteorological effects at an Urban site in China. Atmospheric Research. Vol. 85. p.310-337. 2007

Statistical Character and Transport Pathways of Atmospheric Aerosols in Belgrade

Zoran Mijić, Andreja Stojić, Mirjana Perišić, Slavica Rajšić and Mirjana Tasić

Additional information is available at the end of the chapter

1. Introduction

Clean air is considered to be a basic requirement for human health and well being. Various chemicals are emitted into the air from both, natural and anthropogenic sources. In spite of the introduction of cleaner technologies in industry, energy production and transport, air pollution remains a major health risk and tighter emission controls are being enforced by many governments. Atmospheric particles – aerosols – are some of the key components of the atmosphere. They influence the energy balance of the Earth's surface, visibility, climate and environment as a whole [1-3]. According to World Health Organization (WHO), ozone, particulate matter (PM), heavy metals and some hydrocarbons present the priority pollutants in the troposphere [4]. Public health can also be indirectly affected by deposition of air pollutants in environmental media and uptake by plants and animals, what results in entering of chemicals into the food chain or drinking water, and thereby constituting additional sources of human exposure. A number of epidemiological studies have demonstrated that acute and chronic health effects are related to the inhalable PM_{10} (aerodynamic diameter less than 10 μm) exposure in the urban environment, and some data also seem to indicate possible seasonal effects of the particulate matter on human health [5-10]. This is especially important for urban aerosols, whose variety of size and composition make complete characterization a difficult task. Particulate matter pollution is nowadays one of the problems of the most concern in great cities, not only because of the adverse health effects, but also of the reducing atmospheric visibility and affect to the state of conservation of various cultural heritages [11]. Therefore, the measurement of the levels of atmospheric particulate matter is a key parameter in air quality monitoring throughout the world.

As a result of health and environmental impacts of PM, more rigorous regulations are in force in the USA and European countries. PM standards, issued by European Commission

(EC), have included PM_{10} monitoring and limit values in the Air Quality Directive in 1999 [12]. Directive established in the first stage, annual limit value of 40 µg m^{-3} and 24h limit value of 50 µg m^{-3} (not to be exceeded more than 35 times in a calendar year) to be met by 2005, and in the second stage annual limit value of 20 µg m^{-3} and 24h limit value of 50 µg m^{-3} (not to be exceeded more than 7 times in a calendar year) to be met by 2010. The detailed discussion of these limit values, regulations and relations of new EU standards to US EPA standards can be found elsewhere [13,14].

Most of the trace metals are emitted in particulate form [15-17] and present in almost all aerosol size fractions, but are mainly accumulated in the smaller particles [18]. This has a great effect on the toxicity of metals since the degree of respiratory penetration depends on particle size. Trace metals are persistent and widely dispersed in the environment and interacting with different natural components result in toxic effects on the biosphere [19,20]. The studies of the transport and mobilization of trace metals up to now have attracted attention of many researchers [21-24]. Trace metals are released into the atmosphere by human activities, such as combustion of fossil fuels and wood, high temperature industrial activities and waste incinerations. The combustion of fossil fuels constitutes the principal anthropogenic source for Be, V, Co, Ni, Se, Mo, Sn, Sb, and Hg. It also contributes to anthropogenic release of Cr, Mn, Cu, Zn, and As. High percentages of Ni, Cu, Zn, As, and Cd are emitted from industrial metallurgical processes. Exhaust emissions from gasoline may contain variable quantities of Ni, Cu, Zn, Cd, and Pb [25,26]. In addition to the PM mass concentration limit values, also based on health impact criteria, recent European Union standards set target (Ni, As, Cd) and limit (Pb) values for metals [12]. Environmental technologies may have to be adopted in specific industrial spots to reach the target values. For aimed reduction of PM_{10} levels and their trace metal contents detailed knowledge of sources and their respective contribution to the PM_{10} levels, is required.

One of the main difficulties in air pollution management is to determine the quantitative relationship between ambient air quality and pollutant sources. Various receptor models have been used to identify aerosol sources and estimate their contributions to PM_{10} and trace elements concentrations at receptor sites [27-32]. The combined application of different types of receptor models could possibly solve the limitations of the individual models, by constructing a more robust solution based on their strengths.

Processes in the atmosphere represent a complex problem due to the simultaneous influence of several independent factors. Similarly to other air pollutants, PM_{10} concentrations are random variables influenced by the emission level, meteorological conditions and topography. Each area is specific and the required emission reduction to meet Air Quality Standard (AQS) is different. Information about the frequency distribution of pollutants is useful for developing air pollution control strategies. When the specific probability function of an air pollutant is known, it is easy to predict the required emission reduction, the frequency of exceedance of the AQS, as well as the return period [33]. Many types of probability distributions have been used to fit air pollutant concentrations. These include lognormal distribution [34,35], Weibull distribution [36], gamma distribution [37] and type V Pearson distribution [38]. The Weibull probability model is suitable for presenting air quality data in a "pseudo-lognormal" distribution [39,40]. In the present study, daily average mass concentrations of PM_{10} were taken from 2003 to 2005 in order to estimate the parameters for three theoretical distributions. Lognormal, Weibull and type V Pearson distributions were chosen to fit the PM_{10} data in Belgrade. The distributional parameters were estimated by the maximum likelihood method. Based on the fitted distributions the minimum reduction required in order to achieve AQS was estimated using rollback equation [41-43]. Moreover, since the tail of theoretical distributions diverged in the high concentration region, a two-parameter exponential distribution was used to fit high PM_{10} concentrations and estimate exceedances and return period more precisely [44]. These methods can reasonably predict the return period and exceedances in the succeeding period and can also be used for predicting source emission reduction.

In many countries, the weekly variation in the concentration of atmospheric aerosols and meteorological variables in urban areas has been reported as evidence of anthropogenic influence on the weather system. Of the primary sources road transport is the dominant contributor to high particulate level in the air in most major cities. This study examines the temporal and weekly variations in meteorological variables and PM_{10} mass concentrations in Belgrade. The cumulative frequencies and quantile-quantile plot for Sundays against workdays were constructed. The results showed that Sunday concentrations were significantly reduced from those during the week.

The Mediterranean region and particularly the Balkan Peninsula have been under the influence of Saharan dust transport and deposition over millennia. Identification of the concentration, composition, origin, transport and geographical distribution of PM_{10} in Mediterranean atmosphere has been the subject of research activities since the last two decades as it is heavily affected by two contrasting sources; mineral dust (mainly from Sahara Desert) [45] and various anthropogenic (from industrialized/semi-industrialized countries) emissions. Deviation from lognormal distribution of atmospheric aerosols is a positive indicator of possible transport processes [33]. Since we are particularly interested in transportation of trace metals content in PM_{10} samples, the quantiles for some trace elements were analyzed by the quantile-quantile P-P slope test. This was used for fitting the quantiles of the expected theoretical cumulative probability for normal, lognormal and Weibull distribution with the quantiles of the cumulative probabilities calculated for the experimental dataset [46].

In addition, we investigated the transport pathways and potential sources of PM_{10} concentrations based on backward trajectories and PM_{10} concentration records from 2004 to 2006. Cluster analysis was used to reveal the major pathways for different seasons as well as corresponding statistical analysis related to different clusters. Hybrid receptor models

Potential Source Contribution Function (PSCF) and Concentration Weighted Trajectory (CWT) were used for identification of source regions. The PSCF values can be interpreted as a conditional probability describing the spatial distribution of probable geographical source locations by analysis of trajectories arriving at the sampling site [47]. Since the PSCF method is known to have difficulties in distinguishing strong from moderate sources, the CWT model that determines the relative significance of potential sources has been additionally performed [29,48].

In this review, we report some of the results of the integral monitoring of air quality in Belgrade urban area in order to evaluate the impact of airborne trace metals on the pollution load for the period from 2003 to 2006. Some of the results concerning suspended particle mass and trace metal concentrations in ambient air of the Belgrade will be presented, with the aim to examine elemental associations and to indicate the main sources of trace and other metals in the city. The results of this long-term project of the pollution monitoring could be used as the baseline data for analysis of health risks due to inhalation of suspended aerosols, and to provide scientific evidence for setting up an air pollution control strategy. This information is crucial in environmental quality assessment, and can lead to the determination of a possible exceedance of the critical loads.

2. Experimental methods and procedures

Sampling of particulate matter PM_{10} started in the very urban area of Belgrade (Hs = 117 m, $\varphi = 44^0\,44'\,N$ and $\lambda = 20^0\,27'\,E$) the capital of Serbia, in June 2002 and has continued afterwards. The one sampling site was the platform above the entrance steps to the Faculty of Veterinary Medicine at a height of about 4 m from the ground, 5 m away from a street with heavy traffic and close to the big Autokomanda junction with the main state highway. This point can be considered as traffic-exposed.

The second sampling point was on the roof of the Rector's Office building of Belgrade University, at a height of about 20 m. As this sampling point is in the very city centre, on the rooftop where the airflow is not blocked by any direction, it can be considered as representative for urban-background concentrations. Suspended particles were collected on pre-conditioned and pre-weighed Pure Teflon filters (Whatman, 47 mm diameter, 2 μm pore size) and Teflon-coated Quartz filters using MiniVol air sampler (Airmetrics Co. Inc., 5 l min^{-1} flow rate) provided with PM_{10} cutoff inlet. Particulate matter mass concentration was determined by weighting of the filters using a semi-micro balance (Sartorius, R 160P), with a minimum resolution of 0.01 mg. Loaded and unloaded filters (stored in Petri dishes) were weighed after 48 hours conditioning in a desiccator, in the clean room at a relative humidity of (45-55)% and a temperature of (20 ± 2) ^{0}C. Quality assurance was provided by simultaneous measurements of a set of three "weigh blank" filters that were interspersed within the pre- and post- weighing sessions of each set of sample filters and the mean change in "weigh blank" filter mass between weighing sessions was used to correct the sample filter mass changes. After completion of gravimetric analysis, PM samples were digested in 0.1 N HNO$_3$ on an ultrasonic bath. An extraction procedure with dilute acid was used for the evaluation of elements which can become labile depending on the acidity of the environment [49]. This procedure gives valid information on the extractability of elements, since the soluble components in an aerosol are normally dissolved by contact with water or acidic solution in the actual environment.

The elemental composition (Al, V, Cr, Mn, Fe, Ni, Cu, Zn, Cd, and Pb) of the aerosol samples was measured by the atomic absorption spectroscopy method (AAS). Depending on concentration levels, samples were analyzed for a set of elements by flame (FAAS) (Perkin Elmer AA 200) and graphite furnace atomic absorption spectrometry (GFAAS) using the transversely-heated graphite atomizer (THGA; Perkin Elmer AA 600) with Zeeman-effect background correction. The THGA provided a uniform temperature distribution over the entire tube length, rapid heating and an integrated L'vov platform, which gave an improved signal/interference ratio and high analytical sensitivity. Analyte injection (20 μl) and the atomization were done in five steps controlled by the appropriate software and auto-sampler. For calibration, standard solutions containing all metals of interest were prepared using Merck certified atomic absorption stock standard solutions containing 1000 mg l^{-1} metal in 0.5 N HNO$_3$ and Milli-Q quality deionised water, with no matrix modifier addition. Details on sampling procedures and PM analysis are given in detail elsewhere [50-55].

3. Estimation of emission source reduction

Assuming unchanged spatial distribution of emission sources, meteorological conditions and no reactive species, according to the rollback equation [36], the emission source reduction R (%) required to meet AQS can be estimated as:

$$R = \frac{E\{C_p\} - E\{C\}_s}{E\{C_p\} - C_b} \tag{1}$$

In this equation E{C}$_s$ is the mean (expected) concentration of the distribution when the extreme value equals C$_s$ (the concentration of the AQS), E{C$_p$} is the mean concentration of the actual distribution and C$_b$ the background concentration. If C$_s$ = 125 μg m^{-3}, the PM$_{10}$ daily average concentration is not exceeded more than once per year (P[PM$_{10}$>C$_s$] = 1/365 = 0.00274), then E{C}$_s$ is the expected daily PM$_{10}$ average concentration of a distribution where the probability of a concentration exceeding 125 μg m^{-3} is equal to 0.00274.

Frequency distribution of PM$_{10}$ is a very useful tool to estimate how frequently a critical concentration level is exceeded. In order to obtain the best frequency distribution of PM$_{10}$, three theoretical functions were used: lognormal, Weibull and type V Pearson distributions. The distribution parameters were estimated by the maximum likelihood method. Probability density functions as well as corresponding equations for estimating distribution parameters are presented in Table 1.

Distribu-tion	Probability distribution function	Maximum likelihood estimated parameter
lognormal	$$p_l(x) = \frac{1}{x\sigma_g(2\pi)^{\frac{1}{2}}}\exp\left[-\frac{(\ln x - \ln\mu_g)^2}{2\sigma_g^2}\right]$$ $$x > 0;\ \sigma > 0;\ \mu_g > 0$$	$$\ln\mu_g = \frac{1}{n}\sum_{i=1}^{n}\ln x_i$$ $$(\ln\sigma_g)^2 = \frac{1}{n}\sum_{i=1}^{n}(\ln x_i - \ln\mu_g)^2$$
Weibull	$$p_w(x) = \frac{\lambda}{\sigma}\left(\frac{x}{\sigma}\right)^{\lambda-1}\exp\left[-\left(\frac{x}{\sigma}\right)^{\lambda}\right]$$ $$x \geq 0;\ \ \sigma,\lambda > 0$$	$$\lambda = \left[(\sum_{i=1}^{n}x_i^{\lambda}\ln x_i)\times(\sum_{i=1}^{n}x_i^{\lambda})^{-1} - \frac{1}{n_i}\sum_{i=1}^{n}\ln x_i]\right]^{-1}$$ $$\sigma_w = (\frac{1}{n_i}\sum_{i=1}^{n}x_i^{\lambda})^{\frac{1}{\lambda}}$$
type V Pearson	$$\varphi(x_i, x_i^{eq}) = \frac{x_i^{eq}(\rho_i)^{\rho_i+1}}{\Gamma(\rho_i+1)}e^{-\rho_i\frac{x_i^{eq}}{x_i}}\left(\frac{x_i}{x_i^{eq}}\right)^{-(\rho_i+2)}$$ $$x_i > 0;\ x_i^{eq} > 0;\ \rho_i > 0$$	$$x_i^{eq} = \frac{\rho_i+1}{\rho_i}\frac{1}{\left(\frac{1}{n}\right)\sum_{i=1}^{n}\frac{1}{x_i}}$$ $$\frac{d\ln(\Gamma(\rho_i+1))}{d\rho_i} = \ln(\rho_i x_i^{eq}) - \frac{1}{n}\sum_{i=1}^{n}\ln(x_i)$$

Table 1. Theoretical distributions and equations used for parameters estimation

The appropriateness of each distribution was assessed by the Kolmogorov-Smirnov (K-S) test. The K–S statistic is defined as the maximum difference between the sample cumulative probability and the expected cumulative probability i.e.

$$D = \max|f_n(x) - F(x)| \tag{2}$$

where $f_n(x)$ and $F(x)$ are the expected and observed cumulative frequency functions, respectively. The D value will be compared to the largest theoretical difference, $D_{\alpha/2}$ acceptable for the K-S test at a certain significance level α. If D is less than $D_{\alpha/2}$, one can conclude that the hypothesis that there is no difference between observed and expected distributions has been accepted at the α level of significance. Generally speaking, if D value declines the goodness of fit increases.

The fitted results of three theoretic distributions and the measured data for PM10 are presented in Figure 1. Once the distribution parameters were determined, K-S test could be used to assess which type of distribution is more appropriate for representing the PM10 distribution. From Figure 1 it is clear that the Weibull distribution is inappropriate for representing PM10 distribution (D = 0.073), while type V Pearson distribution is the most suitable one (D = 0.038).

Figure 2 shows the relation between exceeding probability and PM10 concentrations for different distribution functions. It was found that the probabilities of exceeding the AQS (125 μg m⁻³) were 0.12, 0.096 and 0.088 for lognormal, Weibull and type V Pearson distribution, respectively. This means that the number of days exceeding the AQS in the following year would be 43, 35 and 32, respectively. The actual probability for this period was 0.075 (27 days) and it is clear that all three distributions overestimated the exceeding probability although type V Pearson distribution most closely represented the true PM_{10} data.

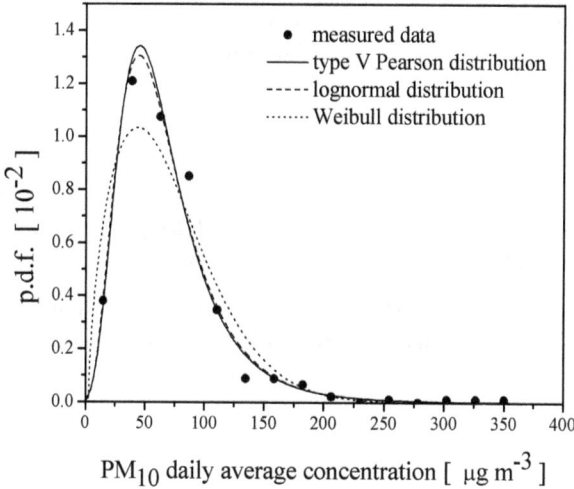

Figure 1. Comparison of the measured PM_{10} data with different probability distribution functions

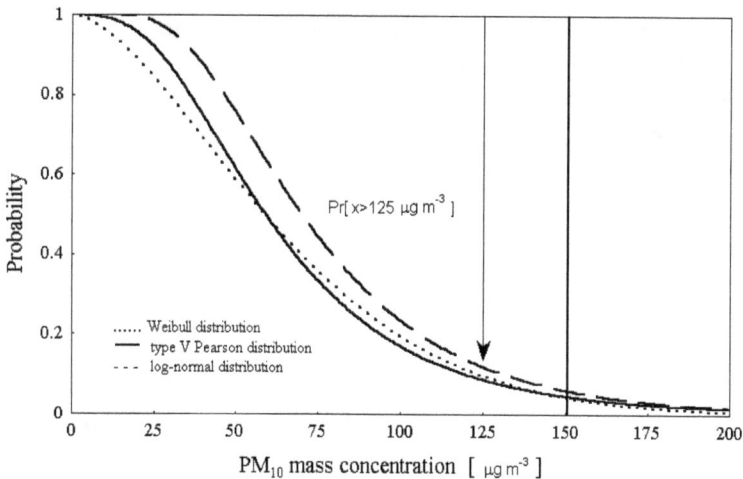

Figure 2. Relation between PM_{10} mass concentration and probability for different distribution functions

After determining the most appropriate distribution function for PM_{10}, the emission source reduction required to meet AQS can be predicted from a rollback equation. The complementary distribution function for calculating the probability of variable x_i exceeding the critical value x for lognormal and type V Pearson distributions can be found in a statistical textbooks. It was found that the value of E{C}s (the average concentration of lognormal distribution where the probability of a concentration exceeding 125 μg m⁻³ equals 0.00274) was 32.1 μg m⁻³. Therefore, the mean PM₁₀ concentration should be reduced from the observed value of 68.4 μg m⁻³ to 32.1 μg m⁻³ and the estimated emission reduction can be calculated from equation 1 being equal to 53.1%. For type V Pearson distribution the estimated source reduction required to meet AQS is 58.6%. Therefore, the source emissions should be controlled much more to reduce PM₁₀ concentrations and meet AQS in the future period [41].

Previous theoretical distributions can give good result for estimating the mean concentration and required reduction of source emission. However, the fitted results of the parent distributions are not accurate enough in the high concentrations region. Therefore, a two-parameter exponential distribution was applied to predict the return period and exceedances of the critical PM₁₀ concentration.

A two-parameter exponential distribution derived from extreme value theory [44] represents the cumulative frequency distribution of high concentrations over a specific percentile.

$$F_L = 1 - e^{-y_n} \tag{3}$$

$$y_n = b_m(x_n - \phi) \tag{4}$$

where y_n, b_m and ϕ are the variate and the parameters of the distribution and x_n is the chosen PM₁₀ concentration exceeding the specific percentile. From the complete data set of PM₁₀ (2003 - 2005), concentrations exceeding 80-th percentile were chosen to fit the two-parameter exponential distribution. The estimated cumulative probability can be calculated from the chosen high PM₁₀ concentration, x_n, as

$$\overline{F}_L(x_n) = \frac{N_1 - r + 1}{N_1 + 1} = P_{rN_1} \tag{5}$$

where N₁ is the size of the chosen high PM₁₀ concentration and P_{rN_1} is the probability of a value that is ranked r out of N₁ values. The relation between variate y_n and P_{rN_1} is:

$$y_n(r) = -\ln(1 - P_{rN_1}) \tag{6}$$

and the parameters b_m and ϕ can be estimated by the least-squares method. In addition, the return period $R(x_c)$, defined as the average number of averaging periods (or observations) between exceedances of a given critical concentration, x_c, can be calculated as:

$$R(x_c) = \frac{1}{(1 - f)(1 - F_L(x_c))} \tag{7}$$

where f is the chosen specific percentile. After determining return period the days of exceedances within one year can be calculated in order to develop air control strategy.

Figure 3 shows the fitted theoretical line of the variate y_n and PM$_{10}$ concentration over 80-th percentile, x_n. The fitted equation for the theoretical two-parameter exponential distribution is $F_L(x_n) = 1 - \exp[-0.016(x - 75.37)]$. From equations (3) and (7) for critical concentration, $x_c = 125$ µg m^{-3}, F_L and the return period can be calculated as: $F_L(x_c) = 1 - \exp[-0.016(125 - 75.37)] = 0.548$ and $R(x_c) = 1/[(1 - 0.8)(1 - 0.548)] = 11$ days. From this calculation predicted exceedances over AQS are estimated to be 33 days in the following year. The prediction of return period for a critical concentration can also be used for estimating source emission reduction. If the AQS of PM$_{10}$ is 125 µg m^{-3} and the allowed exceeding probability is once per year, then the expected return period is 365 days and $F_L(125$ µg m$^{-3}) = 0.986$. In addition, the PM$_{10}$ concentration expected to be equaled or exceeded once per year can be calculated from equation (3). The expected concentration is 337 µg m^{-3} and it can be seen that a reduction of 212 µg m^{-3} (about 63%) needs to be achieved in Belgrade in order to meet AQS (125 µg m^{-3}).

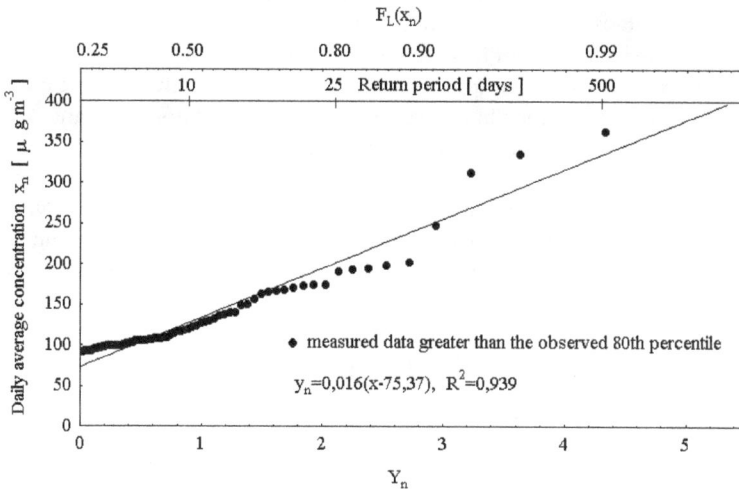

Figure 3. The fitted theoretical line of variate and PM$_{10}$ concentrations over 80-th percentile by type I two-parameter exponential distribution

Although this technique can predict the probability of exceedance of the AQS and return period for the highest concentrations level it does not allow reproducing the time pattern of the estimated concentration levels. Based on the observed time series data set an empirical modelling approach can be applied in order to derive mathematical formulation suitable for the description of the feature of the concentration event [56-58].

4. Estimating the frequency distribution of PM$_{10}$ by the statistics of wind speed

Since the distributions of air pollutant concentrations are influenced by meteorological conditions, such as wind speed, Simpson et al. [59] linked air quality data and wind speed to the Atmospheric Turbulence and Diffusion Laboratory (ATDL) model. The concentration of air pollutant data at cumulative probability, p, is inversely proportional to the wind speeds v at probability (100-p) when the frequency distributions of both data are log-normally distributed and the geometric standard deviation (shape factor) is the same e.i.

$$K = C_p v_{(100-p)} \qquad (8)$$

where K depends on the average area source strength of pollutant and atmospheric stability factor (which varies with atmospheric conditions). This model was used to predict the maximum level of air pollutant. Finally, the frequency distribution of air pollutants can be estimated from the frequency distribution of wind speed [60]. Here we demonstrate the relationship between the frequency distributions of wind speed and daily PM_{10} measured in Belgrade during 2003 year. Three theoretical frequency distributions, (lognormal, Weibull and gamma) were used to fit the measured PM_{10} data. The distributional parameters were estimated by the method of maximum-likelihood and the K–S test was used to determine which type of distribution is appropriate to represent the frequency distribution of PM_{10} and wind speed. The fitted results of PM_{10} are shown in Figure 4 and it can be seen that the lognormal distribution is the most appropriate one to represent the statistical character of PM_{10} (D_{max} = 0.06). The fitted results of lognormal distributions for PM_{10} and wind speed, with corresponding parameters, are presented in Figure 5. If is found that the shape factors of lognormal distributions σ_g are similar, therefore equation 8 can be used to estimate the distribution of PM_{10} from the distribution of wind speed. The average K value can be calculated from the regression line using method of least-squares for the wind speed data and PM_{10} concentrations from the 30-th to 80-th percentiles (the K value is almost constant in that range). The obtain regression is

$$C_{p,PM_{10}} = 141.5 / v_{(100-p)} + 3.64 \qquad (9)$$

with the correlation coefficient R = 0.99 and estimated K = 141.5 µg m^{-2} s^{-1}.

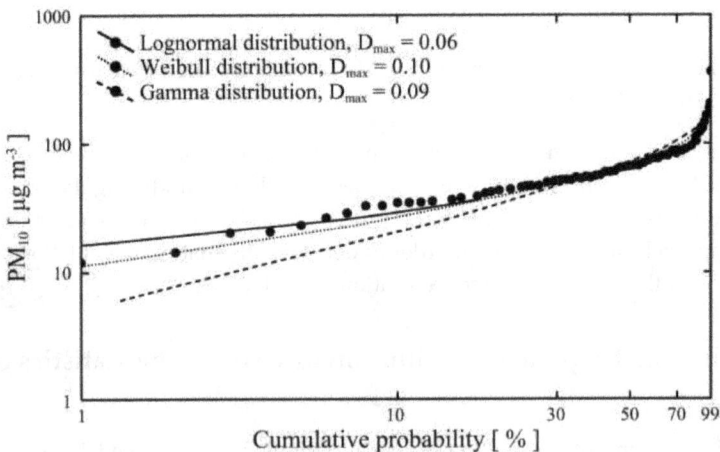

Figure 4. The fitted results of PM_{10} for different distribution functions

When the atmospheric condition is unchanged and the K value is estimated, the concentration of PM10 at different percentiles can be obtained. Observed and estimated frequency distribution of PM10 is shown in Figure 6. The difference between the observed and estimated concentration is significant in the higher cumulative probability range, because the inconsistency of the both, wind speed and PM10 data with lognormal distribution in this region. The result of K-S test shows that the difference between observed and estimated distribution is not significant at the 5% significance level, so the obtain results are acceptable.

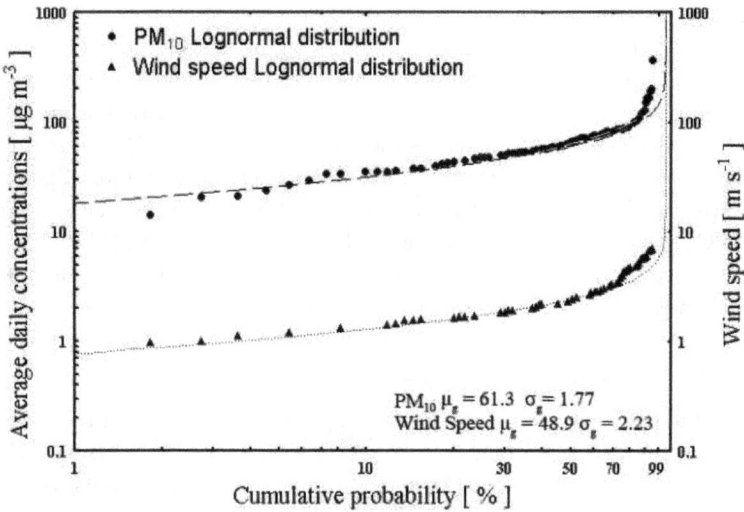

Figure 5. The fitted results of lognormal distributions for PM10 and wind speed

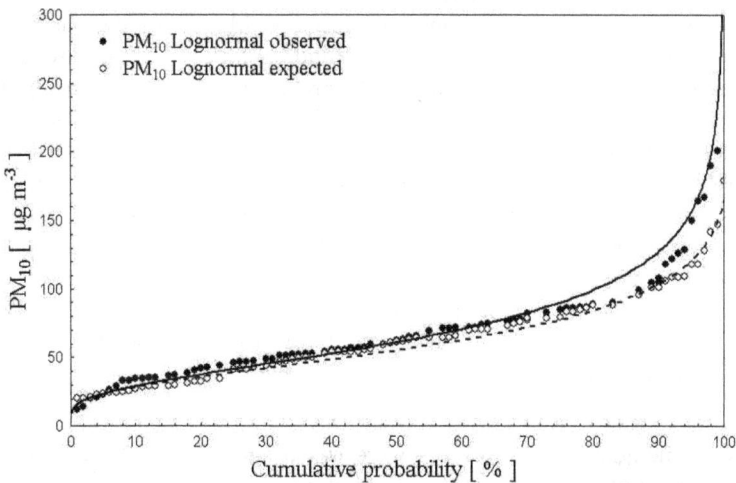

Figure 6. The observed and estimated frequency distributions of PM10

It is clear that there is no universal distribution that can be used to represent the frequency distribution of PM$_{10}$ concentrations. This analysis shows that if the distribution type and shape factors are the same for the wind speed data and air pollutant concentration, there exists simple relationship between two data sets and the frequency distribution of the air pollutant can be reasonably estimated. It can be useful for understanding statistical characteristics of air quality in different region.

5. Weekend effect and PM$_{10}$ temporal variations

Weekly cycles of the mass concentrations of anthropogenic aerosols have been observed in many regions and the intensity of these cycles vary significantly depending on region and season [61,62]. In addition, ozone concentrations tend to be higher on weekends than on weekdays, despite the fact that lower emissions of ozone precursors are expected on weekends. This phenomenon is known as the "weekend effect" [63]. The mechanisms for weekend effect on ozone formation are still not well understood. In a lot of papers that are dealing with weekend effect in detail, the diminution of NOx, CO, PM$_{10}$ and volatile organic compound (VOCs) concentrations can be noticed during the weekend [64,65]. Although the relationship between the variation in aerosol concentration and weather variables has not been clearly understood, many studies have revealed significant weekday–weekend differences in weather variables such as the diurnal temperature range [66], and clouds and precipitation [67,68].

Generally speaking, several source types make significant contribution to particle level in Belgrade: combustion processes, mixes road dust, traffic, metal processing and transport [30,50]. Like in many other major cities with a few millions inhabitants, of the primary sources, fossil fuel combustion and traffic contributions play the most significant role [51].

Since traffic and commercial pattern in urban area show a marked change at weekends, it is expected that particulate concentrations will decrease. Markovic et al. [69] reported the mean daily concentrations of SO$_2$, NO$_2$, PM$_{10}$, CO and O$_3$ during the weekdays and weekend in November/December 2005 case study in Belgrade urban area (Figure 7). The results explicitly show that daily average concentrations of SO$_2$, NO$_2$, PM$_{10}$ and CO show significant weekly cycles with the largest values around midweek and smallest values in weekend. With the exception of ozone, concentrations of airborne pollutants were higher during weekdays. It has been suggested that the weekend condition of lower particle concentrations resulted in less absorption of sunlight, which in turn leaded to enhanced photochemical reaction and enhanced ozone formation [70]. Besides, the higher concentrations of PM$_{10}$, NO$_x$, CO and SO$_2$ during the weekdays influence on the increased deposition of O$_3$ on the particles as well as on the increased consummation of O$_3$ in the reactions with NOx, CO and SO$_2$ what also can have an impact.

In order to investigate differences in PM$_{10}$ mass concentrations between Sundays and weekdays quantile-quantile (Q-Q) plots were constructed [71]. Q-Q plots are useful in detection the distribution differences in data set and allow comparison of concentration distribution of the entire range of values, from the smallest to the largest. If two cumulative frequency diagrams (x and y) are plotted, then, corresponding to any frequency p, there are two quantile values $q_x(p)$ and $q_y(p)$ and Q-Q plot is a scatter plot of $q_y(p)$ versus $q_x(p)$ for various p. Obviously, if x and y are identically distributed, the plot of $q_y(p)$ versus $q_x(p)$ would fall on a straight line with the slope equal to one.

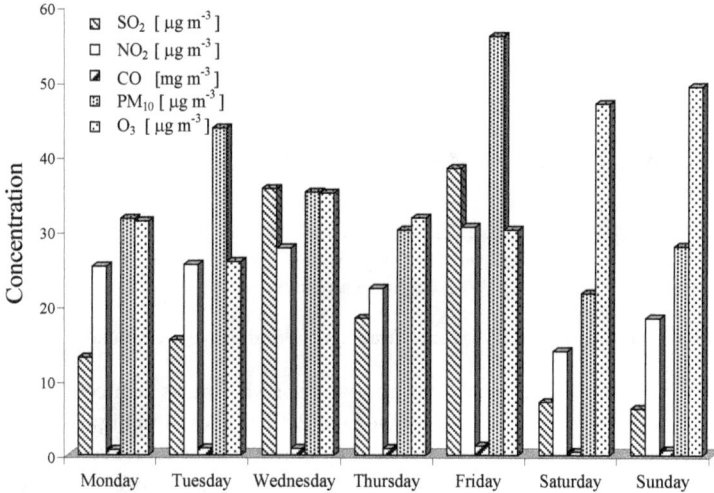

Figure 7. Mean daily concentrations during the weekdays and weekend [63]

The cumulative frequency diagram and quantile-quantile plot for Sundays against weekdays (weekdays were defined as Monday to Friday) were constructed and are shown in Figures 8 and 9. It can be seen that Sunday concentrations were reduced from those during the weekdays. These figures are based on complete PM_{10} data set for the period 2003-2005 in Belgrade urban area continuously recorded by the Public Health Institute of Belgrade.

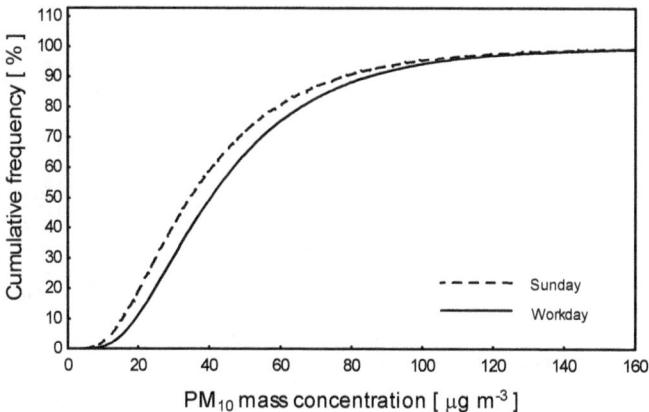

Figure 8. Cumulative frequency diagram of PM_{10} mass concentrations in Belgrade

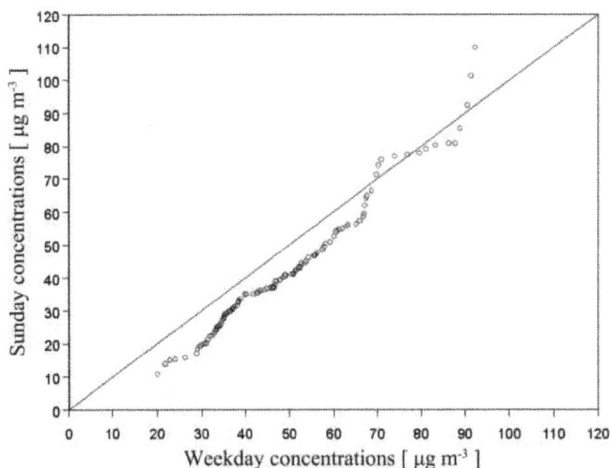

Figure 9. Quantile-quantile plots of PM$_{10}$ concentrations of Sundays versus weekdays

We have analysed mean diurnal variations of PM$_{10}$ mass concentrations and the characteristic pattern based on data set for the year 2005 is shown in Figure 10. It can be seen that the maximum concentrations occur in the morning and late evening which can be significantly related to traffic pattern and anthropogenic influence. Significant seasonal variations of some metals (Ni, V, Al and Fe) have already been reported [30] indicating the impact of fossil fuel combustion and local heating units. Monthly variation of PM$_{10}$ mass concentrations for the year 2005 is presented in Figure 11 showing significant differences during the winter and summer period.

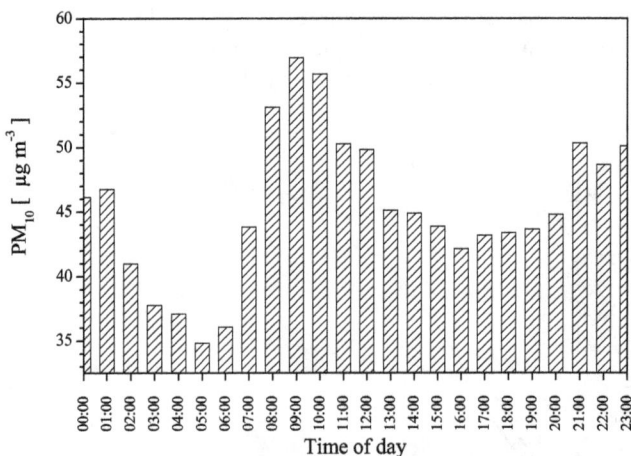

Figure 10. Mean diurnal concentration of PM$_{10}$ in Belgrade, 2005.

Analysis of air quality data in the frequency domain contributes to the understanding of periodic behaviours and yields information about temporal and spatial scales of the underlying mechanisms [72]. The results obtained in many studies show that the frequency analysis of air quality time series is a powerful technique that can provide important information about the nature of the processes behind the measurements. This methodology could be applied to determine representative background concentrations of air pollutants by removing short-term fluctuations associated with influence of local emission sources [73].

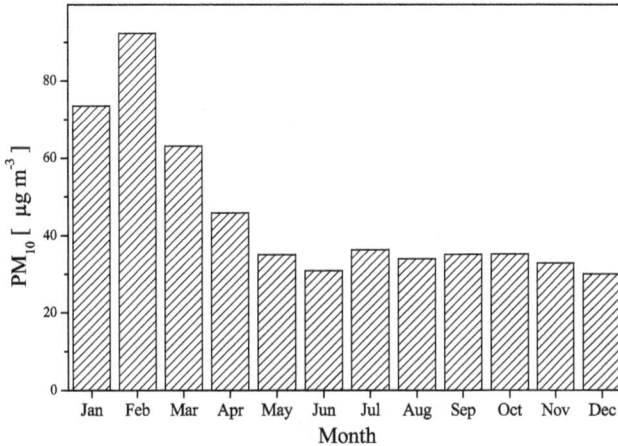

Figure 11. Monthly variations of PM$_{10}$ concentration in Belgrade, 2005

6. Transport analysis

Processes in the atmosphere represent a complex problem due to the simultaneous influence of several independent factors (meteorological conditions, pollutant emission level and topography) and thus the environmental data are random variables that follow natural lognormal distribution [33]. Deviation from lognormal distribution of atmospheric aerosols is a positive indicator of possible transport processes. In this study the quantiles for some trace metals in PM$_{10}$ samples were analyzed by the quantile-quantile P-P slope test. It is used for fitting the quantiles of the expected theoretical cumulative probability for normal, lognormal and Weibull distribution with the quantiles of the cumulative probabilities calculated for the experimental dataset.

Fittings of quantiles of the experimental database with the expected quantiles for theoretical normal, lognormal and Weibull distributions for Al, Mn and V are shown in Figure 12.

The P-P slope test results show that the Weibull distribution is the most appropriate for V and Ni suggesting their possible regional transport. The other metals followed lognormal distribution (except Cu and Zn data which are not successfully described by any of theoretical distribution function) what is a positive signal that an active sources exist in the vicinity. For unknown metal sources, hybrid models that incorporate wind trajectories Potential Source Contribution Function (PSCF) and Concentration Weighted Trajectory (CWT) were used to resolve source locations [29].

Al

Mn

V

a)   b)   c)

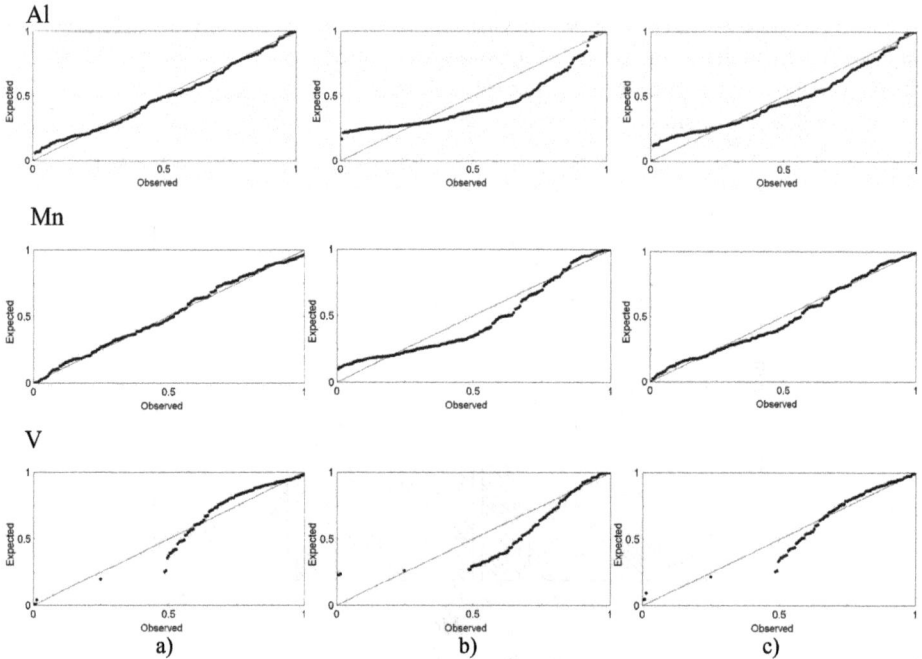

Figure 12. Expected and observed quantiles according to a) lognormal b) normal and c) weibull distribution for Al, Mn and V in PM_{10}

6.1. Potential Source Contribution Function (PSCF)

Air parcel back trajectories, ending at the receptor site, are represented by segment endpoints. Each endpoint has two coordinates (latitude, longitude) representing the central location of an air parcel at a particulate time. To calculate PSCF, the whole geographic region of interest is divided into an array of grid cells whose size is dependent on the geographical scale of the problem so that PSCF will be a function of locations as defined by the cell indices i and j. If a trajectory end point lies at a cell of address (i, j), the trajectory is assumed to collect material emitted in the cell. Once aerosol is incorporated into the air parcel, it can be transported along the trajectory to the receptor site. The PSCF value can be understood as a conditional probability that describes the spatial distribution of probable source locations. The staying time of all trajectories in a single grid cell is n_{ij}, and m_{ij} is the staying time in the same cell that corresponds to the trajectories that arrived at a receptor site with pollutant concentrations higher than a pre-specified criterion value (average PM_{10} and metal concentrations were used in this study). The PSCF value for the ij-th cell is then defined as

$$PSCF_{ij} = \frac{m_{ij}}{n_{ij}} \qquad (10)$$

Cells related to the high values of potential source contribution function are the potential source areas. However, the potential source contribution function maps do not provide an emission inventory of a pollutant but rather show those source areas which emissions can be transported to the measurement site. To remove the large uncertainty caused when a grid cell has small staying time and large PSCF values, usually a weight function $W(n_{ij})$ multiplies the PSCF value to better reflect the uncertainty in the values for these cells [27]. In this study the grid covers area of interest defined by $(35^0 - 50^0)$ N and $(10^0 - 30^0)$ E with cells $0.5^0 \times 0.5^0$ latitude and longitude.

Air masses back trajectories were computed by the HYSPLIT (HYbrid Single Particle Lagrangian Integrated Trajectory) model [74] through interactive READY system [75]. Daily 48h back trajectories, started from the center of Belgrade (44.804^0 N 20.478^0 E) at 12:00 UTC each day, were evaluated for six different heights above the starting point at ground level (350, 500, 750, 1000, 1500 and 2000 m). The total number of days was 280, therefore, the total number of end points was 6x48x280 or 80640. In average, there is 80640/1200 or about 70 end points per cell.

In the present study the weight function $W(n_{ij})$ was defined as:

$$W(n_{ij}) = \begin{cases} 1 & n_{ij} \geq 140 \\ 0.85 & 70 \leq n_{ij} < 140 \\ 0.5 & 40 \leq n_{ij} < 70 \\ 0.25 & n_{ij} < 40 \end{cases} \tag{11}$$

The weighting function reduced the PSCF values when the total number of the endpoints in a particular cell was less than about two times the average value of the end points per each cell.

6.2. Concentration Weighted Trajectory (CWT)

In the current PSCF method, grid cells having the same PSCF values can result from samples of slightly higher concentrations than defined by the criterion or extremely high concentrations. As a result, larger sources can not be distinguished from moderate sources. Therefore, a method of weighting trajectories with associated concentrations (CWT - concentration weighted trajectory) was developed. In this procedure, each grid cell gets a weighted concentration obtained by averaging sample concentrations that have associated trajectories that crossed that grid cell as follows:

$$C_{ij} = \frac{1}{\sum\limits_{l=1}^{M} \tau_{ijl}} \sum\limits_{l=1}^{M} C_l \tau_{ijl} \tag{12}$$

C_{ij} is the average weighted concentration in the grid cell (i,j), C_l is the measured PM concentration observed on arrival of trajectory l, τ_{ijl} is the number of trajectory endpoints in the grid cell (i,j) associated with the C_l sample, and M is the total number of trajectories. Similar to the PSCF method, CWT method also employs the arbitrary weight function to eliminate grid cells with few endpoints. Weighted concentration fields show concentration gradients across potential sources. This method helps to determine the relative significance of potential sources.

For air mass trajectory visualization and statistical analysis, a software application called TrajStat was used in which clustering, PSCF and CWT methods were included [76].

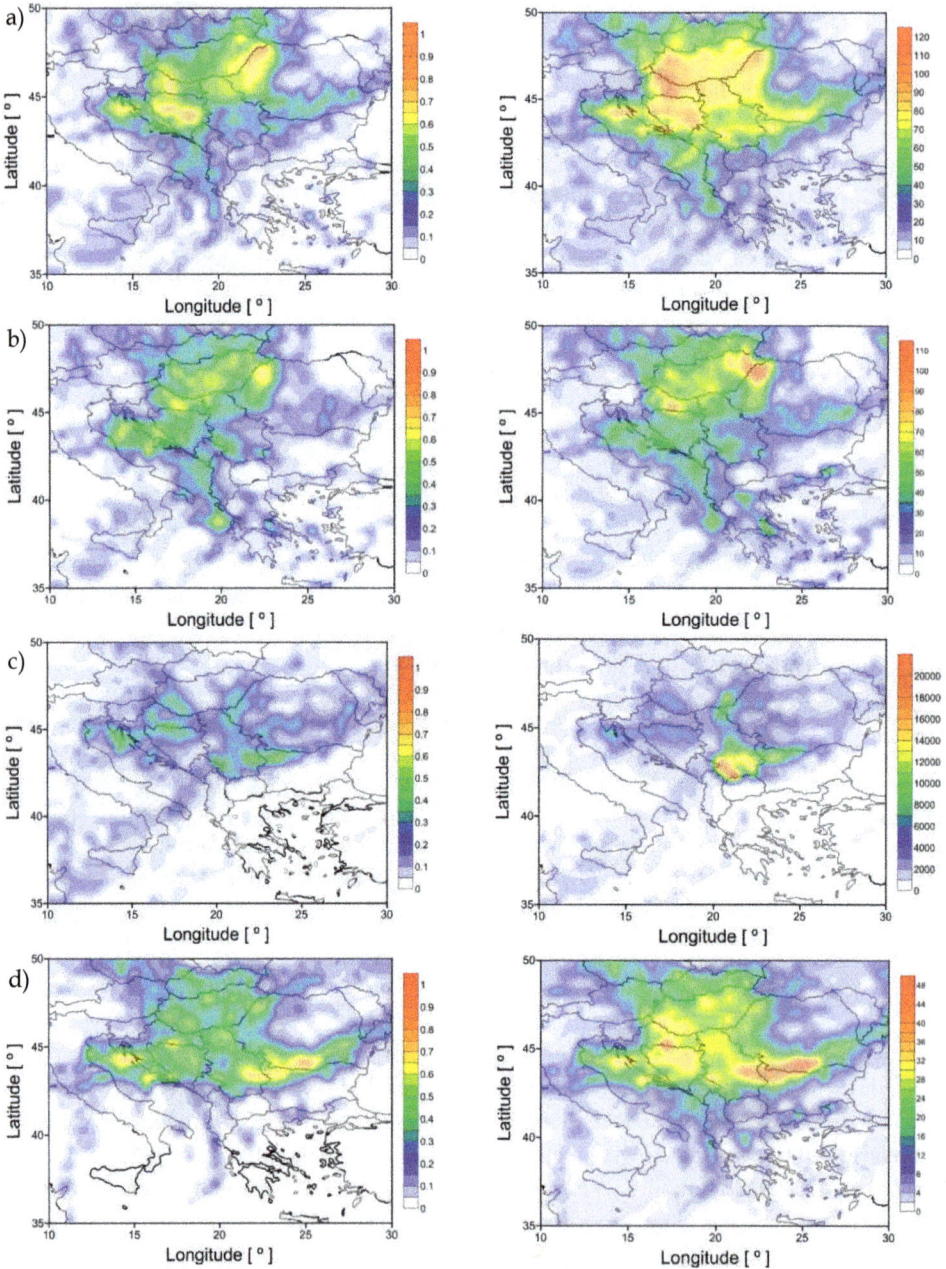

Figure 13. Figure 13. PSCF and CWT maps for a) PM10 (μg m^{-3}) and b) V, c) Al and d) Mn (ng m^{-3}) in PM$_{10}$

6.3. PSCF and CWT Results

Some studies have already reported a long–range transport of PM from western countries over Balkan Peninsula which is sporadically (mostly in spring and summer) associated with African dust outbreaks [77,78]. Based on the trace metals data set contents in PM10 for the period 2003-2005 the possible transport process over Belgrade was investigated. The PSCF and CWT plots for PM10, V, Al and Mn are presented in Figure 13. It can be seen that PSCF plot for PM10 shows high probability area in the northeast and west region. CWT plot indicated that the high concentrations of PM10 are related to the sources located in the west, northwest and northeast direction distinguishing major sources from moderate ones by calculating concentration gradients. The highest PSCF and CWT values for V were similarly distributed in the northeast region. Since V followed Weibull distribution, transportation process is expected and this region the most likely influences receptor site. Aluminum and Mn are metals related to dominant local sources. As can be seen from Figure 13, there are no significant emission region that Al can be transported from, while Mn transportation process can be expected from southeast direction.

6.4. Clustering of modelled backward trajectories

During 2004-2008, 48-hour backward trajectories of air masses over Belgrade were daily calculated at the previously defined six different altitudes. Transport paths of air masses with similar history and origin were obtained by cluster analysis. Since we were concerned on the directions of the trajectories, the angle distance between back trajectories [76] has been used as the cluster model. The angle distance between two backward trajectories was defined as

$$d_{1,2} = \frac{1}{13}\sum_{i=1}^{13}\cos^{-1}\left(\frac{1}{2}\frac{(A_i + B_i - C_i)}{\sqrt{A_i B_i}}\right) \tag{13}$$

where

$$A_i = \left(x_1(i) - x_0\right)^2 + \left(y_1(i) - y_0\right)^2 \tag{14}$$

$$B_i = \left(x_2(i) - x_0\right)^2 + \left(y_2(i) - y_0\right)^2 \tag{15}$$

$$C_i = \left(x_2(i) - x_1(i)\right)^2 + \left(y_2(i) - y_1(i)\right)^2 \tag{16}$$

The variables x_0 and y_0 define the position of the studied site and $x_1(i)$, $y_1(i)$ and $x_2(i)$, $y_2(i)$ are coordinates of i-th segment for trajectories 1 and 2, respectively. In order to investigate possible seasonal variation in transportation process of PM10 over Belgrade, we analysed summer and winter season separately. Cluster analysis resolved the four main directions of air mass flows over Belgrade during both periods. A cluster number can be deduced through visual inspection and comparison of the mean-trajectory maps. The representative trajectories for each cluster group are presented in Figure 14. As can be seen, the four classes were related to west southwest, east, west northwest and north directions. Trajectories within different classes of airflow had distinct effects on the PM_{10} concentrations. The mean PM_{10} concentrations and frequency of air back trajectories occurrences within each cluster group are presented in Figure 15. During the summer period the most frequently arriving direction of air masses is from west northwest direction (40%) and evenly distributed among the others cluster groups. Despite the high frequency of occurrences, the contribution from this direction to PM_{10} concentration measured at the receptor site is the least. High PM_{10} concentrations are associated with the trajectories coming from west southwest and east directions. Also, the highest number of air back trajectories from these directions is associated with the observed PM_{10} concentrations above critical level of 50 µg m^{-3}. During the winter period the most frequently arriving direction of air masses is from west southwest direction (37%) and north direction (26%). Although the frequency of air back trajectories coming from east is the lowest, their contributions to the high PM_{10} concentrations are the most significant. The highest average contribution to the observed PM_{10} is from the cluster group representing the arrival direction from west southwest.

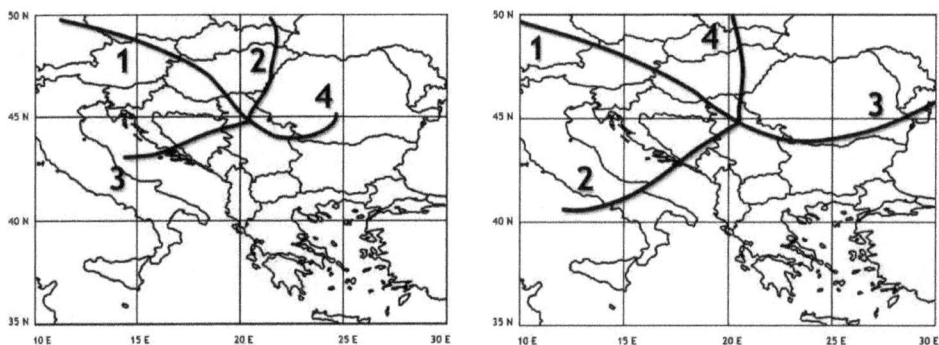

Figure 14. Trajectories representing grouping of 48 hour backward trajectories of air masses over Belgrade into four classes for the summer (left) and winter (right) season

In these directions there are several possible emission sources: coal-fired thermal power plants – "Nikola Tesla A and B", Obrenovac (west southwest); "Kostolac A and B", Kostolac (east southeast); "Kolubara", Veliki Crljeni (south southwest); as well the steel industry complex in Smederevo (east southeast).

Obviously this kind of analysis can reveal the different transport pattern of particulate matter during summer and winter season. Cluster analysis of air back trajectories can be useful tool for investigation of the major transport pathways and their contributions to the high PM_{10} values. Additional analysis can be used to estimate background and long distance transportation of particulate matter [79,80].

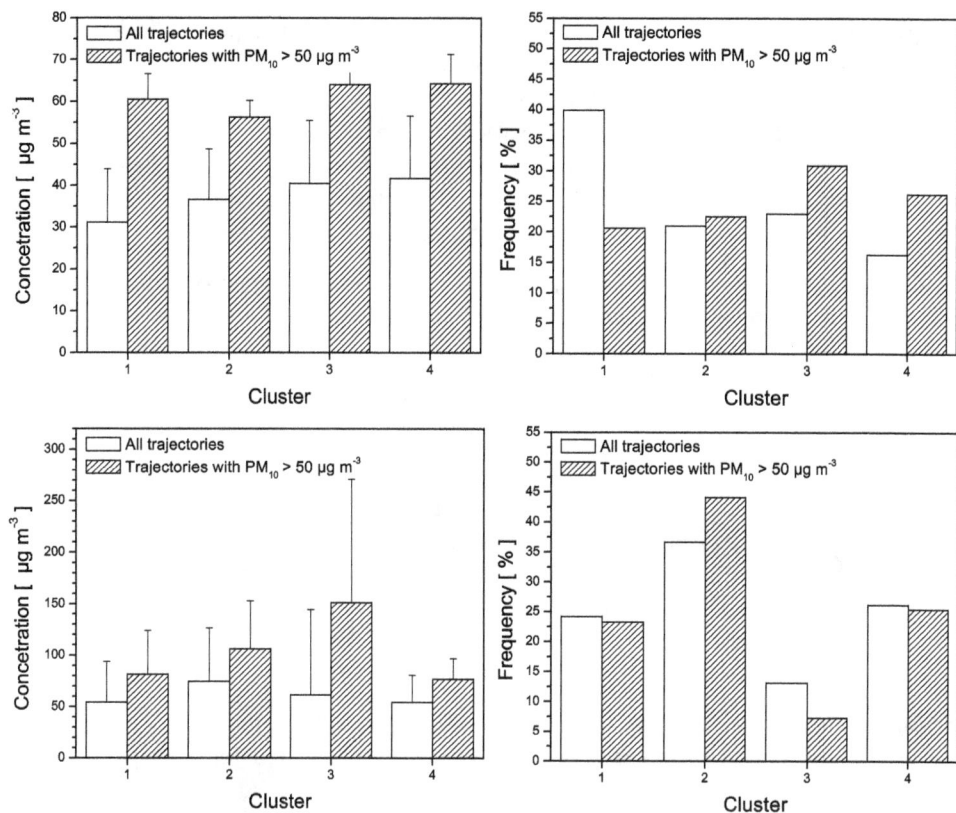

Figure 15. Trajectory frequency and mean PM10 concentrations within four classes for summer (above) and winter (below) period

7. Conclusion

The atmospheric processes are very complex, and pollutant concentrations are inherently random variables because of their dependence on the fluctuation of a variety of meteorological conditions and emission intensity. When sets of air pollution data are available and we aim to predict future trends, we have to know some characteristics of the measured variables, therefore, various statistical characteristics can be determined (maximum and minimum concentration, standard deviation, frequencies of exceedances of some critical level, etc) and assigned to the pollutant concentration. The distribution function is one of the most important characteristic of the random variables. Although there is a priori no reason to expect that "atmospheric" distribution should adhere to a specific probability distribution in some cases, there may be a natural probability distribution which relates emission levels to pollution levels. A great number of papers dealing to this problem have been published and in this study three theoretical distributions (lognormal, Weibull and type V Pearson) were used to fit PM10 concentrations in the Belgrade urban area while two-parameter exponential distribution was used to fit the high percentile PM_{10} daily concentration region. Based on the fitted distributions the minimum reduction required to attain the daily AQS was estimated using the rollback equation. These methods can reasonably predict the return period and exceedances in the succeeding period. The results suggested that the required PM source reduction ranged from 53% to 63% in Belgrade area. Recently measurements indicated that some progress in this direction has already been done and the PM_{10} mass concentrations have been reduced during the last few years.

In addition, it is demonstrated that the distribution of PM_{10} can be successfully estimated from the wind speed data if the corresponding two distributions followed the same type of distribution and have similar shape factors.

Daily and monthly variations of PM_{10} were analyzed and confirmed the existing seasonal variation. Weekend effect has also been observed and the differences between weekdays and weekend were analyzed by Q-Q plot and cumulative distribution function.

The analysis of quantiles of real and theoretical distributions for trace metals in PM_{10} indicated dominant local emission sources for the most of metals. To identify the source locations associated with the high fluxes of trace metal contents in PM_{10} (Al, Mn and V), two trajectory-based models, PSCF and CWT, were used. Both PSCF and CWT resolved similar source locations for those elements. The results indicated that the source areas contributing to receptor site were located mostly in north east and west directions.

In addition, trajectory clustering was further used to investigate the transport pathways during summer and winter season. Four clusters were generated for both periods and corresponding average contributions to PM_{10} mass concentrations. Although the frequency of air back trajectories coming from east was the lowest during the summer period, their contributions to the average PM_{10} concentration were the most significant (41.7 μg m^{-3}) followed by the trajectories coming from west south west (40.4 μg m^{-3}). The highest average contribution to the observed PM_{10} concentration during the winter season was from the cluster group representing the arrival direction from west southwest (74.4 μg m^{-3}). Obviously this kind of analysis can reveal the different transport pattern of particulate matter during summer and winter season.

Nowadays commercial instruments with online measurement are helpful and can provide a large base of measurement. Still, researchers have to analyses these measurements and mathematical models and statistical analysis are some of the useful methods for developing air pollution control strategy related to specific geographic region.

Author details

Zoran Mijić*, Andreja Stojić, Mirjana Perišić, Slavica Rajšić and Mirjana Tasić
Institute of Physics, University of Belgrade, Serbia

---

* Corresponding Author

Acknowledgement

This paper was realized as a part of the projects No III43007 and No III41011 financed by the Ministry of Education and Science of the Republic of Serbia within the framework of integrated and interdisciplinary research for the period 2011-2014. The authors gratefully acknowledge the NOAA Air Resources Laboratory (ARL) for the provision of the HYSPLIT transport and dispersion model and READY website (http://www.arl.noaa.gov/ready.php) used in this publication and the Institute of Public Health of Belgrade, Serbia for providing appropriate data set.

8. References

[1] Anderson L.T, Charlson J.R, Schwartz E.S, Knutti R, Boucher O, Rodhe H, Heintzenberg J (2003) Climate Forcing by Aerosols - A Hazy Picture. Science 300: 1103-1104.

[2] Ramanathan V, Crutzen P.J, Kiehl J.T, Rosenfeld D (2001) Aerosols, Climate, and the Hydrological Cycle. Science 294: 2119-2124.

[3] IPCC (2001) Intergovernmental Panel on Climate Change, Third Assessment Report. Cambridge University Press. Cambridge UK.

[4] World Health Organization WHO (2002) Air Quality Guidelines for Europe.

[5] World Health Organization (WHO) (2003) Health aspects of air pollution with particulate matter, ozone and nitrogen dioxide. Report on a WHO Working Group, Regional Office for Europe; Bonn, Germany, 2003. EUR/03/5042688.

[6] Dockery D. W, Pope III C.A (1994) Acute Respiratory Effects of Particulate Air Pollution. Annu. Rev. Public Health 15: 107-132.

[7] Dockery D.W, Pope III C.A (2006) Health Effects of Fine Particulate Air Pollution: Lines that Connect. J. Air Waste Manage. Assoc. 56: 709-742.

[8] Dockery D.W, Stone P.H (2007) Cardiovascular risks from fine particulate air pollution. N. Engl. J. Med. 356: 511–513.

[9] Schwartz J, Dockery D.W, Neas L.M (1996) Is daily mortality associated specifically with fine particles? J. Air Waste Manage. Assoc. 46: 927-939.

[10] Schwartz J, Ballester F, Saez M, Perez-Hoyos S, Bellido J, Cambra K, Arribas F, Canada A, Perez-Boillos M. J, Sunyer J (2001) The concentration-response relation between air pollution and daily deaths. Environ. Health Perspect. 109: 1001.

[11] Van Grieken R, Delalieux F (2004) In: "Invited Lectures of the 5th Gen. Conf. Balkan Phys. Union, BPU-5", Eds., Serbian Physical Society, Belgrade. 234-246.

[12] EC 1999 Air Quality Directive 1999/30 EC of the European Parliament and of the Council of 22 April 1999 relating to limit values for SO_2, NO_2 and NO_x, particulate matter and lead in ambient air. Off J Eur Communities L163, Brussels.

[13] EC 2008 Directive 2008/50/EC of the European Parliament and of the Council of 21 May 2008 on ambient air quality and cleaner air for Europe.

[14] Brunekreef B, Maynard L.R (2008) A note on the 2008 EU standards for particulate matter. Atmos. Environ. 42: 6425-6430.

[15] Moldovan M, Palacios M.A, Gomez M.M, Morrison G, Rauch S, McLeod C, Ma R, Caroli S, Alimonti A, Petrucci F, Bocca B, Schramel P, Zischka M, Pettersson C, Wass U, Luna M, Saenz J.C, Santamaría J (2002) Environmental risk of particulate and soluble platinum group elements released from gasoline and diesel engine catalytic converters. Sci.Total Environ. 296: 199 – 208.

[16] Popović D, Todorović D, Frontasyeva M, Ajtić J, Tasić M, Rajšić S (2008) Radionuclides and heavy metals in Borovac, Southern Serbia, Environ. Sci. Pollut. Res. 15: 509-520.

[17] Todorović D, Popović D., Rajšić S., Tasić M (2007) Radionuclides and Particulate Matter in Belgrade Air, In, Environmental Research Trends, ed. M.Cato, Nova Science Publishers, New York, pp 271-301.

[18] Espinosa A.J.F, Rodriguez M.T, Barragan de la Rosa F.J, Sanchez C.J (2001) Size distribution of metals in urban aerosols in Seville (Spain). Atmos. Environ. 35: 2595-2601.

[19] Tomašević M, Rajšić S, Đordević D, Tasić M, Krstić J, Novaković V (2004) Heavy metals accumulation in tree leaves from urban areas, Environ. Chem. Lett. 2:151-154.

[20] Tomašević M, Vukmirović Z, Rajšić S, Tasić M, Stevanović B (2008) Contribution to biomonitoring of some trace metals by deciduous tree leaves in urban areas, Environ. Monit. Assess. 137: 393–401.

[21] Nriagy J.O, Pacyna J.M (1988) Quantitative assessment of worldwide contamination of air, water and soil by trace metals. Nature. 333: 134.

[22] Pacyna J.M, Martonova A, Cornille P, Maenhaut W (1989) Modelling of long-range transport of trace elements. A case study. Atmos. Environ. 19: 2109.

[23] Alcamo J, Bartnicki J, Olenrzynski K, Pacyna J (1992) Computing heavy metals in Europe's atmosphere – I. model development and testing. Atmos. Environ. 26A: 3355.

[24] Aničić M, Tomašević M, Tasić M, Rajšić S, Popović A, Frontasyeva M.V, Lierhagen S, Steinnes E (2009) Monitoring of trace element atmospheric deposition using dry and wet moss bags: Accumulation capacity versus exposure time. J. Hazard. Mater. 171: 182-188.

[25] Aničić M, Tasić M, Frontasyeva M.V, Tomašević M, Rajšić S, Mijić Z, Popović A (2009) Active Moss Biomonitoring of Trace Elements with Sphagnum girgensohnii Moss Bags in Relation to Atmospheric Bulk Deposition in Belgrade, Serbia. Environ. Pollut. 157: 673–679.

[26] Samara C, Kouimtzis Th, Tsitouridou R, Kanias G, Simeonov V (2003) Chemical mass balance source apportionment of PM10 in an industrialized urban area of Northern Greece. Atmos. Environ. 37(1): 41-54.

[27] Polissar V.A, Hopke K.P, Poirot L.R (2001) Atmospheric Aerosol over Vermont: Chemical Composition and Sources. Environ. Sci. Technol. 35: 4604-4621.

[28] Song X.-H, Polissar A.V, Hopke P.K (2001) Sources of fine particle composition in the northeastern US. Atmos. Environ. 35: 5277-5286.

[29] Mijić Z, Rajšić S, Žekić A, Perišić M, Stojić A, Tasić M (ed Ashok Kumar) (2010) Air Quality: Characterization and Application of Receptor Models to the Atmospheric Aerosols Research, Sciyo Publisher, Croatia. pp. 143-167.

[30] Mijić Z, Stojić A, Perišić M, Rajšić S, Tasić M, Radenković M, Joksić J (2010) Seasonal variability and source apportionment of metals in the atmospheric deposition in Belgrade. Atmos. Environ. 44: 3630-3637.

[31] Viana M, Kuhlbusch T.A.J, Querol X, Alastuey A, Harrison R.M, Hopke P.K, Winiwarter W, Vallius M, Szidat S, Prevot A.S.H, Hueglin C, Bloemen H, Wahlin P, Vecchi R, Miranda A.I, Kasper-Giebl A, Maenhhaut W, Hitzenberger R (2008) Source apportionment of particulate matter in Europe: A review of methods and results. J. Aerosol Sci. 39: 827-849.

[32] Tasić M, Mijić Z, Rajšić S, Stojić A, Radenković M, Joksić J (2009) Source apportionment of atmospheric bulk deposition in the Belgrade urban area using Positive Matrix factorization, 2nd Int. Workshop on Nonequilibrium Processes in Plasma Physics and Science, IOP Publishing, Journal of Physics: Conference Series 162 012018 doi:10.1088/1742-6596/162/1/012018

[33] Seinfeld J.H, Pandis S.N (1998) Atmospheric Chemistry and Physics From Air Pollution to Climate Change. J. Wiley, New York.

[34] Mage D.T, Ott W.R (1984) An evaluation of the method of fractiles, moments and maximum likelihood for estimating parameters when sampling air quality data from a stationary lognormal distribution. Atmos. Environ. 18:163 –171.

[35] Vukmirovic Z (1989) Lognormal distribution application for air-quality assessment, J. Serb. Chem. Soc. 54: 373-381.

[36] Georgopoulos G. P, Seinfeld H. J (1982) Statistical distributions of air pollutant concentrations. Environ. Sci. Technol. 16: 401A-416A.

[37] Berger A, Melice J.L, Demuth Cl. (1982) Statistical distributions of daily and high atmospheric SO2 concentrations. Atmos. Environ. 16: 2863 –2877.

[38] Morel B, Yeh S, Cifuentes L (1999) Statistical distributions for air pollution applied to the study of the particulate problem in Santiago. Atmos. Environ. 33: 2575-2585.

[39] Vukmirović Z (1990) Expected levels of air pollution in pseudolognormal distribution. Idojaras. 94: 249-258.

[40] Aleksandrapoulou V, Eleftheriadis K, Diapouli E, Torseth K, Lazaridis M, (2011) Assessing PM_{10} source reduction in urban agglomerations for air quality compliance. J. Environ. Monit. 14: 266-278.

[41] Mijić Z, Tasić M, Rajšić S, Novaković V (2009) The statistical character of PM_{10} in Belgrade. Atmos. Res. 92: 420-426.

[42] Lu H (2002) The statistical characters of PM10 concentration in Taiwan area. Atmos. Environ. 36: 491-502.

[43] Lu H (2004) Estimating the emission source reduction of PM10 in central Taiwan. Chemosphere 54: 805-814.

[44] Lu H, Fang G (2003) Predicting the exceedances of a critical PM10 concentration - a case study in Taiwan. Atmos. Environ. 37: 3491-3499.

[45] Z.Vukmirovic, I.Tošic, M.Tasić, S.Rajšic (2004) Analysis of the Saharan dust regional transport. Meteorol. Atmos. Phys. 85: 265-273.

[46] Đorđević D, Mihajlidi-Zelić A, Relić D (2005) Differentiation of the contribution of local resuspension from that of regional and remote sources on trace elements content in the atmospheric aerosol in the Mediterranean area. Atmos. Environ. 39: 6271–6281.

[47] Gao N, Cheng M-D, Hopke P.K (1993) Potential source contribution function analysis and source apportionment of sulphur species measured at Rubidoux, CA during the Southern CaliforniaAir Quality Study, 1987. Anal. Chim. Acta 277: 369-380.

[48] Hsu Y-K, Holsen M.T, Hopke K.P (2003). Comparison of hybrid receptor models to locate PCB sources in Chicago. Atmos. Environ. 37: 545-562.

[49] Kyotani T, Iwatsuki M (2002) Characterization of soluble and insoluble components in PM$_{2.5}$ and PM$_{10}$ fractions of airborne particulate matter in Kofu city, Japan. Atmos. Environ. 36: 639-649.

[50] Rajšić F.S, Tasić D.M, Novaković T.V, Tomašević N.M (2004) First Assessment of the PM$_{10}$ and PM$_{2.5}$ Particulate Level in the Ambient Air of Begrade City. Environ. Sci. Pollut. Res. Int. 11: 158-164.

[51] Rajšić S, Mijić Z, Tasić M, Radenković M, Joksić J (2008) Evaluation of the Levels and Sources of Trace Elements in Urban Particulate Matter. Environ. Chem. Lett. 6: 95-100.

[52] Tasić D.M, Rajšić F.S, Novaković T.V, Mijić R.Z, Tomašević N.M (2005) PM$_{10}$ and PM$_{2.5}$ Mass Concentration Measurements in Belgrade Urban Area. Phys. Scr. T118: 29-30.

[53] Tasić M, Djurić-Stanojevic B, Rajšić S, Mijić Z, Novaković V (2006) Physico-Chemical Characterization of PM$_{10}$ and PM$_{2.5}$ in the Belgrade Urban Area. Acta Chim. Slov. 53: 401-405.

[54] Tasić M, Rajšić S, Tomašević M, Mijić Z, Aničić M, Novaković V, Marković M.D, Marković A.D, Lazić L, Radenković M, Joksić J (2008) Assessment of Air Quality in an Urban Area of Belgrade, Serbia, In: Burcu Ozkaraova Gungor E, editor. Environmental Technologies, New Developments. Vienna, Austria: I-Tech Education and Publishing pp. 209-244.

[55] Tasić M, Rajšić S, Novaković V, Mijić Z (2006) Atmospheric aerosols and their influence on air quality in urban areas. Facta Universitatis, Series Physics, Chemistry and Technology 4: 83-91.

[56] Lonati G, Cernuschi S, Giugliano M (2011) The duration of PM$_{10}$ concentration in a large metropolitan area. Atmos. Environ. 45: 137-146.

[57] Aleksandropoulou V, Eleftheriadis K, Diapouli E, Torsethc K, Lazaridisa M (2012) Assessing PM$_{10}$ source reduction in urban agglomerations for air quality compliance. J. Environ. Monit. 14: 266–278.

[58] Karaca F, Alagha O, Erturk F (2005) Statistical characterization of atmospheric PM$_{10}$ and PM$_{2.5}$ concentrations at a non-impacted suburban site of Istanbul, Turkey. Chemosphere 59: 1183–1190.

[59] Simpson RW, Jakeman A.J, Daly N.J (1984) The relationship between the ATDL and the statistical distributions of wind speed and pollution data. Atmos. Environ. 19: 75 –82.

[60] Lu H, Fang G (2002) Estimating the frequency distributions of PM$_{10}$ and PM$_{2.5}$ by the statistics of wind speed at Sha-Lu at, Taiwan. Sci. Total Environ. 298: 119-130.

[61] Barmpadimos I, Nufer M, Oderbolz D.C, Keller J, Aksoyoglu S, Hueglin C, Baltensperger U, Prévôt A.S.H (2011) The weekly cycle of ambient concentrations and

traffic emissions of coarse (PM10-PM2.5) atmospheric particles. Atmos. Environ. 45: 4580-4590.

[62] Barmpadimos I, Hüglin C, Keller J, Henne S, Prévôt A.S.H (2011) Influence of meteorology on PM10 trends and variability in Switzerland from 1991 to 2008. Atmos. Chem. Phys. 11: 1813–1835.

[63] Marr L. C, Harley R. A (2002a) Spectral analysis of weekday-weekend differences in ambient ozone, nitrogen oxide, and non-methane hydrocarbon time series in California. Atmos. Environ. 36: 2327–2335.

[64] Altshuler S.L, Arcado T.D, Lawson D.R (1995) Weekday Vs weekend ambient ozone concentrations: Discussions and hypotheses with focus on northern California. J. Air Waste Manage. Assoc. 45: 967–972.

[65] Dreher D. B, Harley R.A (1988) A fuel-based inventory for heavy-duty diesel truck emissions. J. Air Waste Manage. Assoc. 48: 352–358.

[66] Gong D.-Y, Ho C.-H, Chen D, Qian Y, Choi Y.-S, Kim J (2007) Weekly cycle of aerosol–meteorology interaction over China. J. Geophys. Res. 112: D22202.

[67] Cerveny R.S, Balling Jr. R.C (1998) Weekly cycles of air pollutants, precipitation and tropical cyclones in the coastal NW Atlantic region. Nature 394: 561–563.

[68] Jin M, Shepherd J.M, King M.D (2005) Urban aerosols and their variations with clouds and rainfall: a case study for New York and Houston. J. Geophys. Res. 110: D10S20.

[69] Marković M.D, Marković A.D, Jovanović A, Lazić L, Mijić Z (2008) Determination of O_3, NO_2, SO_2, CO and PM_{10} measured in Belgrade urban area. Environ. Monit. Assess. 145: 349-359.

[70] Marr L. C, Harley R. A (2002) Modeling the effects of weekday-weekend differences in motor vehicle emissions on photochemical air pollution in Central California. Environ. Sci. Technol. 36: 4099–4106.

[71] Colbeck I (1999) Weekday-Sunday PM_{10} variations. J. Aerosol Sci. 30: S669-S670.

[72] Choi Y.-S, Ho C.-H, Chen D, Noh Y.-H, Song C.-K (2008) Spectral analysis of weekly variation in PM_{10} mass concentration and meteorological conditions over China. Atmos. Environ. 42: 655-666.

[73] Tchepel O, Borrego C (2009) Frequency analysis of air quality time series for traffic related pollutants. J. Environ. Monit. 12: 544-550.

[74] Draxler R.R, Rolph G.D (2010) HYSPLIT (HYbrid Single-Particle Lagrangian Integrated Trajectory) Model access via NOAA ARL READY Website (http://ready.arl.noaa.gov/HYSPLIT.php). NOAA Air Resources Laboratory, Silver Spring, MD.

[75] Rolph G.D (2010) Real-time Environmental Applications and Display sYstem (READY) Website (http://ready.arl.noaa.gov), NOAA Air Resources Laboratory, Silver Spring, MD.

[76] Wang Y.Q, Zhang X.Y, Draxler R (2008) TrajStat: GIS-based software that uses various trajectory statistical analysis methods to identify potential sources from long-term air pollution measurement data. Environ. Modell. Softw. 24: 938-939.

[77] Kubilay N, Nickovic S, Moulin C, Dulac F (2000) An illustration of the transport and deposition of mineral dust onto the eastern Mediterranean. Atmos. Environ. 34: 1293-1303.

[78] Perez N, Pey J, Querol X, Alastuey A, Lopez J.M, Viana M (2008) Portioning of major and trace components in PM_{10}-$PM_{2.5}$-PM_1 at an urban site in Southern Europe. Atmos. Environ. 42: 1677-1691.

[79] Zhu L, Huang X, Shi H, Cai X, Song Y (2011) Transport pathways and potential sources of PM_{10} in Beijing. Atmos. Environ. 45: 594-604.

[80] Bari A, Dutkiewicz V.A, Judd C.D, Wilson L.R, Luttinger D, Husain L (2003) Regional sources of particulate sulfate, SO_2, $PM_{2.5}$, HCl, and HNO_3, in New York, NY. Atmos. Environ. 37 (20): 2837-2844.

Chapter 12

Multi-Physical Modeling for IAQ Monitoring

Giuseppe Petrone, Carla Balocco and Giuliano Cammarata

Additional information is available at the end of the chapter

1. Introduction

Indoor air quality (IAQ) is a term referring to the air quality within buildings, relating to the health and comfort of building occupants. During the past decades, a significant effort was spent in order to understand the phenomenological aspects concerning IAQ and its human perception [1, 2]. IAQ can be affected by microbial contaminants (mold, bacteria), gases (including carbon monoxide, radon, volatile organic compounds), particulates, or any mass or energy stressor that can induce adverse health conditions. Using ventilation to dilute contaminants, filtration, and source control are the primary methods for improving indoor air quality in most buildings. Many researchers investigated both experimentally and numerically on IAQ and thermal comfort in specific applications [3-5], that generally depend on ventilation system, building geometry, pollutant source characteristics, thermal and fluid boundary conditions such as flow rate, locations of supply outlets and return inlets, and diffuser characteristics [6]. Usually, indoor CO_2 concentration is monitored in order to estimate the air quality level and to assess the performance of a mechanical ventilation system [7, 8]. Also, CO_2 concentration is often used to estimate building air change rates. Indoor CO_2 concentrations above about 1000 ppm are generally regarded as indicative of ventilation rates that are unacceptable with respect to body odors. On the other hand, concentrations of CO_2 below 1000 ppm do not always guarantee that the ventilation rate is adequate for removal of air pollutants from other indoor sources [9]. Several class of indoor environments could be considered as interesting subject of studies related to the air quality requirements, such as theatres or auditorium halls, cabins of means of transportation and hospital rooms. In theatre or auditorium halls, people sharing the internal atmosphere represent the main CO_2 source. They do not move significantly and CO_2 production is for the most part related to their breathing. This allows to build up or perform detailed experimental set-up or numerical models in terms of internal CO_2 load. Among the investigations dealing with this subject, a recent paper [10] numerically studied how two ventilation systems with the same air inlet arrangement, but different systems of air extraction, affect the air speed, temperature and CO_2 concentration profile inside an auditorium. It was showed as the lowest rate of air change leads to the increase of temperature. It was found that CO_2 concentration decreases rapidly if the ventilation rate is increased, in this case by the unexpectedly large factor of five. Cheong et al. studied both IAQ [11] and thermal comfort [12] in lecture theatres. The overall results suggested that the ventilation system, in that case a full air-conditioning system, was effective in removing indoor air pollutants and achieving reasonable IAQ. A study proposed by Noh et al. [13] focused on thermal comfort and indoor air quality in a lecture theatre with a 4-way cassette air conditioning and mixing ventilation system. It is showed that increasing the discharge angle from the supply grilles on the cassette unit makes uniformity of thermal comfort worse, but rarely affects IAQ. Recently, Kavgic et al. [14] experimentally monitored and numerically analyzed IAQ and thermal comfort in a theatre hall equipped by a displacement system for ventilation and air conditioning. The paper showed as, for the most part of the monitored period, the environmental parameters were within the standard limits of thermal comfort and IAQ. Otherwise, the calculated ventilation rates showed that the theatre was mostly over-ventilated, with negative consequences for its energy consumption. As above introduced, the bio-effluents diffusion in indoor environments is a very actual issue of interest because of the potential risk of infections transmission between people sharing the same atmosphere. This issue takes also top relevance when considering indoor environment characterized by very high occupant density, such as the cabins of transportation means. In order to avoid high concentration regions of any air pollutant inside the cabins, environmental control system is devoted to dilute the contaminant concentration by introducing fresh air inside. Since cabins have usually more complex geometry and lower outside air supply rate per person as compared to buildings, it is very challenging to design comfortable and healthy cabin environment for transportation means. Among the experimental works concerning this topic, Wang et al. [15] conduced several trials in order to measure the local mean age of air and the ventilation effectiveness factor (VEF) at the breathing level of the passengers. Carbon dioxide was used as the trace gas to determine the local mean age of air and the VEF. The air velocity profiles measured using a volumetric particle tracking velocimetry (VPTV) system were used to generate the airflow patterns and investigate the underlying mechanism affecting the local mean age of air and the VEF. Their results mainly show as the local mean age of air was affected not only by the velocity magnitude, but also by the local airflow patterns such as jet and recirculation. Except for the recirculation regions, the higher mean air velocity corresponded to a lower local mean age of air, indicating a higher net air exchange rate. In the strong air recirculation regions, the higher air velocity did not lead to the lower local mean age of air. This was because the "old" air was kept in the recirculation area, and no sufficient freshly supplied air was introduced. The findings of Wang et al. well underlines the importance of carrying-out velocity and pressure distribution maps that are the real important primary indexes in order to investigate on IAQ. Both in buildings and transportation means applications it is recently also arising interest in evaluating air quality and thermal comfort conditions depending on layout-out adopted for air distribution systems [16]. The most common schemes used in HVAC applications are the Mixing Air Distribution (MAD), the Personalized Air Distribution (PAD) and the Under Floor Displacement system (UFD). In the MAD system, ventilating air is supplied at the ceiling level with a high velocity and then mixes with the air inside the hall. Recovery grids are located at the floor level. The PAD system [17-19] supplies fresh air directly to the breathing area of a person. As it is applied in some movie theatres, diffusers are located on the chair back-side in order to supply fresh air to the person seated on the rear row [20]. In this layout, air is usually recovered at the ceiling level. The functional principle of the UFD system [21] is otherwise based on the buoyancy flow induced close to the human body because of metabolic heat dissipation. Air is supplied at the floor level, close to the heat source. The buoyancy driven upstream flow joins the recovery grids located at the ceiling [22]. In this context it appears more and more interesting to strike a balance among the different scheme based on comparative analyses between different ventilating system layouts in assuring the best air quality and thermal comfort in a chosen application [23]. Let now consider the IAQ in hospitals: because of the risk of infections transmission, the IAQ really assumes a crucial role. This includes both the Isolation Room (IR) and the Operating Theatre (OT). Different engineering standards for the design of operating theatres and Heating Ventilation and Air Conditioning (HVAC) plant system are adopted by several countries. Guidelines and advice are available regarding air quality and how to provide it, as well as other interventions that may reduce the incidence of postoperative infections. The HVAC and IAQ impact on virus and bacteria load, but also aerosol diffusion in the OT has been widely investigated [24, 25]. The risk of virus dispersal and post-operation infections depends mainly on airflow behavior and direction changes caused by people moving or doors opening. In recent years several studies have focused on computational models using fluid dynamics approaches to investigated airflow patterns and the related spreading of infection in isolation and hospitalization rooms and in particular in OTs for different ventilation systems (for example operating under open or closed-door conditions) [26]. The main attention of these papers has been the evaluation of the effects of negative pressure in isolation rooms accommodating patients with highly infectious diseases. Opening the door causes the dispersion of infectious air out of the isolation room. An isolation room held at negative pressure to reduce aerosol escape and a high air-change rate to allow rapid removal of aerosols can eliminate transmission of infectious aerosols to those outside the room. Tunga and collaborators [27] found that an air velocity above 0.2 m/s via a doorway effectively prevents the spread of airborne contaminants out of the isolation room with an open door. In a recent paper the change of an OT from positive to negative pressure environment [28] has been analyzed. Results show that the dispersion pattern of bacteria in the negative pressure theatre were as good as, if not better than, those in the original positive pressure design. Several studies focused on the air flow patterns in the OT, directed to turbulent versus displacement ventilation and laminar air flow systems [29] The laminar airflow (LAF) of OTs is not a strictly accurate description, as it does not fulfill the aerodynamic conditions for genuine laminar flow. Unidirectional or linear airflow is a better alternative, reflecting the actual pattern of air movement in parallel lines. The advantage of using LAF over the turbulent counterpart is its ability to minimize infection by mobilizing a relatively uniform and large volume flow of clean air. After passing through the three stages of filtration (with a HEPA filter as the final stage), the conditioned air enters the OT through a large supply diffuser that usually occupies a substantial ceiling area and moves towards the surgical area making only a single transit. When the room air moves in a single direction at a sufficient velocity, flow due to heat or movement are abolished and the re-entrainments of particles into the operative field are stopped [30]. When a solid object is encountered, the air flows round the object and the laminar-flow pattern is distorted only in the immediate surroundings of the object. Contaminants are flushed out as soon as they are liberated without migration to other areas. The LAF has two main configurations: vertical and horizontal flow. The selection depends on several factors, including the plane of movements employed in the surgical process, the rate of particle emission, the degree of cleanliness required and its cost-effectiveness. A recent article [31] shows the contamination diffusion in an ISO5 class operating room with vertical LAF when the door is open by a computational fluid dynamic (CFD) simulation. In this last paper, the influence of the door-opening procedure was ignored since the door of the operating room is a sliding one and in particular the effect of people crossing with and without a stretcher is disregarded. Some authors used numerical [32] or experimental [33] approaches to investigate the effects of one moving person or the movement of a sliding door on the air distribution, including air velocity and pressure field but also CO_2 contaminant distribution, within generic ventilated or specific negative pressure isolation rooms [34]. Among the numerical based studies, simulation of object or person movements has been handled directly or indirectly [35]. Direct simulation includes the real movement of the real "object" inside the solution domain and requires a moving mesh approach in order to be realized. One of the most recent methods used for this kind of application is the so-called Arbitrary Lagrangian-Eulerian (ALE) method, in certain cases allowing resolution of the limit of small distortions allowed in the numerical model by the traditional Lagrangian approach. However, the direct simulation of the moving object requires very expensive computational cost. They are related on the one hand to the degrees of freedom increasing for a chosen number on mesh nodes, and on the other to the remeshing procedure, often needed during computations. The basic principle of the indirect simulation of a moving object in surrounding air consists otherwise in keeping into account the effects of the object's movement on the air flow. Two indirect numerical procedures were recently tested and applied by Brohus et al. [36] in order to simulate the effect of the surgical staff movements on contaminant concentration in an operating room. In this framework, this chapter presents a collection of results obtained by several numerical models built-up in order to investigate on IAQ basing on real-world scenarios and events. In particular, they refer to critical applications, characterized by high people density (movie theatre hall, cabins of means of transportation) or contamination risk (hospital rooms). The case studies presented in the following are introduced by a brief description of the applied mathematical and numerical models.

2. Numerical modelling

2.1. Continuous equations

Numerical models presented in this chapter have been built in COMSOL Multiphysics v.3.5a, a commercial FEM based software widely used for solving multi-physical problems governed by coupled ordinary and partial differential equations. The multi-physical problems involve at least solution of mass, momentum, energy conservation laws. Under assumptions of Newtonian fluid and uncompressible turbulent flow, the Reynolds Averaged Navier-Stokes equations read as in following:

$$\rho\frac{\partial \mathbf{U}}{\partial t} + \rho(\mathbf{U}\cdot\nabla)\mathbf{U} = \nabla\cdot\left[-p\mathbf{I} + \left(\mu + \mu_T\right)\left(\nabla\mathbf{U} + \left(\nabla\mathbf{U}\right)^T\right)\right] + F \tag{1}$$

$$\nabla\cdot\mathbf{U} = 0 \tag{2}$$

$$\rho\frac{\partial k}{\partial t} + \rho\mathbf{U}\cdot\nabla k = \nabla\cdot\left[\left(\mu + \frac{\mu_T}{\sigma_k}\right)\nabla k\right] + \frac{1}{2}\mu_T\left[\nabla\mathbf{U} + \left(\nabla\mathbf{U}\right)^T\right]^2 - \rho\varepsilon \tag{3}$$

$$\rho\frac{\partial \varepsilon}{\partial t} + \rho\mathbf{U}\cdot\nabla\varepsilon = \nabla\cdot\left[\left(\mu + \frac{\mu_T}{\sigma_\varepsilon}\right)\nabla\varepsilon\right] + \frac{1}{2}C_{\varepsilon 1}\frac{\varepsilon}{k}\mu_T\left[\nabla\mathbf{U} + \left(\nabla\mathbf{U}\right)^T\right]^2 - \rho C_{\varepsilon 2}\frac{\varepsilon^2}{k} \tag{4}$$

$$\rho C_p\frac{\partial T}{\partial t} + \rho C_p\mathbf{U}\cdot\nabla T = \nabla\cdot\left(\lambda\nabla T\right) + Q \tag{5}$$

Where $\mu_T = \rho C_\mu\, k^2/\varepsilon$ represents the turbulent viscosity. Values adopted for constants appearing in the above equations are determined from experimental data and reported in Table 1 [37].

C_μ	$C_{\varepsilon 1}$	$C_{\varepsilon 2}$	σ_k	σ_ε
0.09	1.44	1.92	1.0	1.3

Table 1. Numerical values of constants appearing in the adopted mathematical model.

Conservation laws (1-5) used to solve thermal and fluid-dynamical analyses have been coupled in same of applications presented in the following with two specific transport-diffusion equations, whereof the dependent variables are the temperature (solved both in fluid and solid domains), the carbon dioxide concentration and the mean age of air [38, 39], that quantifies the average lifetime of air at a particular location and so gives an indication of the air "freshness":

$$\frac{\partial CO_2}{\partial t} + \mathbf{U}\cdot\nabla CO_2 = D\nabla^2 CO_2 \tag{6}$$

$$\rho\frac{\partial \tau}{\partial t} + \rho\mathbf{U}\cdot\nabla\tau = \Gamma\nabla^2\tau \tag{7}$$

In the previous equations D and Γ are the diffusion coefficients of carbon dioxide in air and the diffusion coefficient of τ in the air mixture, respectively. Boundary conditions considered for models presented in the following are summarized in Table 2. Logarithmic wall functions were applied in the near wall flow, that has been considered parallel to the wall and being in a wall offset δ_w equal to one half of the boundary mesh element dimension. The equivalent wall offset in viscous unit is defined as $\delta_w^+ = \delta_w \rho U_\tau / \mu$, being U_τ the frictional velocity. Assumed values for the constant C^+ is 5.5, while the Karman's constant value was set equal to 0.42. Under the general assumption of fully turbulent flow, a turbulent length scale of 0.01 m and a 5% of turbulent intensity were applied at the generic air inlet sections.

Equation	Boundary condition	
	Inlet	Symmetry
		$\mathbf{n} \cdot \mathbf{U} = 0$
	$\mathbf{U} = -U_{in}\mathbf{n}$	$\mathbf{t} \cdot \left[-p\mathbf{I} + \left(\mu + \mu_T \right)\left(\nabla \mathbf{U} + \left(\nabla \mathbf{U} \right)^T \right) \right]\mathbf{n} = 0$
	Output	Open
(1-2)	$\left[\left(\mu + \mu_T \right)\left(\nabla \mathbf{U} + \left(\nabla \mathbf{U} \right)^T \right) \right]\mathbf{n} = 0$	$\left[-p\mathbf{I} + \left(\mu + \mu_T \right)\left(\nabla \mathbf{U} + \left(\nabla \mathbf{U} \right)^T \right) \right]\mathbf{n} = 0$
	$p = p_{out}$	
	Wall	
	$\mathbf{n} \cdot \mathbf{U} = 0$	
	$\left[\left(\mu + \mu_T \right)\left(\nabla \mathbf{U} + \left(\nabla \mathbf{U} \right)^T \right) \right]\mathbf{n} = \left[\rho C_\mu^{0.25} k^{0.5} / \left(\ln\left(\delta_w^+ \right)/ \kappa + C^+ \right) \right]\mathbf{U}$	
	Inlet	Symmetry
(3)	$k = 3/2\left(I_T U_{in} \right)^2$	$\mathbf{n} \cdot \left[\left(\mu + \mu_k/\sigma_k \right)\nabla k - \rho \mathbf{U}k \right] = 0$
	Output / Open	Wall
	$\mathbf{n} \cdot \nabla k = 0$	$\mathbf{n} \cdot \nabla k = 0$
	Inlet	Symmetry
(4)	$\varepsilon = C_\mu^{0.75}\left(3/2\left(I_T U_{in} \right)^2 \right)^{1.5} / L_T$	$\mathbf{n} \cdot \left[\left(\mu + \mu_T/\sigma_\varepsilon \right)\nabla \varepsilon - \rho \mathbf{U}\varepsilon \right] = 0$
	Output / Open	Wall
	$\mathbf{n} \cdot \nabla \varepsilon = 0$	$\varepsilon = \rho C_\mu k^2 / \left(\kappa \delta_w^+ \mu \right)$
	Inlet	Symmetry / Insulation
(5)	$T = T_{in}$	$\mathbf{n} \cdot \left(\lambda \nabla T \right) = 0$
	Output	Dispersion
	$\mathbf{n} \cdot \left(\lambda \nabla T \right) = 0$	$\mathbf{n} \cdot \left(\lambda \nabla T \right) = h_{conv}\left(T_{ext} - T \right)$
	Inlet	Interface (fluid/solid)
(6)	$CO_2 = CO_{2in}$	$\mathbf{n} \cdot \left(D\nabla CO_2 + CO_2\mathbf{U} \right) = 0$
	Output/Open	Walls
	$\mathbf{n} \cdot \left(D\nabla CO_2 \right) = 0$	$\mathbf{n} \cdot \left(D\nabla CO_2 + CO_2\mathbf{U} \right) = 0$
	Inlet	Interface (fluid/solid)
(7)	$\tau = 0$	$\mathbf{n} \cdot \left(\Gamma \nabla \tau + \tau \mathbf{U} \right) = 0$
	Output/Open	Walls
	$\mathbf{n} \cdot \left(\Gamma \nabla \tau \right) = 0$	$\mathbf{n} \cdot \left(\Gamma \nabla \tau + \tau \mathbf{U} \right) = 0$

Table 2. Numerical values of constants appearing in the adopted mathematical model.

2.2. Numerical methods

Continuous equations have been spatially discretized by a Finite Element approach based on the Galerkin method on non-uniform and non-structured computational grids made of tetrahedral Lagrange second order elements. For each application presented below, influence of spatial discretization has been preliminary studied in order to assure mesh-independent results. In order to prevent rising and propagation of numerical instabilities, an artificial streamline diffusion technique, based on the Galerkin Least-Squared (GLS) method, was employed in simulations. It should be noted that the GLS is a consistent method, which means that it does not perturb the original transport equation. Steady solutions were carried-out by applying an iterative dumped Newton-Raphson scheme [40], classically based on the discretized PDE linearization by a first-order Taylor expansion. An iterative approximation $s^{(k+1)}$ of the steady state s_0 is performed by solving the following system:

$$\begin{cases} J_E\left[s^{(k)}\right]\delta s^{(k+1)} = -E\left[s^{(k)}\right] \\ \delta s^{(k+1)} = s^{(k)} + \tilde{\lambda}\ \delta s^{(k+1)} \end{cases} \tag{8}$$

where $E\left[s^{(k)}\right]$ is the residual at the k-step, $0 < \tilde{\lambda} \leq 1$ is a dumping factor used in calculating the perturbation in the Newton system and the Jacobian operator J_E consists in the difference between perturbed and not perturbed equations linearized at the specific linearization state $s^{(k)}$. Once $s^{(k+1)}$ evaluated, the relative error $\Lambda^{(k+1)}$ is computed as:

$$J_E\left[s^{(k)}\right]\Lambda^{(k+1)} = -E\left[s^{(k+1)}\right] \tag{9}$$

If $\Lambda^{(k+1)} > \Lambda^{(k)}$, the dumping factor value is reduced and the approximate solution $s^{(k+1)}$ is computed again, until the relative error is less than its value in the previous step or it underflows a given minimum value, that means solution is not converged. If a successful approximated solution is taken, algorithm proceeds to a next Newton iteration. The iterative procedure ends when the chosen converge criterion is satisfied. Time integration of governing equations has been otherwise performed applying an Implicit Differential-Algebraic (IDA) solver [41], which uses variable-order and variable-step-size Backward Differentiation Formulas (BDF). Because the time-marching scheme is implicit, a nonlinear system of equations must be solved each time step. The previously discussed solving procedure, based on the Newton algorithm, has been exploited to solve this nonlinear system of equations. Algebraic systems of equations coming from differential operator discretization were solved both by a PARDISO package, a direct, high-performance, robust, memory efficient package for solving large sparse unsymmetrical linear systems of equations on shared-memory and distributed-memory multiprocessors.

3. Case studies

3.1. Movie theatre hall: Influence of the air-distribution layout on IAQ and thermal comfort

This case study concerns a movie theatre hall, whose geometry is outlined in Figure 1. The mean goal of this analysis consists in investigating the IAQ depending on the adopted layout for ventilation air distribution inside the indoor environment: Mixing Air Distribution (MAD), Personalized Air Distribution (PAD) and Under Floor Displacement (UFD). The internal dimensions of the hall are 18.5 m in length, 10 m in width and a maximum height of 9 m. The hall disposes of 10 rows of seats and 12 seats per row. Exploiting the room symmetry, a two-dimensional approximation was retained in simulations in order to save computational resources. Geometrical elements have been designed in order to simulate chairs (green) and human bodies (gray) as numerical domains in the considered transversal mid-section of the hall. The main region of the geometry is filled by air (cyan). Specific geometrical boundaries, located at the ceiling level, at the chair rear-side and at the floor level (close to the occupant's feet), have been represented to simulate diffusers and recovery grids for ventilating air. Some details concerning inlet/outflow sections for each ventilation layout analyzed are reported in enlargements of Figure 1. Results are mainly presented in by pictures reporting fluid-dynamical and thermal fields and CO_2 computed levels. Firstly, global fluid-dynamical and thermal results related to the steady solution of the considered systems are presented. Then, air quality is investigated by analyzing the results coming from the transient analyses simulating the human breathing. Finally, a brief discussion of results concerning their comparison with some data available in literature is reported.

3.1.1. Fluid-dynamics and thermal behavior in steady conditions

In Figure 2 air velocity vectors are presented in a color-scaled map identifying their magnitude, for MAD, PAD and UFD system, respectively. Figure 3 reports the corresponding thermal fields. The significant differences in air flow patterns can be appreciated from a distribution system to another.

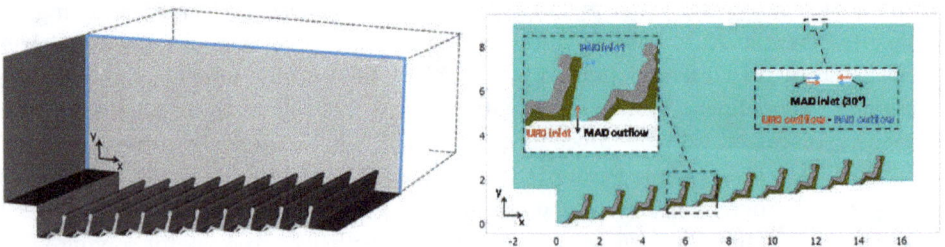

Figure 1. Geometry of the physical system and enlargements reporting details concerning the inlet/outflow sections for each ventilation layout.

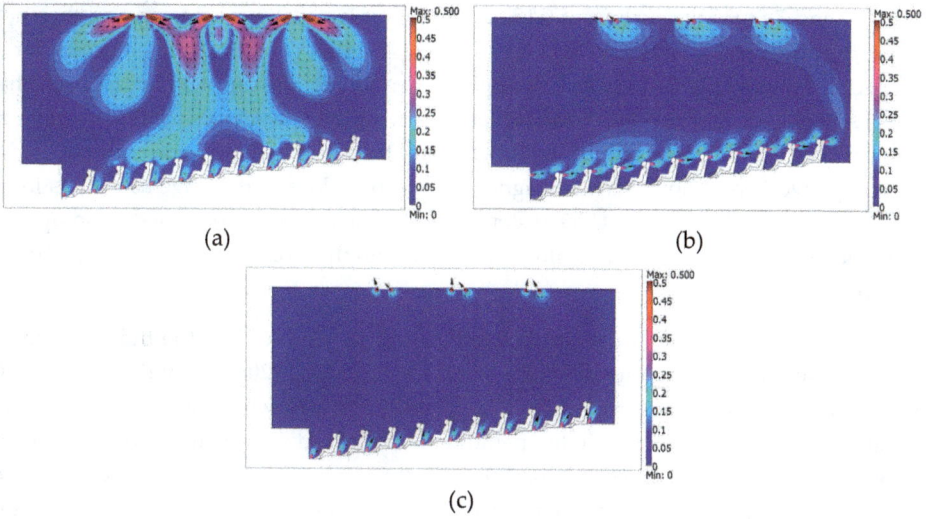

Figure 2. Velocity vectors and velocity magnitude [m/s] for MAD (a), PAD (b) and UFD (c) systems.

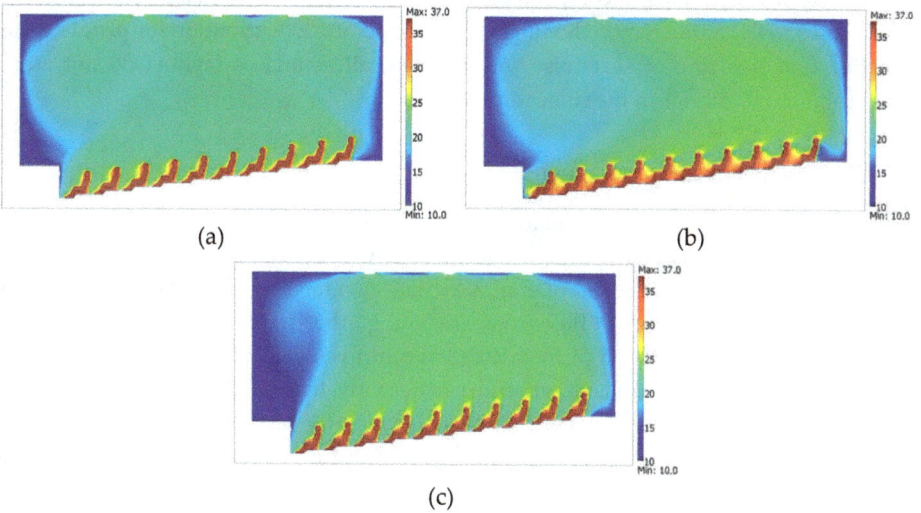

Figure 3. Temperature [°C] distribution for MAD (a), PAD (b) and UFD (c) systems.

Diffusers at the ceiling level determinate in the MAD system an important air circulation in the whole hall, characterized by two main cells located in the headroom close to the stage and the back side of the theatre hall. Otherwise, the intermediate rows of seats appear more subjected by the ventilating effect. Magnitude of air velocity close to some of people heads (rows 3-4 and 7-8) reaches values of 0.15-0.18 m/s, that is usually considered as a threshold inducing a disagreeable draught effect if exceeded [42]. That condition instead occurs for all the spectators when the PAD system is applied. Ventilating air, supplied by the back chair diffusers, flows too fast (up to 0.2 m/s) in proximity of the spectators' heads. The main air stream is blown to the rear wall of the theatre hall, then air flows up and it is recovered at the ceiling level. This allows almost stagnant conditions close to the stage as well as in the central portion of the hall. In UFD system, velocity of air in the occupied zone appears lower, globally assuring a better comfort condition from this point of view. The stage zone is stagnant.

Thermal fields well highlight the convective heat transfer induced by the fluid flow in the different cases. The average temperature of air for the MAD (20.7 °C) and UFD (20.8 °C) systems appears slightly lower than the PAD one (21.2 °C). However, when PAD system is simulated a high gradient of air temperature is remarked in proximity of each seated person. This is due to the stagnant conditions occurring in region between the chair back and the person legs: very slow velocity of air determinates an "ineffectiveness" in thermal dissipation of the metabolic heat. In that condition the PAD system appears the worse one referring to spectators thermal comfort. However, it is to notice as three-dimensional simulations are needed in order to clarify this aspect. In UFD system thermal plumes upon people heads are evident. In comparison with the MAD, the displacement ventilating system allows to get a much more homogeneous temperature distribution for people seated in the first and late row of seats also. As outlined, all simulated layouts do not assure thermal comfort conditions in the stage region.

3.1.2. Transient analyses and air quality

Transient analyses simulating human breathing have been carried out in order to investigate on air quality inside the movie theatre. CO_2 concentration in indoor air has been considered as an index of overall ventilation adequacy. The human breathing determinates periodical mass flux rate incoming the indoor environment. As a consequence, CO_2 concentration of fresh air (350 ppm) is locally modified. Depending on ventilating layout, the air streams transport differently the bio-effluent inside the hall. Figure 4 report the evaluated maps of CO_2 concentration for MAD, PAD and UFD close to one of occupants. Snapshots refer to 4 time-steps during a breathing period, that was considered occurring in 4 seconds.

Those pictures well elucidate the effect of the periodic function applied for breathing simulation. By adopting the local CO_2 level as index of effectiveness for the air distribution system, the best one seems to be the PAD layout. In fact, it assures a good dilution of the bio-effluent breathed out by the spectators, determining lower concentration of it close to the occupants' faces. On the other hand, the UFD system is characterized by almost stagnant condition in that region, so that extended zones presenting high levels of CO_2 are detected.

The MAD system determinates intermediate conditions from the previous ones. As previously introduced, in order to investigate on the average lifetime of air at a particular location, and giving an indication of the air "freshness", the additional transport-diffusion equation (7) has been also integrated during transient analyses. The "mean age of air" τ [s] has been computed for the studied ventilating layouts. As attended, starting from an arbitrary initial condition, each ventilating system assures, once a given time elapsed, the mean age of air becomes constant. The time range needed to assure that condition is shown to be different from a system to another. The UFD system achieves a τ stationary behavior exceeding 300 seconds about, while that threshold increases for MAD (1000 seconds about) and PAD (1400 seconds about) systems. This different behavior is also highlighted by Figure 5, where the computed mean age of air, evaluated for t=3600 seconds, is plotted in color-scaled maps overall the air volume for the investigated systems. In a MAD layout, air supplied by the ceiling located diffusers has to cover a long path before getting the occupied zone of the theatre hall. As a consequence, audience receive air 100-250 seconds "old" depending on specific place (spectators seated in the intermediate rows take advantage from this point of view). Different conditions occur when supplying air by PAD and UFD systems. Ventilating air is directly introduced by diffusers in the occupied zone, so that the mean age of air close to the people's face results globally lower than 50 seconds for the PAD system (except the first and the second row) and lower than 30 seconds for the UFD system.

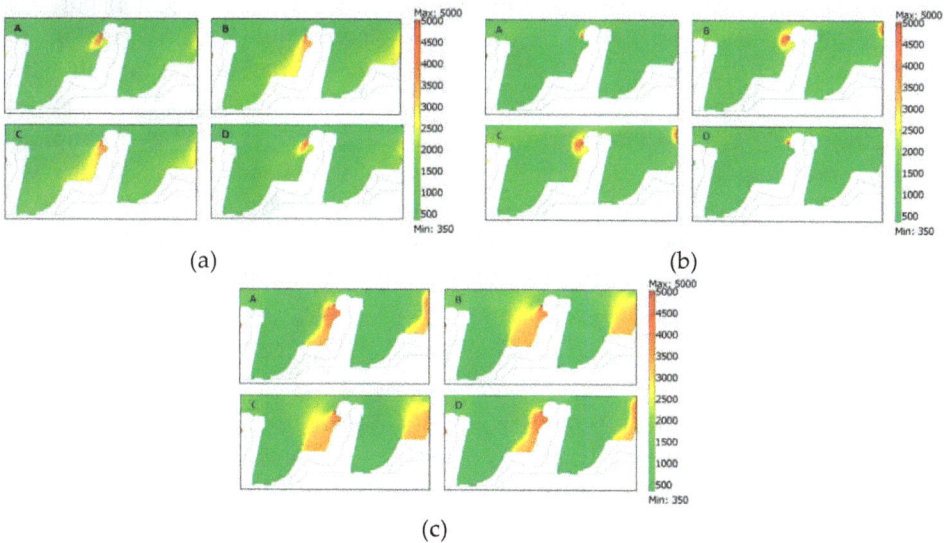

Figure 4. CO_2 concentration map [ppm] close to a person at 4 time steps of a breathing period (t=0 s (A), t=1 s (B), t=2 s (C), t=3 s (D)) for MAD (a), PAD (b) and UFD (c) systems.

(a)   (b)

(c)

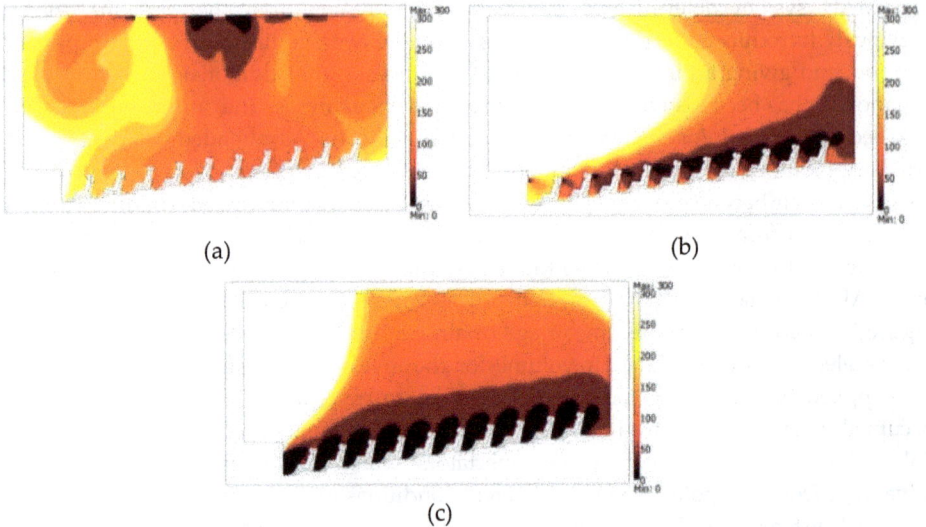

Figure 5. Distribution of the "mean age of air" for MAD (a), PAD (b) and UFD (c) systems.

3.2. Cabins of transportation means: influence of the air-distribution layout on IAQ and thermal comfort

This second case study deals with an investigation on bio-effluents transport and diffusion in ventilated cabins of transportation means and thermal comfort for passengers. As for the previously discussed application, different layouts for ventilation system (MAD, UFD, PAD) are preliminary analyzed in order to strike a balance between air quality degradation and comfort conditions for passengers. Firstly, a simplified geometry consisting in a 2D representation of 5 rows of seats standing inside a cabin was considered (Figure 6), then a more detailed 3D system was investigated in order to further optimize thermal and fluid-dynamical conditions for passengers. Depending on the air distribution system, small differences can be remarked in model geometries: as an instance, the geometrical elements used as inlet and outlet sections for fluid flowing the control volume. Geometrical elements are designed also in order to represent seated human occupants inside the cabin.

Physical properties of fluid are considered constant, except for density in the buoyancy term of the Navier-Stokes equations. Periodic functions has been used as boundary conditions to simulate the human breathing of occupants during time and the relative mass rate of carbon dioxide introduced in the cabin. Referring to the inlet velocity function, it is evaluated considering: the mass rate of air inhaled by a standard person every breathe, the air density, the surface of the nose holes and the breathing frequency. The carbon dioxide mass rate incoming in the control volume is as well computed following the same analytic procedure. In this case the concentration flux is evaluated considering the CO_2 molecular mass and the rate of CO_2 contained in air breathed out. In order to simulate more real conditions it is supposed that passengers breathe not in phase each other. The phase displacement is imposed in 0.2 second for each passenger.

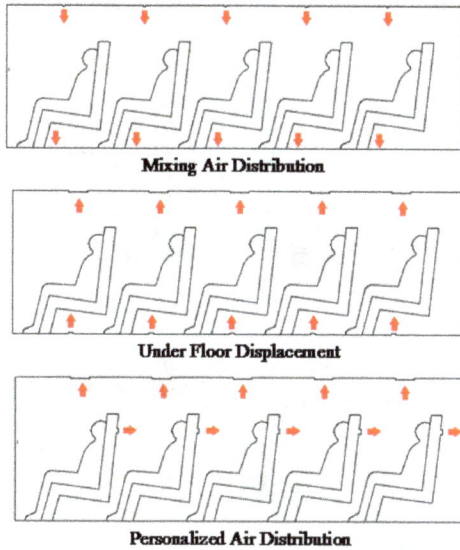

Figure 6. Outline of the cabin for the different layouts of the ventilating system.

3.2.1. Simplified geometry

The obtained results for the simplified 2D geometry are presented in this subsection. Figure 7 shows the velocity field at time instant t=120 [s] for the MAD, UFD and PAD air distribution systems. The significant differences in flow patterns can be appreciated from a distribution system to another. It is to remark the relative difference in the air velocity magnitude occurring close to the passenger faces. While the MAD and UFD systems assure magnitude of velocity lower than 0.15 m/s, the PAD system application determinates values comprised between 0.3-0.4 m/s. This represents the threshold value of induced discomfort in passengers due to a potential air draft perception. In Figure 8 CO_2 concentration maps are plotted for each air distribution system and at different time steps. Because of it is assumed a breathing frequency of 0.25 Hz, images (captured in the range 60-63 seconds) describe the concentration of bio-effluent during a complete breathing act close to the passenger's face. It can be observed as, from the air quality point of view, the best air distribution system appears the PAD one. In fact, it assures a good dilution of the bio-effluent breathed out by the passengers, determining very low concentration of it close to the occupant's nose. On the other hand, the UFD system is characterized by almost stagnant condition in that region, so that high levels of CO_2 are detected. The MAD system determinates intermediate conditions from the previous ones.

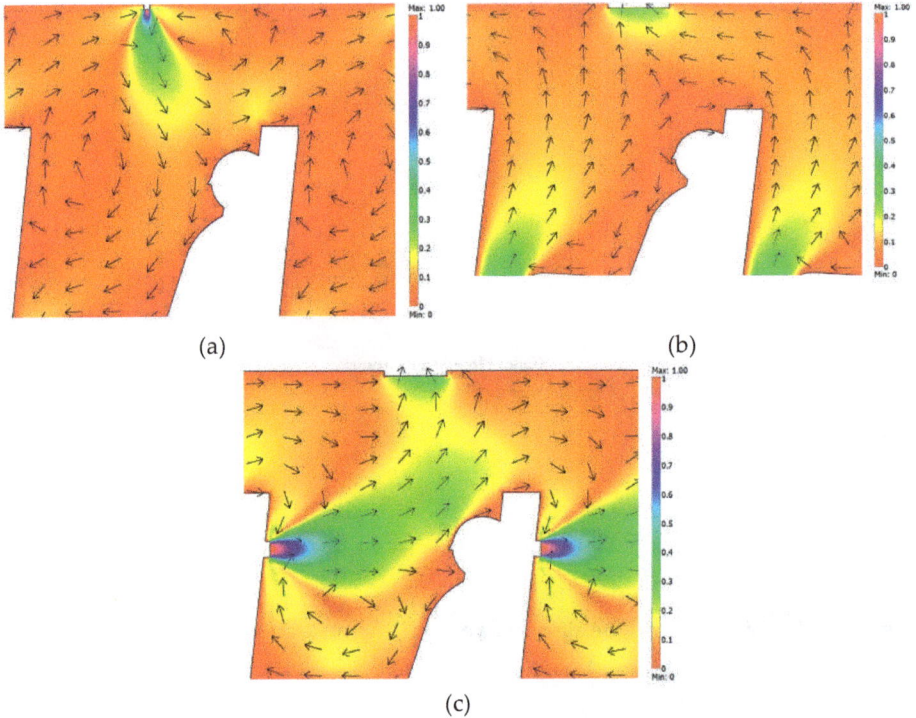

Figure 7. Velocity field [m/s]: MAD (a), UFD (b) and PAD (c) system.

Focalizing now the attention on the potential contamination risk inside the cabin, Figure 9 shows the tracing obtained by monitoring the path of a particle introduced, at the initial time of simulation, close to the nose of the passenger seated in the third row. This kind of post-processing allows to well understand the transport effect on a small mass generated by the fluid flow. Some remarks need to be pointed out. For each air distribution system, the particle path follows the streamlines of air flow. In MAD system, fresh air coming from the cabin ceiling blows the particle down as far as the recovery grids arranged on the floor. The time needed is 23 about seconds. In UFD system, fresh air coming from the bottom push up the particle as far as the grids, this time located on the roof. The time needed is about 24 seconds. In the PAD system, fresh air blown by the seat in front of the breathing passenger let his bio-effluent flow toward the passenger lodged in the rear row. The particle is then blown toward the outlet section by the rear air jet. The time needed is about 10 seconds. Results mainly show as from the comfort condition the most appropriate system is the UFD system. In fact it assure the lower velocity level close to the passenger's face. From the air quality point of view, the PAD system represent instead the best choice because it allows very low level of stagnant bio-effluent close to the passenger's nose. Anyway, referring to the contamination risk inside the cabin, this system is detected to be the most critical because it allows particle breathed out by a passenger to be potentially inhaled by another. Globally it appears that in absence of relevant challenges to be pursued in the most recent UFD and PAD systems, the classical MAD represent the better compromise between opposite requirements.

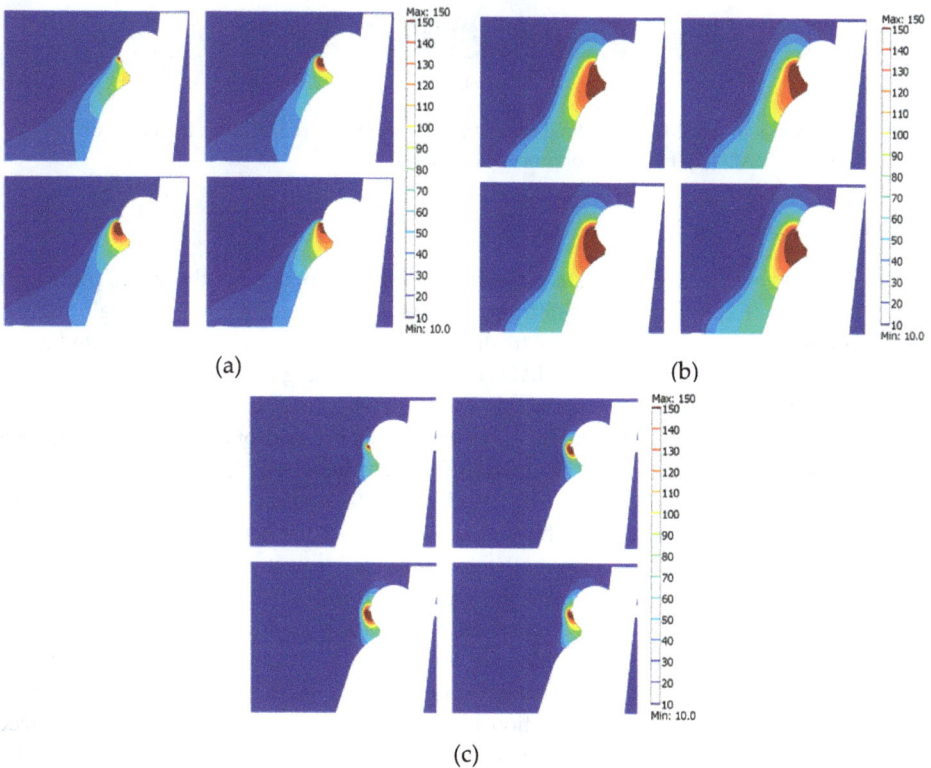

Figure 8. CO_2 levels [mol/m^3] in a breathing cycle for MAD (a), UFD (b) and PAD (c) system.

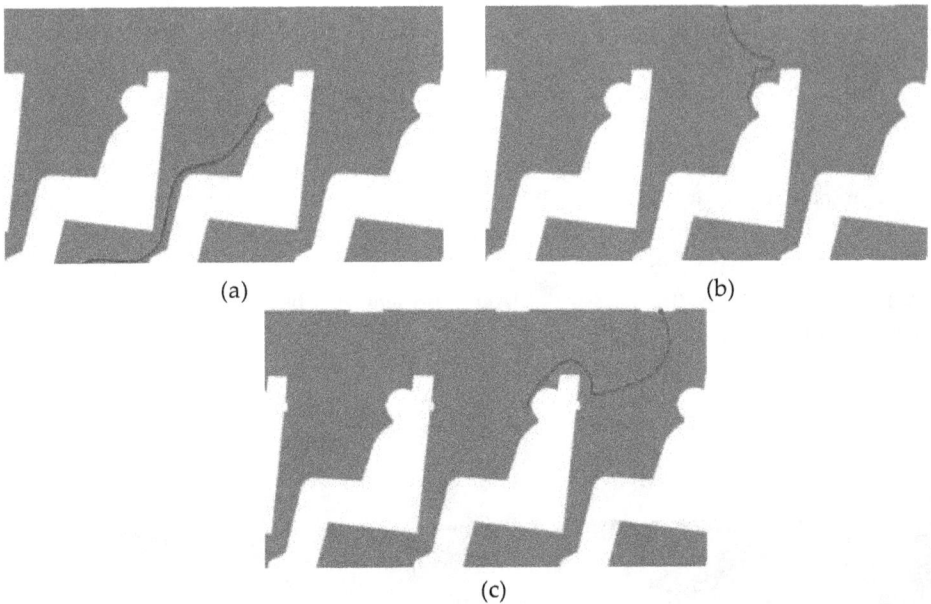

(a)   (b)

(c)

Figure 9. Particle tracing for MAD ((a), final time of processing t=23[s]), UFD ((b), final time of processing t=24[s]), PAD ((a), final time of processing t=10[s]) .

3.2.2. Detailed geometry

The detailed 3D geometry of the system consists in a shorter portion of the cabin, containing two rows of seats. Exploiting the longitudinal symmetry of the cabin, just one half of it has been reproduced. Two inlet sections for fresh air are considered: the first one, located at the roof and representative of a ceiling MAD, is made of rectangular slots positioned in the horizontal channel developing along the cabin aisle; the second one, representative of a UFD ventilation system, is located in correspondence of the lateral side of each seat. Results have been carried-out for the two operating layouts: firstly a pure ceiling system for air inlet is studied, then a second scheme has been considered: inlet sections for conditioning air are also arranged in correspondence of the lateral side of each passenger seat. This layout is representative of a combination between the standard ceiling system and a displacement air distribution system. Both configurations are studied for winter conditions, referring to an external temperature of 5°C. Figure 10 presents isosurfaces of velocity for the first analysed configuration. Ventilation air is provided by four rectangular slots located at the roof in correspondence of the aisle of the bus cabin. The interest was firstly focused on verifying passengers comfort referring to the motion field of ventilation air. In fact it is recognized that values of air velocity exceeding 0.8 m/s could be responsible of discomfort condition for human occupants of a ventilated ambient. Figure 11 shows distribution of air velocity and velocity vectors plotted in a transversal section of the cabin chosen as indicative for verifying comfort of passenger (the surface crosses passengers head). From analysis of Figure 11 it can be verified as the ceiling system for air distribution does not involve in discomfort for passengers. The inlet velocity for incoming air was chosen at value 2.5 m/s, that well fits with reasonable values usually adopted for this kind of ventilation system . The above discussed consideration is also confirmed by results reported in Figure 12. That figure reports the velocity field in a horizontal transversal section of the cabin. Maximum value of air velocity is reached in a zone not occupied, such as the cabin aisle. From a fluid-dynamical point of view, the second element that could determinate disagreeable condition for occupants is the turbulence level. In Figure 13 streamlines of the vorticity function for the obtained air flow have been reported. From analysis of that finding it is possible to deduce as high turbulence levels only occur in the top portion of the cabin space, so that passengers do not result affected by this potential cause of discomfort. Let now examine the ceiling air distribution system from a thermal point of view. Heating air is introduced from the lateral side of the passenger's seat close to the cabin aisle. On the other hand, the higher level of heat dissipation occurs on the opposite side, close to lateral wall of the bus cabin. This wall obviously confines with the external ambient. Moreover, it is partially made of the window glass characterized by a high thermal conductivity. These assumptions let guess to results obtained from thermal analysis of this inlet layout for heating air.

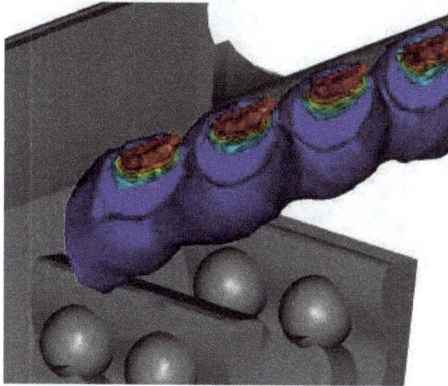

Figure 10. Isosurfaces of velocity.

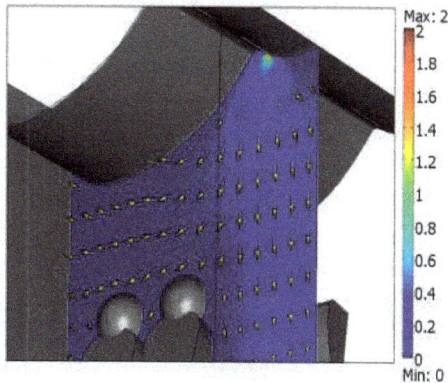

Figure 11. Motion field of air.

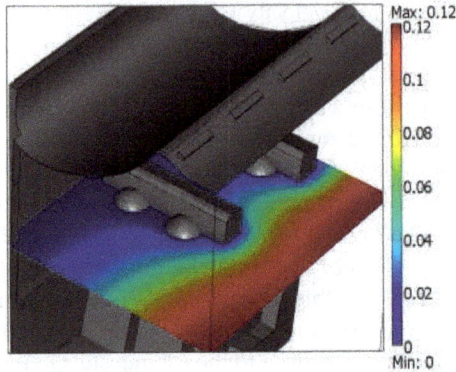

Figure 12. Velocity field.

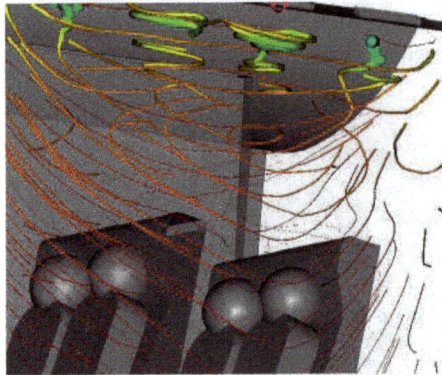

Figure 13. Streamlines of the vorticity function.

In Figure 14 temperature distribution has been plotted on contours of solid elements of the numerical model. Results reported in that figure well show the ineffectiveness of the ceiling air distribution system from a thermal point of view. As previously mentioned, thermal comfort highly depends on uniform distribution of the air temperature surrounding the human body. From Figure 14 is clearly observable as passengers seated close to the window are submitted to strong temperature gradient in horizontal direction. The obtained thermal field does not assure thermal comfort for passengers. Because of that reason it was simulated a second scheme for heating air incoming. As yet mentioned, this second outline of the numerical model takes its inspiration from the displacement air conditioning systems. In order to test this layout of inlet section location, heating air was considered to enter the computational domain in trough out the lateral bottom side of each passengers seat. Figure 15 presents the obtained results. As observable thermal field at solid-fluid interfaces of the numerical model show an almost uniform distribution of temperature in correspondence of the human occupants. That clearly means better comfort conditions for passengers.

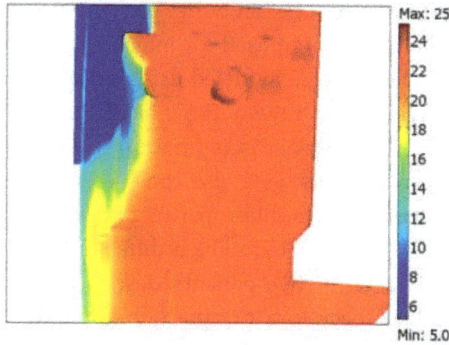

Figure 14. Temperature field at solid-fluid interfaces (MAD).

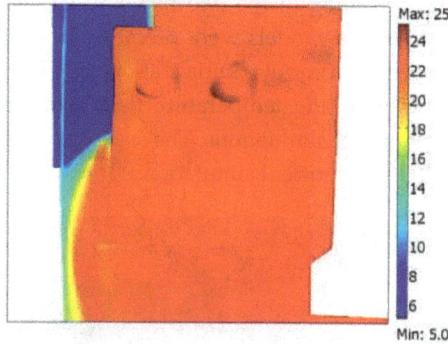

Figure 15. Temperature field at solid-fluid interfaces (MAD + UFD).

3.3. Hospital Isolation Room (IR): Influence of bio-effluent impulsively introduced by patients coughing

The third case study presented in this chapter concerns IAQ in a hospital IR. Analyses take into account the effect of patients' coughing on the indoor environment. Transient simulations were performed to investigate the efficiency of the existing HVAC plant with a Variable Air Volume (VAV) primary air system. Solid modelling of the room, taking into account thermo-physical properties of building materials, architectural features and furnishing arrangement of the room, inlet turbulence high induction air diffuser, the return air diffusers and two patients lying on two parallel beds was carried out. The 3D model of the room studied is provided in Figure 16.

The HVAC primary air system, designed for hospitalization infected patients, provides 6 m³/h of inlet air and 7 m³/h of extraction air. The room is ventilated by a commercial "turbulence" high induction air diffuser, located in the center of the ceiling and in the middle area between the two beds. The exhaust air is expelled by three return air diffusers located on the ceiling, the door of the toilet and the one adjacent to the corridor. The exhaust vents of the ventilation system were set at an outlet pressure value in order to maintain the negative pressure within the room as imposed for infectious patients. The heads of the two patients lying on two parallel beds were modeled using solid geometry to take into account their different positions. In particular three outflow head surfaces were modeled and used to simulate the inlet surface of the bio-aerosols (mouth) related to the different position of each patient. The heads of the patients were also considered as two heat sources. Building thermal performances and thermo-physical properties of building envelope were taken into account. Three scenarios were investigated, which differ in the direction of coughing and breathing for the same position of the two patients, described by the "supine" position, the "face-to-face" position and the "back-to-back" position. Time dependent simulations were based on governing equations solution for the three scenarios with the aim of investigating the temporal patterns of the ventilation flow and the particles tracing and diffusion in the conditions of coughing and breathing of the two patients. The transient simulations with thermal and fluid-dynamic coupled models were preceded by a long enough run without cough/sneeze events in order to compute initialization states for transient simulations. Airflow pattern of steady solution and temperature field in the room were computed to define initial conditions for transient simulations. The sneeze and/or cough events were then considered at the beginning of the transient simulation.

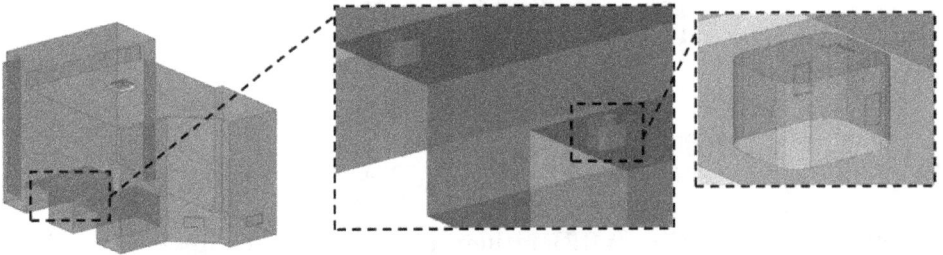

Figure 16. Outline of the geometrical system.

Then, considering each relative position between the patients, three events (coughing and/or breathing) were investigated: the first event (model_1) concerns the first patient coughing three times when the second is breathing; the second event (model_2) concerns the first patient coughing three times when immediately after the second coughs three times, the third (model_3) concerns both patients coughing, at the same time, three times. In particular for all the models the door was kept closed, so that particle diffusion would be influenced by ventilation airflow patterns alone. The "impulse" functions (volumetric and mass flow due to coughing) for the simulation of coughing events were repeated over time, with a time interval as "peak-to-peak" equal to 2 seconds. In model_2, the time phase displacement between coughing events of the two patients was taken as 6 seconds.

These conditions remain throughout the entire simulation of 60 seconds when the ventilation system operating and the coughing/breathing events happen. A peak of inlet velocity value due to the cough was set at 28 m/s, which is the maximum experimentally determined connected to the transient flow due to the impulse that has a length of about one second. The inlet/outlet breathe velocity was considered constant 0.9 m/s. A particle tracing and diffusion model, connected to cough events, was developed to simulate the dispersal of bacteria-carrying droplets in the isolation room equipped with the existing ventilation system. An analysis of the region of droplet fallout and the dilution time of bacteria diffusion of coughed gas in the isolation room was performed.

The analysis of transient simulation results concerning particle path and distance, and then particle tracing combined with their concentration, provided evidence of the formation of zones that should be checked by microclimatic and contaminant control. Figure 17 reports, for model_2 (the first patient coughing three times when immediately after the second coughs three times) studied in the scenario called as "supine position", an isosurface of aerosol concentration during first patient coughing. Figure 18 illustrated the same computed data for scenario "face-to-face". For "supine position" of the patients the effluent concentration in the form of aerosol achieves the higher values in the time interval of 2-4 seconds and then decreases with time because the continuous dilution provided by the air flow inlet. For "face to face" position the effluent concentration, achieves the higher values in the same time interval, but during the transient air flow displacement a local effluent stagnation zone is produced between the two beds and small recirculation flows are generated near the patients due to lower displacement efficiency of the plant. For "overleaf to overleaf" position this effluent stagnation reaches intermediates values obtained from other two models. Particles tracing for the above cited analyses are also reported in figures 19-20 for "supine" and "overleaf to overleaf" scenarios . The particle tracing obtained from the models for the several scenarios simulations highlights the position during time of the particles due to cough with a diameter of 1E-5 m; the obtained trajectories are referred to those particles which position at the initial instant overlap with the surface of the mouth of the first coughing patient and the second breathing. Concerning the particle path and distance, for the "supine" position, during the first 10 seconds the particles go through about 1.8-1.9 meters along the orthogonal direction to the surface of the inlet air diffuser; in the final part of the transient analysis the particles trajectory, initially dependent on the inertial effect of the air throw, is influenced by the air ventilation outflow. For "face-to-face" position, during the first 10 seconds the particles go through about 1.6-1.7 meters along the orthogonal direction to the surface of the inlet air diffuser, then the particles flow due to the patient coughing can reach the face of the second patient. For the "overleaf-to-overleaf" position, in about four seconds the particles go through 1.4 meters along the orthogonal direction to the surface of the inlet air diffuser and then collide against the wall. Globally it can be observed as the potentially influenza viruses aerosolized could maintain the biological properties 24 hours if the humidity is 20%. The inactivation rate is about 1.69 day^{-1} on steel and 0.58 day^{-1} on linen, cotton and fabric fomites. From simulation results can be deduced as the stagnation zones could represent a source of infection when the rooms is used for different patients during the day.

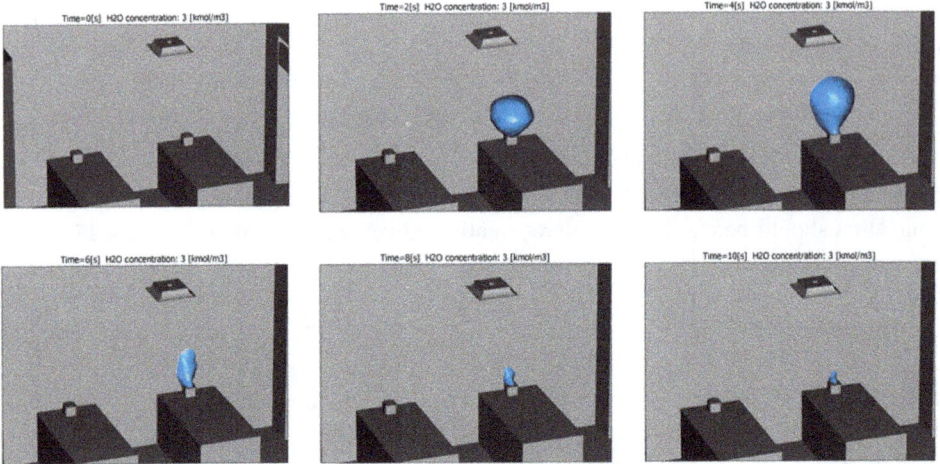

Figure 17. Isosurface of aerosol concentration for 1st patient coughing (model_2,"supine" scenario).

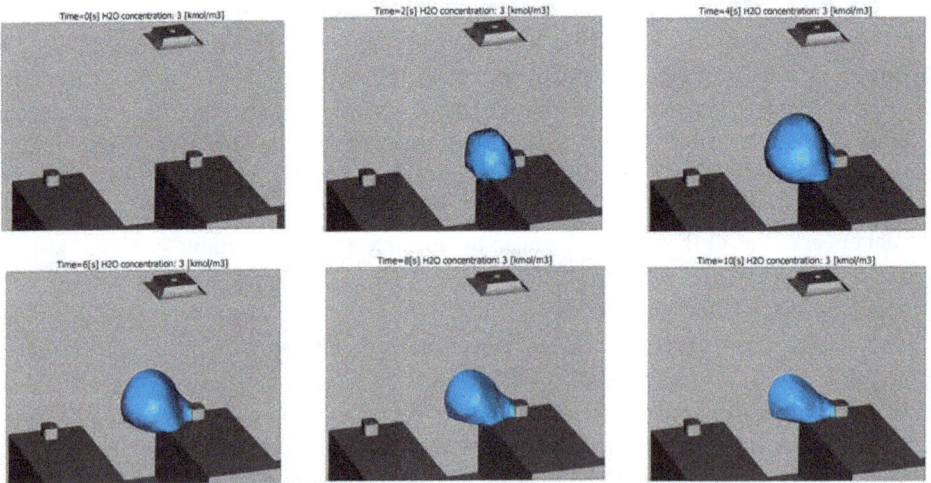

Figure 18. Isosurface of aerosol concentration for 1st patient coughing (model_2,"face-to-face" scenario).

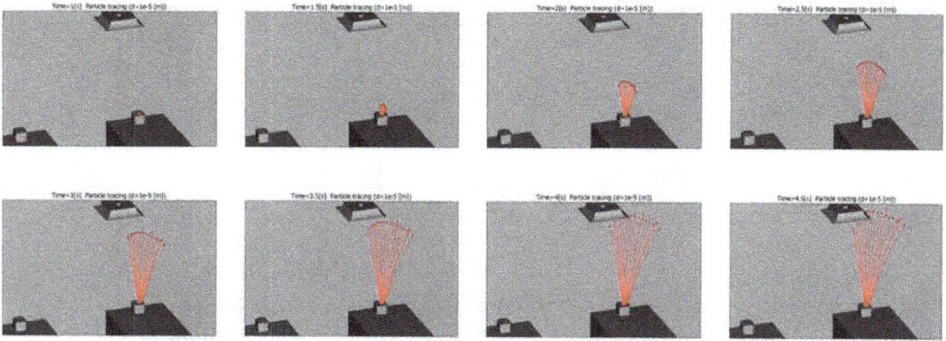

Figure 19. Particle tracing for aerosol during first patient coughing (model_2, "supine" scenario).

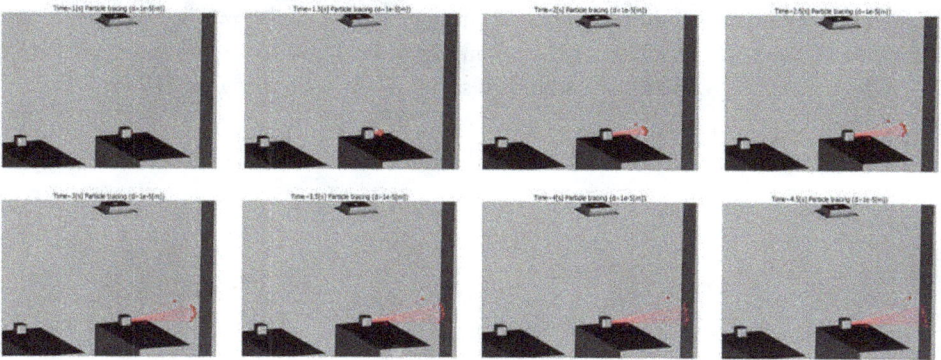

Figure 20. Particle tracing for aerosol during first patient coughing (model_2, "back-to-back" scenario).

3.4. Hospital Operating Theatre (OT): Influence of door opening/closing and people moving

The last presented application is devoted to investigate on IAQ in a OT keeping into account effects due to the opening/closing of a sliding door and the medical staff movement when crossing the open door-space. Transient simulations were performed for three cases:

a. Opening and closing of the sliding door (Case A);

b. Opening of the sliding door, one person crossing, closing of the sliding door (Case B);

c. Opening of the sliding door, two persons with a stretcher crossing, closing of the sliding door (Case C).

The physical and architectural model of the OT was based on a typical and standard layout of the ISO5 class with ultra clean air filters system. The operating room is 6.3 m wide, 6.3 m long with a total volume of 119.07 m^3. We referred to a HVAC plant systems that meets the most important standard requirements. The functional and operating parameters of the HVAC system with primary air unit and a ceiling unidirectional diffuser are provided in Table 3.

Air ceiling unidirectional inlet diffuser		
Effective surface (> 90% of the floor area)	36	[m²]
ACH (Air Change per Hour)	200	[1/h]
Minimum air change flow rate	1500	[m³/h]
Mean inlet velocity adopted	0.184	[m/s]
Minimum over-pressure differential	5	[Pa]

Table 3. Functional and operating parameters of HVAC system.

The internal air temperature of the OT was set at 20 °C and that of the surrounding zones at 26 °C as suggested. Taking into account literature suggested values [33] for a standard ISO5 class operating room, furniture, basic safety lighting systems, surgical lighting surgical staff position were disregarded but thermal power provided were considered: the total thermal power released by 8 persons was taken a 920 W, for general lighting 400 W and for surgical equipment 1 kW. The sliding door clear opening of the OT is 900 mm x 2100 mm with a thickness of 40 mm, representative of a door made by an aluminum frame and a galvanized steel sheet panel. The geometry of the studied system is outlined in Figure 21.

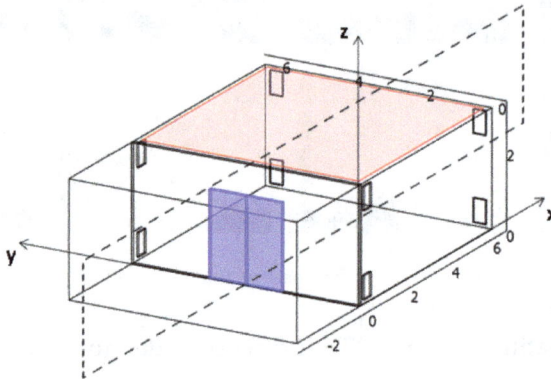

Figure 21. Geometry of the outlined OT and representation of the "zx" symmetry plane (dashed lines) considered for studying one half of the system.

The operating room is accessible from a 3 m breadth adjacent corridor by a two-panels sliding door, represented in blue in Figure 21. As introduced, the considered room is equipped by specific air vents and recovery grids in order to assure an unidirectional flow inside the hall. The air diffuser covers almost all the ceiling surface, extending for 6 m x 6 m over the room (pink surface in Figure 21). The recovery grids are located at the top and at the bottom corners of the room. Exploiting the geometrical symmetry of the system with respect to the mid zx-plane, one half of the presented system was considered in computations. In order to validate this assumption some preliminary tests were carried-out by considering the full geometry of the system. No important differences in test results have been remarked between the full model and the one-half model of the considered OT.

3.4.1. Implemented procedure for "moving objects" simulation

The adopted procedure in order to simulate the solid objects moving in the fluid domain is mainly based on the definition of specific source terms in the above reported governing equations (1-5), assuming assigned values in the portions of the computational domains where the solid objects are located at a chosen time. Moving of interfaces describing the position of the solid object, that are not explicitly designed in the geometrical models, is driven by some logical functions preliminarily defined, that allow simulating of the dynamics associated with the sliding door opening/closing and persons walking across the OT. The procedure mainly consists in defining some logical functions assuming binary values that identify the portions of domain where solid objects are located (binary value 1) or not (binary value 0) at the initial time. The binary value assumed by the logical functions depends on assigned geometrical coordinates for each object. In those regions fluid-dynamical properties and source terms assume specific values determining rest conditions for fluid. Time-dependent functions allow modification of the geometrical coordinates identifying the position of the "solid" objects during time, so that a prescribed motion law can be assigned to the moving objects. Let us first consider the door as example of the applied procedure. The following expressions have been used to define its spatial coordinates:

```
x1_door = x10_door
x2_door = x20_door
y1_door = y10_door
y2_door = y20_door-func_dy(t)
z1_door = z10_door
z2_door = z20_door
```

A graphic representation is given in Figure 22 to better elucidate the meaning of the above reported expressions. In this figure the solid objects (door outlined in pink and persons/litter in blue) are geometrically represented only for the sake of clarity, these geometries in fact are not really outlined in the numerical models used for computations. The above expressions assume constant or variable values in time. The time-function func_dy(t) denotes the motion law assigned to the door in the y direction (constant sliding velocity equals 0.66 m/s). When func_dy assumes the value "0" the door is closed, otherwise, when the assumed value is "1" the door is completely open (the door width being 1 meter). The following logical functions have been then defined in order to identify the coordinates' range, along each axis, identifying the door position:

```
x_door = if ((x>x1_door)and(x<x2_door)) then 1 else 0
y_door = if ((y>y1_door)and(y<y2_door)) then 1 else 0
z_doo = if ((z>z1_door)and(z<z2_door)) then 1 else 0
```

Position of the solid object "door" analytically corresponds to value "1" assumed by the following expression:

```
door = x_door*y_door*z_door
```

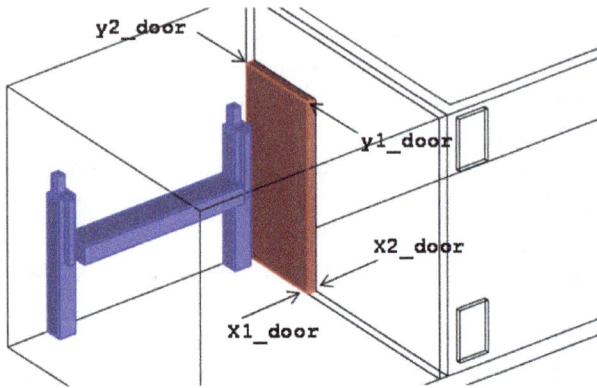

Figure 22. Expressions used to characterise positions of the "solid objects".

The same procedure is applied for defining position of the other "solid objects" (head, harms and body for each person and litter) during time. In this case, a second motion law (func_dx(t)) is defined in order to simulate the person and stretcher movements along the x-direction throughout the free door-space once the door is open. The constant advancing velocity chosen for the person is 0.84 m/s. Time evolution of the functions applied to define the passage of the "solid objects" for the different simulated case studies reported in this paper are graphically reported in Figure 23. Due to its formulation, the present numerical procedure can be applied to prismatic solid objects only. Once the solid object position is defined during time, some further functions could be used for assigning physical properties to these portions of the numerical domain, as given below:

```
eta_dom = if((door>0)or(P#1>0)or(P#2>0)or(L>0)) then eta_solid else
eta_fluid
rho_dom = if((door>0)or(P#1>0)or(P#2>0)or(L>0)) then rho_solid else
rho_fluid
k_dom   = if((door>0)or(P#1>0)or(P#2>0)or(L>0))  then  k_solid  else
k_fluid
Cp_dom = if((door>0)or(P#1>0)or(P#2>0)or(L>0))  then  Cp_solid else
Cp_fluid
```

where expressions P#1, P#2 and L relate to person 1, person 2 and stretcher and, in analogy with expression door used for the door, they identify the people and stretcher position where their value is "1". It should be noticed that this procedure makes it possible to use average values of the physical properties for all the solid objects only. This does not represent an inconvenience for viscosity in the momentum equations, as its value is chosen arbitrarily high (1E+5) in order to contribute to motion enabling.

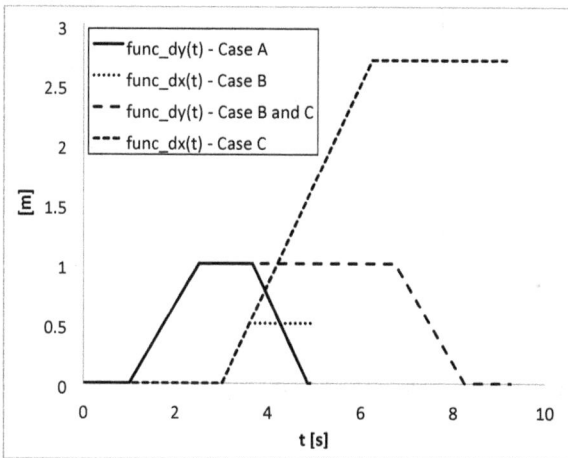

Figure 23. Functions defining the "solid objects" dynamics.

Otherwise, the use of an average value of density, thermal conductivity and thermal capacity in the energy equation should be verified by specific test cases. During the present study we compared steady solutions obtained by solving a model based on a classical implementation of the solid objects (partially opened position for the door) and a model based on the discussed here modelling procedure. Comparison has shown very low difference in temperature profiles, especially in airflow behaviour, that is the main point of modelling method proposed in this study. Numerical values used for physical properties in "fluid" and "solid" domains are listed in Table 4, where values used for the partition wall facing the corridor are also given.

	S.I Unit	"Fluid"	"Solid"	Wall
ϱ	[kg/m³]	1.2	900	600
η	[Pa s]	5E-5	1E+5	-
k	[W/(m K)]	0.62	0.6	0.16
C_p	[J/(kg K)]	1004	2000	840

Table 4. Numerical values of constants used in the adopted mathematical model.

Therefore, some special sources terms have been implemented in the governing equations. In order to enable motion in the solid objects, we defined:

```
S_dom = if((door>0)or(P1>0)or(P2>0)or(L>0)) then 0 else 1
F_dom = (C*(1-S_dom)^2)/(S_dom^3+ES)
```

Being C=1.6E+6 and ES=1E-4 constant values [43]. Note that F_dom becomes null where S_dom=1 (volume of fluid). In order to enable thermal transport in solid objects, in the energy equation we defined the components of thermal transport vector as follows:

```
u_dom = if((door>0)or(P#1>0)or(P#2>0)or(L>0))then 0 else u
v_dom = if((door>0)or(P#1>0)or(P#2>0)or(L>0))then 0 else v
```

```
w_dom = if((door>0)or(P#1>0)or(P#2>0)or(L>0))then 0 else w
```

being u, v ,w the components of the velocity vector. Finally, in order to consider the metabolic heat produced by people, we considered the following source term in the energy equation:

```
Q_dom = if((P#1>0)or(P#2>0))then Q else 0
```

Metabolic rate was assumed to be 2 Met (walking person), that corresponds to a specific heat source Q_dom = 1300 W/m³.

3.4.2. Transient results

Figure 24 presents the velocity field and the velocity vectors in a horizontal slice of the OT (1.5 meters from the floor) during the door opening in the Case A.

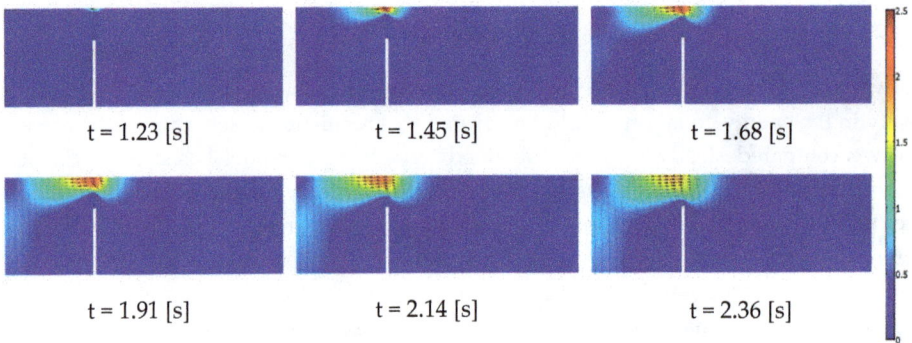

Figure 24. Velocity field [m/s] and velocity vectors in a horizontal plane (z = 1.5) at several time instants during the door opening (Case A).

Due to the pressure level inside the OT, a flow rate of air from the room to the corridor is highlighted when the door opens. The maximum magnitude of the flow velocity is reached for a partial opening position of the door. As a matter of fact, Figure 25 shows, for several time instants, the x-velocity profiles along the y-direction in correspondence of the door-space (line represented in red in the figure, lying at 1 meter from the floor). The negative values assumed by the x-velocity component depend by the chosen orientation of the Cartesian system of coordinates, and well underline the air flow direction when the door is opening. This effect, caused by the mixing between the air flow inside the room with that at the ceiling supply diffuser and the outside flow, is confirmed by recent studies [32]. The effect of temporal or extended period of door opening but also of surgical staff movements, that can obstruct the ventilation system efficiency, widely explained in the above cited papers, can be compared with our simulation results connected to relative velocity and pressure difference variations. In particular our obtained results can be strongly compared with those concerning the work of Brohus et al. [38] on the influence of person movements on contaminant transport in an operating room evaluated by smoke visualization and CFD simulations. As a matter of fact the effects we found on the velocity distribution and pressure variations due to persons and sliding door movements through the LAF field are very similar. A similar way of representation used in Figure 25 is applied in Figure 26-27 to report results obtained for Case B and Case C. Figure 26 refers to Case B and presents the velocity magnitude computed along a line lying along the x-direction (red line in figure, y=3.05, z=1.5) for several time instant during the person crossing of the open door. As can be appreciated, velocity achieves null value in the x-range corresponding to the person position during the door-space crossing. Similar considerations can be pointed-out by looking at Figure 27, that refers to Case C simulation. In this case the x-range where velocity magnitude is zero is much wider, because of the presence of two persons and a transported stretcher crossing the door-space. In confirmation, Figure 28 presents for the Case C the velocity magnitude evaluated along a x-directed line (red in the figure) lying at a distance from the floor higher than the stretcher one. In that figure two zero x-ranges can be appreciated, that correspond to the position, at each time step, of the two persons. Figure 29 collects some pictures referring to Case C, reporting the velocity fields in a longitudinal section of the OT (y=3.05) during the persons/stretcher crossing of the door-space. They highlight the "solid objects" moving across the door-space, from the corridor to the OT. The air distribution is significantly perturbed by the persons/stretcher incoming inside the OT, that involves in local airstream from the OT to the corridor, reaching values of velocity magnitude up to 1.5 m/s.

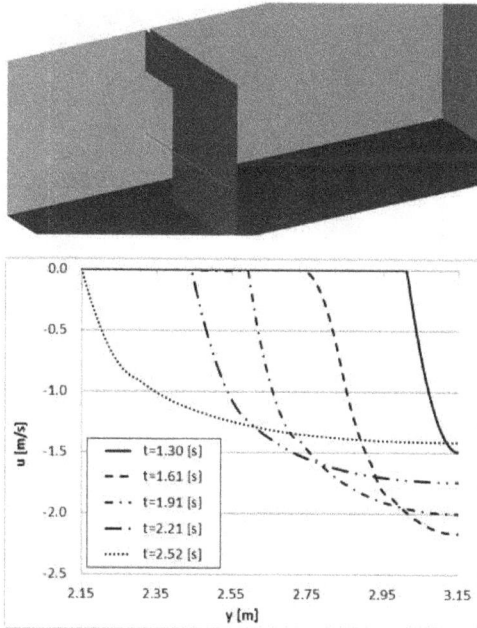

Figure 25. Velocity component u along a y-lying (z = 1 and x = 0) for several time instants during the door opening (Case A).

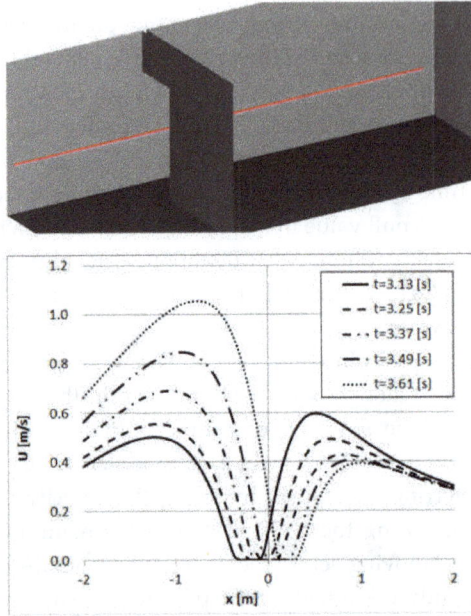

Figure 26. Velocity magnitude along a x-lying (z = 1 and y = 3.05) for several time instants during the person crossing (Case B).

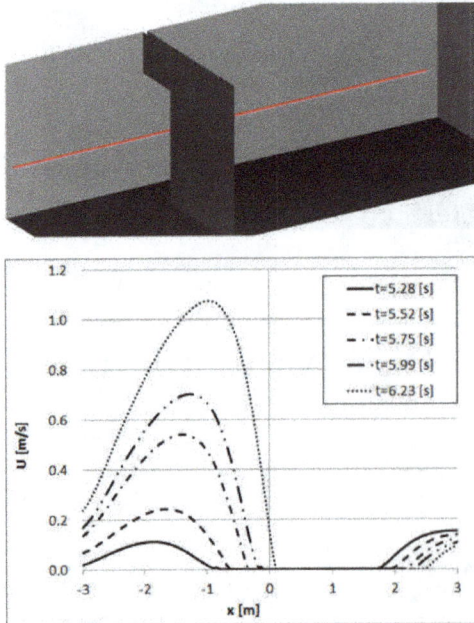

Figure 27. Velocity magnitude along a x-lying (z = 1 and y = 3.05) for several time instants during the persons and stretcher crossing (Case C).

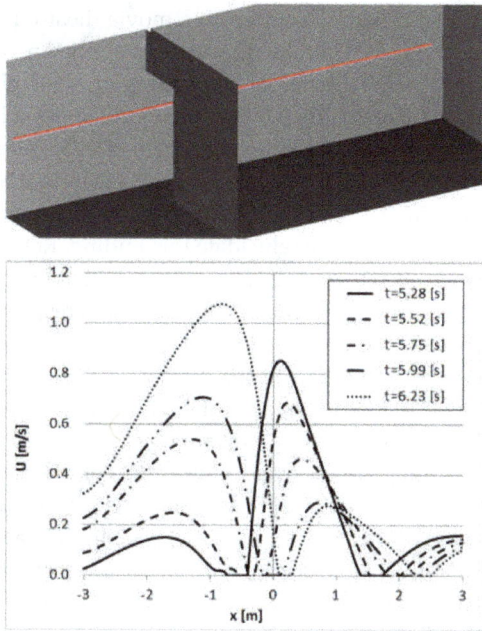

Figure 28. Velocity magnitude along a x-lying (z = 1.5 and y = 3.05) for several time instants during the persons and stretcher crossing (Case C).

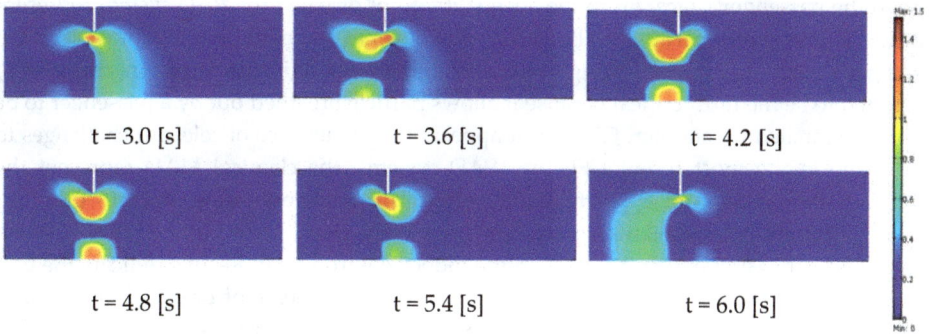

Figure 29. Velocity field [m/s] in a vertical plane (y = 3.05) at several time instants during the persons/stretcher crossing (Case C).

4. Conclusion

This chapter presents several case studies concerning multi-physical simulation for IAQ monitoring in internal environments with high crowding level and/or risk of contamination. A brief description on mathematical and numerical schemes adopted for carrying-out simulation results leads up to the chapter main body, where four real-world applications are presented and discussed. The first one is devoted to investigate on the influence of the air-distribution layout on IAQ and thermal comfort in a movie theatre hall. From results some interesting items can be remarked. When MAD system is applied a potential draught effect could be perceived by some of the spectators, depending on their place. That condition is increased for PAD system. Otherwise, the UFD system guarantees a better level of comfort from this point of view. From thermal analyses it appears that UFD system assures the most homogeneous thermal field, while PAD system generates disagreeable conditions related to high temperature gradients in the occupied zone. Authors are carrying out some three-dimensional simulations in order to better elucidate this finding. Referring to carbon dioxide level in the occupied zone, the PAD system is otherwise detected to be better than MAD and UFD ones, assuring a higher dilution of the bio-effluent with respect to the other systems. Anyway, the air stream determined by the personalized ventilation allows air breathed out by a person to be potentially transferred to the person seated in the rear row. This item could become critical because of a potential risk of contamination. From the air "freshness" point of view, the MAD is surely the worst system among the studied ones. The diffusers location, at the ceiling level, determinates very high values of the mean age of air supplied to spectators with respect to PAD and UFD layouts. The aim of the second case study presented in very similar to the that refers to the first one. In this last case we investigated on IAQ inside cabins of means of transportation depending on the applied layout for ventilation air distribution. The goal was to strike a balance between air quality degradation and comfort conditions for passengers standing in an aircraft cabin potentially equipped by three kinds of air distribution system. Results mainly show as from the comfort condition the most appropriate system is the UFD system. In fact it assure the lower velocity level close to the passenger's face. From the air quality point of view, the PAD system represent instead the best choice because it allows very low level of stagnant bio-effluent close to the passenger's nose. Anyway, referring to the contamination risk inside the cabin, this system is detected to be the most critical because it allows particle breathed out by a passenger to be potentially inhaled by another. Globally it appears that in absence of relevant challenges to be pursued in the most recent UFD and PAD systems, the classical MAD represent the better compromise between opposite requirements. The third application concerns the effect of patients coughing in a hospital isolation room and the performance of the HVAC system in diluting the bio-effluents introduced in the indoor environment. Results indicate the best conditions for the high induction air inlet diffuser and the scheme of pressures imposed in the room to provide the effective means of controlling flows of virus-loaded droplets. Our findings stress that the position of the air-supply inlets in the ceiling and the exhaust vents at the opposite side of the room and in the ceiling, provide an up-draft effect and infection control efficiency. When the air-supply inlet is located in the middle of the ceiling of the room, and the exhaust vents are positioned in the wall behind the patient beds, the coughed particles and gas are contained at the side of the patient in the region of the exhaust vents. This can minimize the region of coughed gas diffusion and droplet fallout. The high induction air inlet diffuser and the scheme of pressures imposed in the room provide the effective means of controlling droplet flows containing viruses. This type of analysis allows to predict air recirculation zones which can host pathogens. Indeed our analysis of the particle tracing and path distribution combined with their concentration, was effective in identifying examples of such zones between the two beds, around the ceiling surface and on the surface of the bed lamps. The last case study is devoted to IAQ investigation in a hospital operating room, taking into account the sliding door and the medical staff movements. The study provides a simplified indirect numerical method to simulate the influence of solid objects movement, that exploits analytical expressions to dynamically track the solid-fluid interfaces in the computational domain. Special source terms were implemented in the governing equations in order to enable air-flow in domain's portions where solid objects are located at a chosen time. The used turbulence method was preliminary validated by comparing numerical results with experimental data available in literature for similar cases. Results, obtained for the three cases studied, highlight a strong modification of the air velocity field inside the OT, due to opening/closing door and staff movements. Air leakages through the door clear-space were estimated for each configuration. The air-flow rates, coming out the OT under transient conditions, were also correlated to the HVAC energy overload. An important reduction of the overpressure assured by the HVAC system inside the studied room was clearly evaluated, highlighting that pressure reaches values very close to the minimum threshold limits recommended by the most important international standards concerning the OT climate conditions. We are persuaded that the proposed numerical approach, based on FE multi-physical analysis in order to investigate on IAQ, can be reputed as a powerful and reliable tool to be used for the comprehension and analysis of the airflow and pressure distribution in several class of applications. For that reason, we believe that the introduction of the practice of performing multi-physical simulations on real cases will be an effective, relatively low cost, means of prediction and control.

Author details

Giuseppe Petrone* and Giuliano Cammarata
Department of Industrial Engineering, University of Catania, Catania, Italy

Carla Balocco
Energy Engineering Department, University of Firenze, Firenze, Italy

5. References

[1] Fanger PO (1970). Thermal comfort. Copenhagen: Danish Technical Press.
[2] Fanger PO (1988a). Introduction of the olf and the decipol units to quantify air pollution perceived by humans indoors and outdoors. Energy and Buildings, 12: 1-6.
[3] Awbi HB (2003). Ventilation of buildings. London: Spon Press, Taylor & Francis Group.
[4] Alamari F, Butler DJG, Grigg PF, Shaw MR (1998). Chilled ceilings and displacement ventilation. Renewable Energy, 15: 300-305.
[5] Fredriksson J, Sandberg M, Moshfegh B (2001). Experimental investigation of the velocity field and airflow pattern generated by cooling ceiling beams. Building and Environment, 36: 891-899.

* Corresponding Author

[6] Zhang L, Chow TT, Fog KF, Tsang CF, Qiuwang W (2005). Comparison of performances of displacement and mixing ventilations. Part II: indoor air quality. International Journal of Refrigeration, 28: 288-305.

[7] Fleming WS (1986). Indoor air quality, infiltration, and ventilation in residential buildings. In Proceedings of: ASHRAE Conference IAQ'86: Managing Indoor Air for Health and Energy Conservation, Atlanta, GA, USA; pp. 192–207.

[8] Persily A (1997), Evaluating building IAQ and ventilation with indoor carbon dioxide. ASHRAE Trans, 103: 193–204.

[9] Daisey JM, Angell WJ, Apte MG (2003). Indoor air quality, ventilation and health symptoms in schools: an analysis of existing information. Indoor Air, 13: 53-64.

[10] Papakonstantinou KA, Kiranoudis CT, Markatos NC (2002). Numerical simulation of CO_2 dispersion in an auditorium. Energy and Buildings, 200: 245–250.

[11] Cheong KWD, Lau HYT (2003). Development and application of an indoor air quality audit to an air-conditioned tertiary institutional building in the tropics. Building and Environment, 38: 605-616.

[12] Cheong KWD, Djunaedy E, Chua YL, Tham KW, Sekhar SC, Wong NH, Ullah NB (2003), Thermal comfort study of an air-conditioned lecture theatre in tropics. Building and Environment, 38: 63–73.

[13] Noh KC, Jang JS, Oh MD (2007). Thermal comfort and indoor air quality in the lecture room with 4-way cassette air-conditioner and mixing ventilation system. Building and Environment, 42: 689-698.

[14] Kavgic M, Mumovic D, Stevanovic Z, Young A (2008). Analysis of thermal comfort and indoor air quality in a mechanically ventilated theatre. Energy and Buildings, 40: 1334-1343.

[15] Wang A, Zhang Y, Sun Y, Wang X (2008) Experimental study of ventilation effectiveness and air velocity distribution in an aircraft cabin mockup, Building and Environment, 43, 337-343.

[16] Petrone G, Cammarata L, Cammarata G (2011). A multi-physical simulation on the IAQ in a movie theatre equipped by different ventilating systems, Building Simulation, 4: 21-31.

[17] Arens E, Xu T, Miura K, Zhang H, Fountain M, Bauman F (1998). A study of occupant cooling by personally controlled air movement. Energy and Buildings, 27: 45–49.

[18] Demetriou DW, Khalifa HE (2009). Evaluation of distributed environmental control systems for improving IAQ and reducing energy consumption in office buildings. Building Simulation, 2: 197-214.

[19] Niu J, Gao N, Phoebe M, Huigang Z (2007). Experimental study on a chair-based personalized ventilation system. Building and Environment, 42: 913-925.

[20] Bauman F (2003). Underfloor Air Distribution (UFAD) Design Guide. American Society of Heating, Refrigerating, and Air Conditioning Engineers Research Project RP-1064. Atlanta.

[21] Zhang T, Chen QY (2007). Novel air distribution systems for commercial aircraft cabins. Buildings and Environment, 42: 1675-1684.

[22] Tang J, Li Y, Eames I, Chan P, Ridgway G (2006). Factors involved in the aerosol transmission of infection and control of ventilation in healthcare premises. Journal of Hospital Infection, 64: 100-114.

[23] Zhao B, Yang C, Chen C, Feng C, Yang X, Sun L, Gong W, Yu L (2009). How Many Airborne Particles Emitted from a Nurse will Reach the Breathing Zone/Body Surface of the Patient in ISO Class-5 Single-Bed Hospital Protective Environments? A Numerical Analysis. Aerosol Science and Technology, 43: 990–1005.

[24] Talon D, Schoenleber T, Bertrand X, Vichard P (2006). Performances of different types of airflow system in operating theatre. Annales de Chirurgie 131: 316-321.

[25] Balocco C, Lio P (2011). Assessing ventilation system performance in isolation rooms. Energy and Buildings, 43:246-252.

[26] Balocco C (2011). Hospital ventilation simulation for the study of potential exposure to contaminants. Building Simulation, 4: 5-20.

[27] Tunga YC, Hu SC, Tsai TI, Chang IL (2009). An experimental study on ventilation efficiency of isolation room. Building and Environment, 44: 271-279.

[28] Chow TT, Kwan A, Lin Z, Bai W (2006b). Conversion of operating theatre from positive to negative pressure environment, Journal of Hospital Infection, 64: 371-378.

[29] Menazadeh F, Manning AP (2002). Comparison of Operating Room ventilation systems in the protection of surgical site. ASHRAE Transactions, 108: 1-13.

[30] Chow TT, Yang XY (2004). Ventilation performance in operating theatres against airborne infection: review of research activities and practical guidance. Journal of Hospital Infection, 56: 85–92.

[31] Dong S, Tu G, Cao R, Yu Z (2009). Numerical Study on Effects of Door-Opening on Airflow Patterns and Dynamic Cross-Contamination in an ISO Class 5 Operating Room. Trans. Tianjin University, 15: 210-215.

[32] Shih YC, Chiu CC, Wang O (2007). Dynamic airflow simulation within an isolation room. Building and Environment, 42: 3194–3209.

[33] Matzumoto H, Ohba Y (2004). The Influence of a Moving Object on Air Distribution in Displacement Ventilated Rooms, Journal of Asian Architecture and Building Engineering, 3: 71-75

[34] Halvonovà B, Melikov AK (2010). Performance of "ductless" personalized ventilation in conjunction with displacement ventilation: Impact of disturbances due to walking person(s). Building and Environment, 45: 427–436.

[35] Balocco C, Petrone G, Cammarata G (2012). Assessing the effects of sliding doors on an operating theatre climate. Building Simulation, 5:73-83.

[36] Brohus H, Balling KD, Jeppesen D (2006). Influence of movements on contaminant transport in an operating room. Indoor Air, 16: 356-372.

[37] Wilcox DC (1998), Turbulence Modelling for CFD, DCW Industries Inc.

[38] Abanto J, Rarrero D, Reggio M, Ozell B (2004). Air flow modelling in a computer room. Building and Environment, 39: 1393-1402.

[39] Zhang L, Chow TT, Fog KF, Tsang CF, Qiuwang W (2005). Comparison of performances of displacement and mixing ventilations. Part II: indoor air quality. International Journal of Refrigeration, 28: 288-305.

[40] Deuflhard P (1974). A modified Newton method for the solution of ill-conditioned systems of nonlinear equations with application to multiple shooting. Numerical Mathematics, 22: 289-315.

[41] Hindmarsh AC, Brown PN, Grant KE, Lee SL, Serban R, Shumaker DE, Woodward CS (2005). SUNDIALS: Suite of Nonlinear and Differential/Algebraic Equation Solvers. ACM Transactions on Mathematical Software, 31: 363-396.

[42] Fanger PO, Melikov A, Hanzawa H, Ring J (1988b). Air turbulence and sensation of draught. Energy and Buildings, 12: 21-39.

[43] Debabrata P, Yogendra KJ (2001). Melting in a side heated tall enclosure by a uniformly dissipating heat source. International Journal of Heat and Mass Transfer, 44: 375-387.

Chapter 13

Air-Polluted with Particulate Matters from Livestock Buildings

Ehab Mostafa

Additional information is available at the end of the chapter

1. Introduction

Livestock production has a harmful effect on the environment during the breeding period, such as the emittance of dust, odour and ammonia into the surrounding environment through the ventilation system as well as its harmful influence on the animals and workers inside these animal houses.

Airborne dust is normally considered to be one of the contaminants in livestock buildings. Particle matter reduces the air quality within the livestock buildings compromising the health of farmers and animals, (Hinz et al., 2007).

Commercial livestock production facilities are always associated with some level of airborne particles. High concentrations of airborne particles could affect the external environment, production efficiency, health and welfare of humans and animals, (Banhazi and Seedorf, 2007).

The improvement of the farm animal health is an important goal to ensure proper livestock production. Apart from management factors the internal environmental conditions play a key role for ensuring the well-being of intensively housed livestock and farm workers. Livestock farmers are exposed to dust concentrations inside their animal houses that are a factor of 10 to 200 times higher than those of the outside air, (Aarnink and Ellen, 2007).

The ventilation system of a building discharges dust particles into the environment, considering the high dilution rate with the outside air, the following discussion focuses on the dust level and control inside the livestock building. Requirements for good management and ventilation in animal husbandry systems ensure that the quality of indoor air is acceptable for animals' and humans' health, (Haeussermann et al., 2007).

2. Harmful effect of the dust on human and animal health

It is generally assumed that dust particles are capable of transporting different chemical compounds and microorganisms from one livestock building to the other, or from a livestock building to the farmhouse and to the neighboring houses. This may cause increased risks of airborne infections of animals and malodour problems. Farmers in animal houses are exposed to gases and a complex aerosol of bacteria, fungi, endotoxin and organic dust, which are linked to the development of respiratory diseases in farmers' lungs, (Takai et al., 2002).

2.1. Air quality requirements

According to the federal pollution control law people, animals, plants, soil, water, atmosphere and other cultural assets need to be protected from harmful environmental effects. This is determined in the administrative regulation (TA-Luft, 2002) (technical instructions for cleaning the air) by specifying limits for the emission mass flow and the mass concentration of harmful substances in concrete.

According to the statutory mandate, it is a goal of (TA-Luft, 2002) to provide authorities with up to date information on nationwide guidelines in order to carry out an evaluation of the emissions and immissions especially within licensed facilities. In order to indicate the values in (TA-Luft, 2002) the terms emission and immission are defined with the pertinent defaults using standardized evaluation criteria. "Emissions are defined within these administrative rules as those of air pollution" (TA-Luft, 2002).

The emissions are thus indicated on the one hand as the mass of the emitted substances or groups of substances related to the volume and mass concentrations. The mass indication of emitted substances or groups of substances is related to the unit time (emission mass flow). The dust contained in the exhaust emissions should not exceed a 20 mg/m³ mass concentration or 0.20 kg / h of mass flow. There are other values in the MAC list (Maximum Acceptable Concentration) published from the senate committee of the German research council, (DFG, 2006). The "maximum workplace concentration value" in (GefStoffV, 1999) defines the value of substances permissible in workplace atmospheres, in order not to affect the health of workers within an eight hour daily work schedule. This value differentiates between two groups of dust, the respirable (< 5 μm) and the alveolar dust (< 1.1 μm). The respirable group may not exceed a concentration of 4 mg/m³ and for the alveolar group the limit value is 1.5 mg/m³, (DFG, 2004). In case of non-compliance with these limits in animal barns protective arrangements should be employed for the staff such as breathing masks, (Scheuermann, 2004). Under Danish conditions a consistent relationship between environmental exposure in livestock buildings, lung function changes and/or respiratory symptoms in workers and identified exposure-response thresholds for workers on the basis of exposure response thresholds for poultry and swine confinement buildings has been observed by Pedersen et al. (2000). The limit recommendations for humans are 2.4 mg/m³ of total dust, 0.23 mg/m³ of respirable dust with a total of 800 EU/m³ (EU = endotoxin unit) and 7 ppm of ammonia (Pedersen et al. 2000).

2.2. Particle influence on the respiratory system of animals and humans

Keder (2007) reported that the particles suspended in the air enter the human body by breathing. These particles include natural materials such as bacteria, viruses, pollen, sea salt, road dust, and anthropogenic emissions. The hazard caused by these particles depends on their chemical composition as well as where they deposit within human respiratory system. Hence, understanding the deposition of aerosol in the human respiratory system is critical to human health, so that the deposition of "bad" aerosol must be reduced. The respiratory system works essentially as a filter. The viscous surface of the airway wall almost guarantees the deposition without re-entrainment when a particle is in contact with it. The most important mechanisms are impacting, settling, diffusion and interception. A particle entering the respiratory system is subject to all the deposition mechanisms previously mentioned. The actual deposition efficiency of a given particle size has been determined experimentally. Several models have been developed to predict the deposition. Over 700.000 people in the United States are exposed to hazardous levels of swine dust each year, and over 60 % of these suffer from various respiratory disorders including organic toxic dust syndrome, chronic bronchitis, hypersensitivity pneumonitis and occupational asthma (Rosentrater 2004). There are significantly higher prevalences of chronic cough, chronic phlegm, chronic bronchitis and chest tightness in poultry workers than in control workers (Eugenija et al. 1995 & Iversen et al. 2000).

2.3. Transportation of harmful substance inside animal buildings

Particulate matter can be considered as a good media to absorb odour and other harmful gases such as ammonia. It can then transfer inside the animal buildings and to the environment by ventilation. The airborne dust is one of the primary means by which disease-causing organisms are spread throughout a poultry house. Reductions in airborne dust levels have been associated with even greater reductions in airborne bacteria. Poultry (meat and eggs) contaminated with Salmonella continue to be important vehicles for Salmonella infections in humans. Pathogens, such as Salmonella can be introduced into the food chain at any point. Airborne transmission of Salmonella is a major factor for the spread of Salmonella from bird to bird and hatching eggs in breeder houses. It has been shown to be a major factor in the spread of disease in hatching cabinets (Mitchell et al. 2004). A significant proportion (15 to 23 %) of airborne ammonia in enclosed livestock facilities is associated with dust particles (Reynolds et al. 1998). The dust particles in swine buildings may be responsible for a considerable portion of odourant emissions from the buildings and odour perceptions by downwind neighbours of swine farms. Therefore, controlling the odour will require a reduction of dust emissions from buildings (Robert 2001). Ammonia and odours may be absorbed by the dust particles. Viable bacteria and viruses carried into the air by dust particles may have a greater ability to survive (Takai et al. 1998).

3. Definition of the particle

The Dust is defined according to ISO (1995) as small solid particles conventionally taken as those particles below 75 μm in diameter which settle out under their own weight but may remain suspended for sometime. On the other side, IUPAC (1990) defined it as a small, dry, solid particles projected into the air by natural forces such as wind, volcanic eruption and by mechanical or man-made processes such as crushing, grinding, milling, drilling, demolition, shovelling, conveying, screening, bagging, and sweeping. Dust particles are usually in the size range from about 1 to 100 µm in diameter, and settle slowly under the influence of gravity.

4. Particles characterization

4.1. Chemical properties of dust particles

Dust is analyzed according to its chemical composition into inorganic and organic (viable and non-viable) components. The chemical composition of airborne and the settled dust from different sources have nearly the same concentrations of dry matter, ash, N, P, K, Cl and Na (Pedersen et al. 2000). From the chemical analysis of the airborne and settled dust in broiler houses and pig rooms, the dust from broiler houses was higher in its chemical composition than that from the pig barns (Ellen et al. 2000).

The dust from poultry houses contains the highest amounts of protein. This is caused firstly by the relatively high protein content in the feed which is usually between 20 and 25 %. Secondly the other proportion of up to 45 % comes probably from feathers and claw abrasion. Also in the pig house the dust percentage of about 20 % seems to come from the skin and the hair of the animals (Hartung and Saleh, 2007).

4.2. Biological properties of dust particles

The particles, especially large particles, act as carriers of other air pollutants such as bacteria, viruses, odour and gases (Zhang et al. 2005). The dust within livestock buildings has viable microorganisms, fungi and absorbed toxic gases (Wang et al. 2000). There are larger amounts of airborne microorganisms in alternative housing systems of poultry houses. These high concentrations of viable fungi in the multiple level systems may be caused by using wood shavings in the bedding area that might have been contaminated with fungal spores. The dust and microorganisms with different admixture are abundant in the air of livestock houses (Bakutis et al. 2004). The dust can carry and promote large aggregations of microorganisms including viruses and bacteria (both gram-positive and gram-negative), especially Salmonella, Staphylococcus, Micrococcus, Endotoxin, and Rotavirus.

4.3. Physical properties of dust particles

The dust particles are subjected to a variety of physical processes according to their density, size and shape. The most important physical effects are sedimentation, agglomeration, aerodynamics, adsorption and resuspension (Rosenthal et al. 2007). There is a strong influence of the particle size, density, surface and shape on the distribution of airborne particles (Schmitt-Pauksztat 2006).

5. Sources of the dust inside the livestock buildings

The dust particles in animal housing may originate from feed (80 to 90 %), litter (55 to 68 %), animal surfaces (2 to 12 %), feces (2 to 8 %) and from structural elements in the house such as the walls and floor (Hartung and Saleh 2007 & Seedorf and Hartung 2000). It is also primarily generated from feed grains, dried fecal materials and excrements, animal dander (skin cells and hair), feathers, insect parts, mold, pollen, grain mites, mineral ash, and dead micro-organisms which are comprised of viable organic compounds, fungi, gram-negative bacteria, endotoxins, absorbed toxic gases and other hazardous agents (Wang et al. 1999a & Jay et al. 1994). The amount of dust released is proportional both to the number of animals and to their weights. This fact indicates that a considerable part of the dust can be generated from the animals themselves (Gustafsson 1997). A small amount of particles could also enter the animal house with the incoming ventilation air.

6. Indoor dust concentration and emission rate in livestock buildings

The concentration of both airborne inhalable and respirable fractions was overall higher in pig and poultry buildings than in cattle houses. Dust concentrations and emissions were affected significantly by several things such as housing type, the season of year and day/night time (Takai et al. 1998). The inhalable and respirable dust concentrations in poultry buildings were higher in winter than in summer season (Takai et al. 1998). The low level of dust concentration in warmer seasons was related to the high humidity and high exhausted ventilation (Zhu et al. 2005). The mean inhalable dust emission rates were higher in summer season than in winter season. The highest concentration of the total dust and respirable fraction in the laying hens houses were during June. This result can be explained by the fact that during this period the birds molted thus, increasing the amount of dust. Moreover, the increased ventilation due to the higher temperatures on the one hand helped extracting dust from the unit but on the other stirred up previously deposited dust. This indicates the enormous importance of the systematic general cleaning of the unit (Guarino et al. 1999). The diurnal change has a big influence on the concentration of airborne dust in a commercial animal houses. There is a variation in number of different sized particles during the day with constant ventilation rate in a building. A clearly increase in the dust particle number during daytime when the animal activity is higher than at night-time (Gustafsson 1997). The highest dust concentrations were measured at 5:00 O'clock and the lowest dust concentrations were found at 14:00 O'clock (Hessel and Van den Weghe 2007). The dust levels in livestock houses were consistently low during the night-time hours and high during the afternoon is correlated to animal activity, operation of feed delivery equipment and worker activity (Nannen and Büscher 2007 b). There is a high variation in the pattern of spatial dust distribution in mechanically ventilated animal buildings. Thus, the ventilation systems have direct effects on the spatial dust concentration whereas the increase of the ventilation rate will not necessarily reduce the overall dust level effectivel because the dust production rate will increase with increasing ventilation (Pedersen et al. 2000). A comparative study of respirable dust concentration between European countries and the state of Texas in the USA has been presented by Redwine et al. (2002). The dust concentration in European countries ranges in the literature from 0.4 to 9.7 mg.m^{-3}, on the other side, it ranges from 0.1 to 0.3 mg.m^{-3} in Texas, which is slightly less than comparable data from European studies. These results return to the warm climate of Texas as mentioned by the authors. This warmer climate requires a higher ventilation rate and the use of evaporative cooling systems. A higher ventilation rate may dilute the dust concentration and the evaporative coolers may suppress dust emission rates by maintaining a higher relative humidity in livestock buildings. The number of different sized particles in the animal houses have indicated that increased ventilation rate mainly reduces the number of particles larger than 1.0 μm and had only a limited effect on the number of particles smaller than 1.0 μm (Gustafsson 1997).

7. Quantifications of dust concentration

7.1. Planning and preparation of the measurements

According to (VDI 2066, 1975) placement of the equipment and accessibility of the test points affect the dust determination. In new installations the requirements of the measuring sections and test points must be considered in the planning stage. These requirements are as following:

1. The flow in the measuring sections should be as undisturbed as possible.
2. The measurement cross section should be placed within a straight measuring section and have an inlet and outlet free of any interference.
3. The length of the inlet and outlet sections should be at least three times the hydraulic diameter of the measuring cross section.
4. The test place should be easily accessible by the measuring staff and for transport of the instruments.
5. The test place should be protected against external effects (rain, wind, heat, etc.) and it must comply with the accident prevention regulations.

7.2. Dust concentration measuring methods

The determination of dust concentrations with the help of filters has been explained by VDI 2066 (1975) and VDI 2463 (1999). Different procedures for measuring the particles in gases or liquids such as the Coulter Counter have been explained by Cox and Wathes (1995).

Schmitt-Pauksztat (2006) explained the different procedures to measure the dust concentration such as:

1. Aerodynamic procedures
   - Elutriator
   - Inertia impactor
   - Particle size analysis
2. Optical procedures
   - Mie theory

- Laser particle counter
- White light method
- Influence of the particles form and structure

Gustafsson (1997) measured the dust concentration using the following methods:

1. Gravimetic measurements of the amount of total dust (mg/m³) with 37 mm diameter Millipore filters at a flow rate of 1.9 l/min.
2. Gravimetric measurements of the amount of respirable dust (mg/m³) with a millipore filter after separation of larger particles with a SKC cyclone.
3. Counting the number of different sized particles with a Rion optical particle counter.
4. Weighing the settled dust on 0.230 m² settling plates.
5. Measuring the ventilation rate with an Alnor hot wire anemometer in the exhaust air ducts.

The particulate matter (PM) in the ventilation exhaust air could be measured by using a tapered element oscillating microbalance (TEOM). The instrument draws aerosol through an exchangeable filter attached to a hollow tapered oscillating glass rod at a constant flow rate. The real-time PM concentration is based on a sample flow rate coupled with gains in mass on the filter measured by its effect on the oscillation frequency. Each TEOM system consists of controller and sensor units, figure (1). The sensor unit contains a mass transducer and is heated to 50 °C to minimize moisture effects. The PM10 sample inlet is attached to the sensor unit and can be replaced with PM2.5 inlets. Sample flow is split isokinetically into a main flow passing through the filter and a bypass flow each controlled by a mass flow controller (Lim et al. 2003 and Kosch et al. 2005).

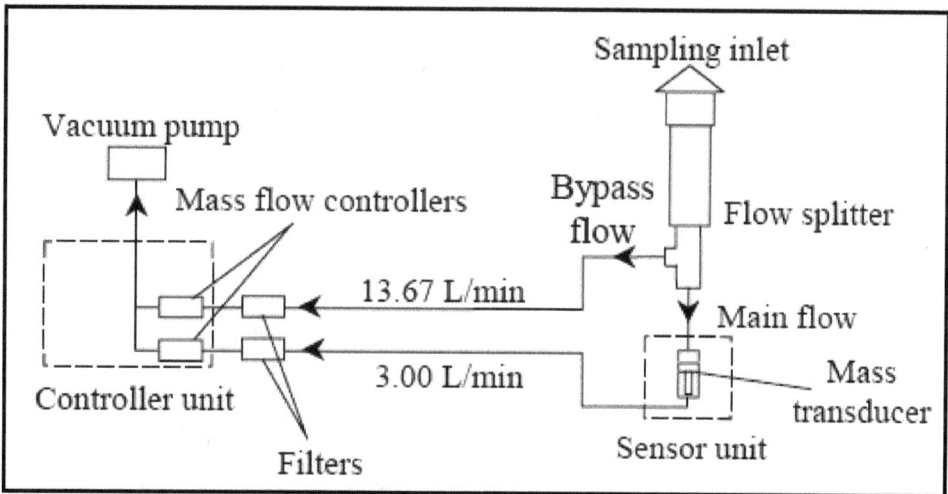

Figure 1. Schematic layout diagram of the tapered element oscillating microbalance (Lim et al., 2003)

A multi-point dust sampler has been developed by Wang et al. (2000) and Wang et al. (1999a) to measure the spatial dust distribution at different ventilation rates in a mechanically ventilated airspace using an array of critical venturi orifices for controlling the airflow rate at each sampling point, figure (2). It consists of a commercially available vacuum pump, a pressure monitor, a pressure regulator, an array of filter holders, filters, critical venturi orifices and sampling heads. When air is drawn through the sampling head and the filter the volumetric flow rate remains constant for all venturi orifices even though the pressure may vary as long as the pressure across the venturi orifices is higher than the critical pressure drop. Since the critical pressure drop of the venturi was below 11 kPa, the pump operated at a sufficiently high vacuum (approximately 35 kPa) and a constant flow through the filters was maintained. This multi-point sampler was used in this study to measure the dust mass concentration in a cross-section of the ventilated airspace.

The technical set-up and measuring method of Optical Aerosol Spectrometers (OAS) including the device characteristics has been explained by (Mölter and Schmidt 2007). The set-up principle of an OAS in forward scattering is presented in figure (3).

During forward scattering, the light scattered by particles as shown in figure (4) towards 180º is collected by the light source with a light sensitive detector, e.g. a photomultiplier. At the 90º scattered light detection the photomultiplier is attached orthogonally to the image plane.

The height of the scattered light impulse is a measure for the particle diameter, while the number of impulses supplies the information on the concentration since the volume flow is known. With the help of a lens system the light is focused on the desired measuring volume size. Before the receiver optics a light collector in forward scattering must be installed. This protects the light detector against direct irradiation. Due to diffraction actions of the light and of the scattered light the light collector leads to an ambiguous calibration curve also when using white light. However, a source of white light in connection with a 90º scattered light detection secures a clear calibration curve for many refractive indices.

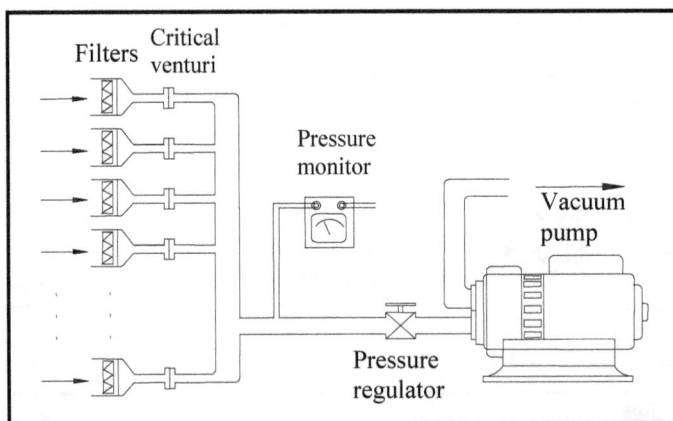

Figure 2. A schematic diagram of the multi-point dust sampler (Wang et al., 2000 and Wang et al., 1999a)

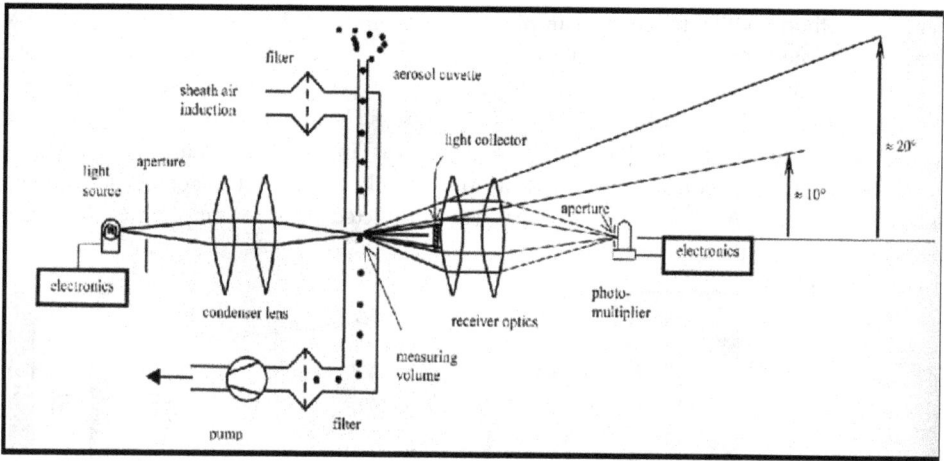

Figure 3. Optical aerosol spectrometer (Mölter and Schmidt, 2007)

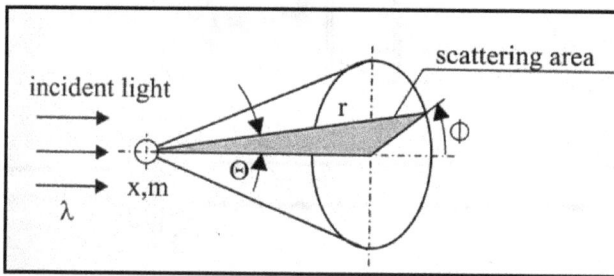

Figure 4. Principle of incident light scattering (Mölter and Schmidt, 2007)

The measurement of the emission rates of particulate matter from mechanically ventilated livestock buildings in the laboratory, using a test chamber and at the exhaust duct, using three air sampling methods (Predicala and Maghirang 2004):

1. **Low-volume traverse under isokinetic conditions.**

This method used a sampling head with a 14 mm probe inlet diameter and a 37 mm filter assembly, as shown in figure 5a. The sampling head was attached to a 0.80 m long rigid tube which was connected by flexible tubing to a flow meter with a flow control and a vacuum pump. The sampling flow rate was adjusted to isokinetic condition. Isokinetic sampling was achieved by varying the sampling flow rate to match the air velocity at the inlet plane of the sampler with the air stream velocity outside the sampler. The required sampling flow rates for isokinetic sampling were determined by conducting a velocity traverse at the sampling plane prior to sampling.

2. **Fixed sampling at specific locations within the duct cross-section.**

This method used a 14 mm sampler and an Institute for Occupational Medicine (IOM) sampler, as shown in figure 5b. The 14 mm sampler was similar to that in the low-vol traverse method, while the IOM sampler was a commercially available inhalable PM sampler, operated under either isokinetic conditions or at the recommended flow rate of 2 l/min (sub-isokinetic sampling). The IOM sampler was typically used to assess occupational worker exposure in livestock buildings; thus, its possible application for measuring PM concentrations to determine PM emission rates was investigated. Sampling was sub-isokinetic when the actual sampling flow rate was lower than the required isokinetic sampling flow rate. A velocity traverse was also conducted prior to sampling to determine the required isokinetic sampling flow rate.

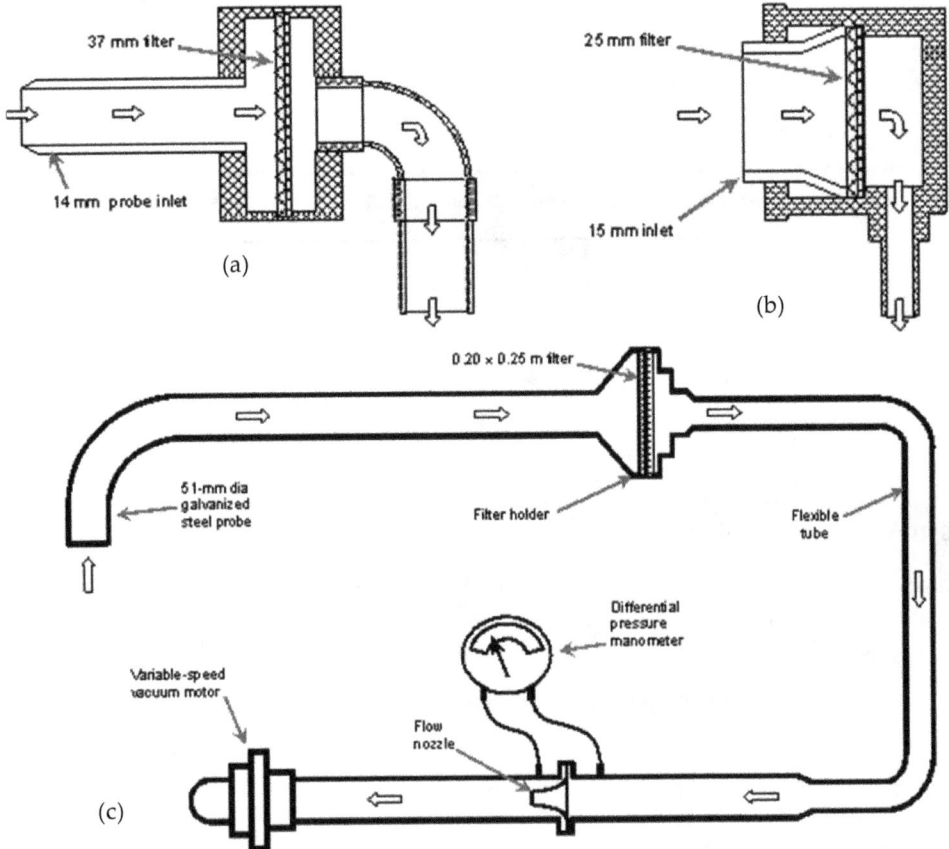

a) The low-volume traverse method, b) The fixed sampling method, c) The high-volume sampling train

Figure 5. Schematic diagram of three air sampling methods (Predicala and Maghirang, 2004)

3. Hi-volume traverse under isokinetic conditions.

This method is considered as the reference method for this study. The sampling train consisted of a 51 mm diameter probe, a 0.20 x 0.25 m filter holder, a flow nozzle and a variable-speed vacuum motor presented in figure 5 c. Similar to the low-vol traverse method PM was also extracted isokinetically at specified sampling locations within the sampling plane. The sampling flow rate indicated by the differential pressure across the flow nozzle was adjusted by varying the speed of the vacuum motor. After sampling, the probe and the front part of the filter holder were rinsed with acetone to collect the PM deposited along the probe and filter holder walls. The acetone was allowed to evaporate and the residual PM was added to the PM mass collected on the filter. The sampling duration at each traverse point was determined by preliminary tests so the total collected PM mass was at least 100 mg.

The low-volume traverse and fixed sampling under isokinetic conditions agreed well with the high-volume traverse (mean difference ranging from 7 % to 14 %). Methods involving room sampling, fixed sampling at exhaust and high-volume traverse at exhaust were also compared in a swine finishing barn. Room sampling overestimated concentrations at the exhaust by an average of 30 % and PM concentration from fixed sampling did not differ significantly ($p > 0.05$) compared to the high-volume traverse method. It appears that fixed sampling under isokinetic conditions can be used as an alternative to the high-volume PM traverse method to accurately measure PM concentrations at the exhaust from which the PM emission rate can be determined.

8. Means for reduction of dust in/from animal houses

Several methods for dust control such methods include spraying or sprinkling oil or oil-soap solution in the airspace accelerating dust sedimentation onto the floor by investigating the air ionization systems and separating dust from the air stream with air cleaning devices and ventilation. There are number of mechanical methods for dust control. These methods include fiber filters, water or oil scrubbers, electrostatic precipitators and traditional cyclones (more particles smaller than 10 microns can't be separated by the conventional cyclones because of the strong turbulence associated with the high pressure typically higher than 500 Pa) but these methods may be associated with the ventilation system in the barns (Zhang et al. 2001). The different methods used for reducing the indoor concentration and dust emission rate as following:

8.1. Dust suppression with spraying oil and /or water

Indoor dust concentration inside animal houses could be reduced by spraying a mixture of oil and water. This method proved to be very effective to reduce dust at relatively low costs. The main effect of oil/water spraying is preventing dust on surfaces to become airborne. With a good design dust reduction could be reach up to 90 %. Designing this system, the following items are important to be considered:

- Oil concentration should be at least 20 %. With this concentration the relative humidity inside the animal house slightly increased (< 2 %).
- Oil drops should be bigger than 150 µm to descend to the floor at a fast speed to increase efficiency. Furthermore, small droplets might affect the respiratory health of animals and humans when the small droplets are inhaled.

- Generally, all kinds of vegetable oils can be used although some remarks have to be made:
  - It is not necessary to use purified oil however the oil should be free of particles.
  - Oil with a strong odour is less suitable because of possible effects on animal behaviour.
  - Oil should contain a low concentration of Iodine.
- The dust binding effect of the oil is long lasting (some days). Frequent spraying is needed (Aarnink and Ellen 2007, and Pedersen et al. 2000).

The oil applications could be on the feeding materials for reducing dust concentrations by livestock industries. A variety of vegetable oils including canola, corn, sunflower, flax, soybean and rapeseed oils along with mineral oils have been used to control dust from feed sources and building floors. Soybean oil reduced dust counts by as much as 99 % following 0.5, 1.0 and 2.0 % additions to dry feed (Ullman et al. 2004, Pedersen et al. 2000 and Takai 2007). Although oil sprayed on birds is not recommended and application would be an incompatible practice with broiler rearing. Due to high bird density oil sprinkling may still hold promise as an effective dust control technique. The technical parameters regarding the spraying of oil-water mixtures on surfaces in pig buildings to enable consistent dust reduction efficiency with the least possible oil application rate could be concluded as following (Takai 2007):

1. Number of treatments within the range of 1 and 14 per day does not have an influence on dust reduction efficiency.
2. Oil concentration in the oil-water mixture should be higher than 20 %.
3. Droplet diameter should be greater than 150 µm.
4. Further development of methods to prevent plugging in the spray system is desired.

An ultra sonic sprayers unit (USSU) as shown in figure 6 has been used to reduce the dust concentration in an enclosed experimental layer and floor feeding broiler house. For laying hens 1 and 2 % solutions of emulsified canola oil (weight base) were sprayed by (USSU) once a day after feeding. However for the floor feeding broiler house 2 % solutions of emulsified canola oil (weight base) were sprayed by (USSU) every hour for 10 minutes (75 g were sprayed) or when a dust concentration detector detected a threshold concentration which was 5.0×10^8 particles/m³ with less than 5 µm in aerodynamic diameter. Spraying 2 % solution of emulsified canola oil with the ultra sonic sprayer unit in the enclosed layer house reduced the concentration of dust with $0.5 \leq$ aerodynamic diameter < 2 µm and with $10 \leq d <$ 30 µm to 58 by 51 %, respectively. On the other side, 1 % oil spraying in the layer house reduced the dust concentration to about 20 %. This spraying method could reduce the dust concentration to a daily average of 47 % in the floor feeding broiler room, but concentration itself was 100 times higher than in the layer house (Atsuo 2002).

Showering water on the floor surfaces in the walking alleys reduced the total dust concentration with 9% average and spraying salt solution (KCl) in the air with nozzles reduced the total dust concentration by 41%. Spraying water droplets gave different results depending on the type of nozzles used. The use of high pressure nozzles (ultrasound nozzles) which created droplets in the size range of 5-10 μm resulted in a significant increase of both total and respirable dust (Gustafsson 1997). An aerosol application unit has been used to distribute an oil mixture-emulsion under high pressure inside a pig barn. The oil mixture contained different types of essential oils (to reduce airborne germs and fungi) and a carrier oil. By operating the aerosol application unit every 30 minutes, it was possible to obtain an almost continuous indoor air treatment within the barn. In comparison with the reference pig barn (same building, different compartment; ceteris paribus conditions), an average indoor concentrations were reduced to an average by 59 % for total dust and 54 % for PM_{10}. Emissions reduced to 68% for total dust and 65 % for PM_{10} (Hölscher 2006).

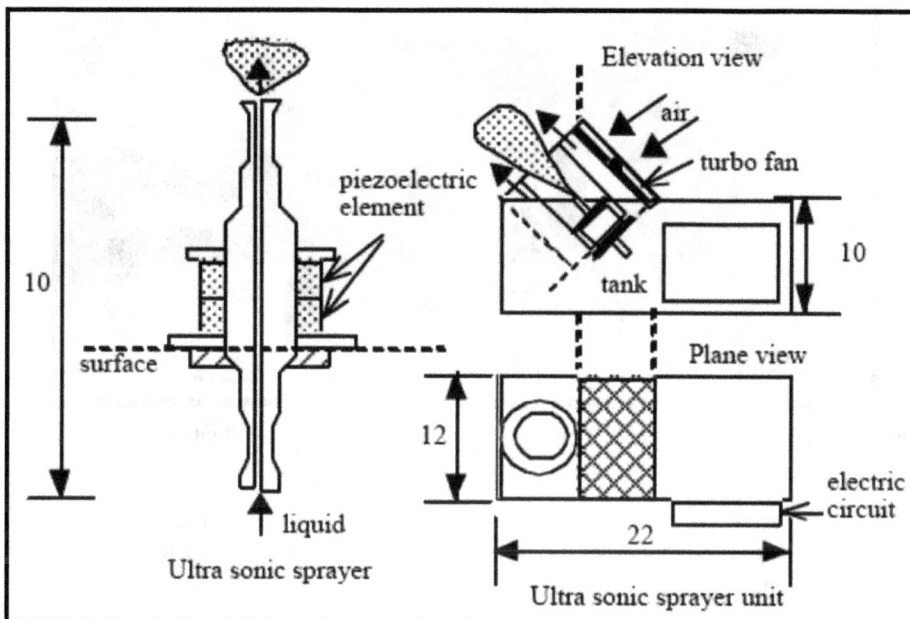

Figure 6. Ultra sonic sprayer unit (Atsuo, 2002)
(All units in cm)

8.2. De-dusters

There are two aerodynamic uniflow de-dusters (a cyclone type particle separator & gas remover with airflow capacity 188 l/s and 1,880 l/s) have been developed with low pressure requirement and high particle separation efficiency. This development is based on fluid dynamics, particle mechanics and sensitivity analysis.

- The small model de-duster employs a set of turbine-type vane guides, an involute separation chamber and a flow converging section to minimize turbulence and reduce the pressure loss. As shown in figure 7, dusty air is drawn from the air inlet passing through a set of vanes to establish a spiral flow pattern. The air then passes through the involute chamber and converges at the exit section above the dust bunker. Particles are collected in the dust bunker and clean air is exhausted through the blower. This device, as shown in figure 2.30, unlike the conventional cyclones, can remove respirable particles at pressures of 50 Pa.

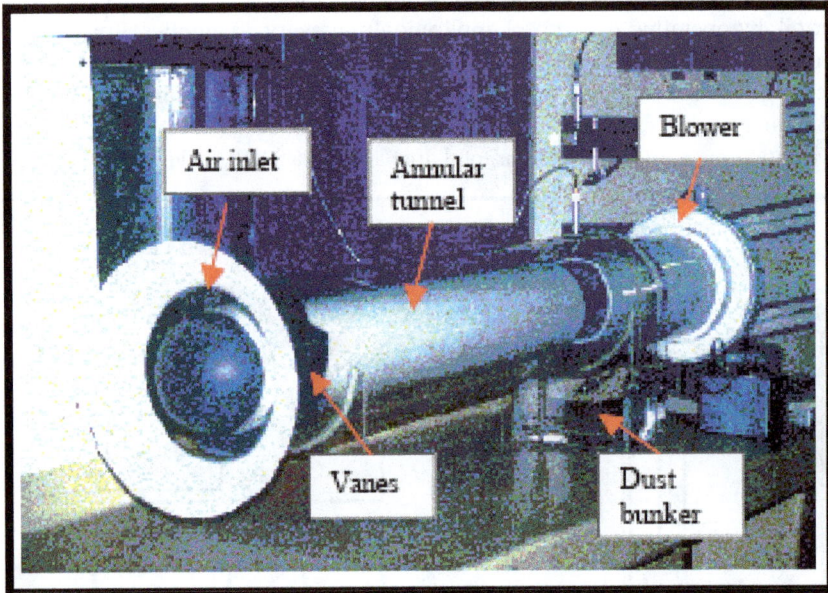

Figure 7. The prototype of the uniflow deduster fabricated based on the sensitivity analysis (Zhang et al., 2001)

- The large model de-duster contains three concentric de-dusters. The outer cylinder of the smaller de-duster serves as the inner cylinder of the bigger de-duster. Thus, the total cross sectional area is increased to allow air delivery and the volume of the unit is minimized as shown in figure 8. The fan speed can be varied via a frequency controller so that the performance at different airflow rates can be evaluated (Zhang et al. 2001). An automatic dust flushing system was developed to periodically clean the dust in the dust bunker. The new design is aimed at reducing dust emissions for exhaust fans with large air flow rates. The dust mass concentration was measured at the inlet and the outlet of the de-duster using filter collectors during 24-h periods. The results showed that the dust mass removal efficiency was 91 % at the 60 % power level. The dust reduction efficiency was 89 % at 100 % power level. From the study of dust concentration reduction through cleaning the air using aerodynamic de-dusters, the ratio of air flow rate through the de-duster to the ventilation room is 32 % with a dust removal efficiency of 85 %. The large flow rate for the de-duster is required to improve the room air cleaning efficiency (Wang et al. 1999a). The effectiveness of air cleaning devices on dust concentration is dependent not only on the airflow through the device but also on the ventilation rate in the building (Gustafsson 1997). Considering the mass balance of the dust it is obvious that the air cleaning equipment needs large airflow capacities if the dust concentration in the air is to be affected. The airflow through an air cleaner has the same influence on the dust concentration as an equally large increase in the ventilation rate in the building. The particle separation efficiencies of this de-duster were 50, 77 and 90 % for particles diameter of about 4, larger than 7, and larger than 10 μm respectively. In terms of mass concentration measured using mass samplers, the particles separation efficiency was 85 %. Because most of the dust mass is attributed to the larger particles, the number separation and mass separation efficiency agreed very well (Gustafsson 1997).

Figure 8. The large de-duster prototype (Zhang et al., 2001)

8.3. Ionization

Ionization is the physical process of converting an atom or molecule into an ion by adding or removing charged particles such as electrons or other ions. This process works slightly differently depending on whether an ion with a positive or a negative electric charge is being produced. A positively charged ion is produced when an electron bonded to an atom (or molecule) absorbs enough energy to escape from the electric potential barrier that originally confined it, thus breaking the bond and freeing it to move. The amount of energy required is called the ionization potential. A negatively charged ion is produced when a free electron collides with an atom and is subsequently caught inside the electric potential barrier releasing any excess energy. An electrostatic space charge system was shown to remove up to 91 % of artificially generated dust and 52 % of dust generated by mature White Leghorn chickens in a caged layer room. An apparatus consisting of 2 negatively charged needles located 0.25 m above the floor and a positively charged aluminium collector plate (0.76 m high by 1.4 m long) located in front of the door, charged at 12 and 8 kV, respectively, was tested at a livestock facility. Ionization was approximately 6 times greater at dust removal than gravity alone. Relative humidity had no apparent impact on reductions in dust concentrations (Ullman et al. 2004). The ionization in pig houses has resulted in a 20–30 % decline in dust concentration (Gustafsson 1997). Electrostatic collectors are devices that impart electric charges to dust particles and then push them out of the air stream using electromagnetic force. They typically exhibit low operating costs and high removal efficiencies. The electrostatic ionization could produce airborne swine dust removal rates of up to six times greater than gravitational sedimentation alone. The ability of the electrostatic precipitator system in figure 9 to remove airborne particles has been tested. This electrostatic precipitator consisted of a discharge electrode which was constructed from a single strand of stainless steel wire and a grounded collection electrode pipe positioned 17.8 cm below the wire. The discharge wire and the collection pipe were supported by PVC end plates. Additionally, an ionization guard was located above the wire to direct electrons and charged dust particles down toward the collection electrode. The entire unit was 3.05 m in overall length. To charge the precipitator and provide negative ionization at the discharge wire (which imparts electrical charges to passing dust particles). The electrode wire was connected to a -20 kV, 50 mA, and rectified a.c. power supply unit (Rosentrater 2004).

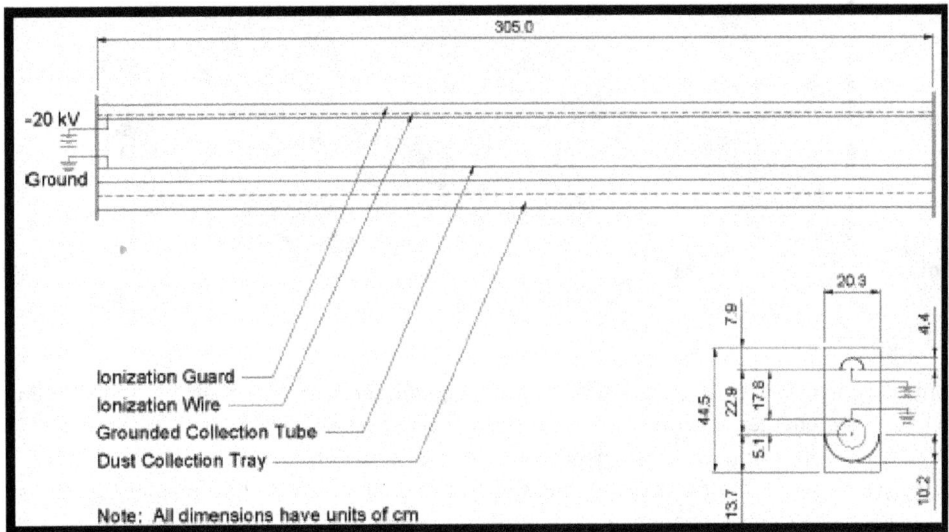

Figure 9. Schematic of electrostatic precipitator unit (Rosentrater, 2005)

An electrostatic space charge system (ESCS) has been used by Mitchell et al. (2004) to demonstrate the effectiveness of this system in the breeder/layer farm environment for reducing airborne dust in a several month long study. The system as shown in figure 10 used ceiling fans to distribute negatively charged air throughout the room and to move negatively charged dust downward toward the grounded litter where most of it would be captured. The dust concentration was reduced by an average of 61 % over a period of 23 weeks.

Mitchell and Baumgartner (2007) used previous ESCS for reducing the dust particle concentration in poultry production houses and a hatchery. The effectiveness of ESCS for PM_{10} dust reduction ranged from 78 % in commercial poultry hatchers to 47 % in commercial caged layer houses. The effects of airflow on the distribution for negative air ions using three types of ion generators has been studied by Mitchell (1997) as shown in figure 11, which have potential for dust reduction in animal house or hatchery applications. All of the devices used limited current power supplies which restricted the current to 2 mA or less for safety and the ozone output was limited to less than 0.1 ppm. The first type was a self-contained Ceiling Ionizer that was designed to hang from the ceiling in the middle of a room where a space charge was desired. The second was a Room-Ionizer-System (RIS) consisting of a metallic bar with external power supplies operated at –8 kVDC, or –15 kVDC. Both of these devices require an external air moving device. The third type was an ionizer which is designed to be used inline in a duct or with a self-contained air source to charge clean outside air prior to injecting it into a treatment area. This device will be referred to as the IDI (in-duct-ionizer). These devices could be effectively used to reduce dust and microorganisms in a variety of applications with air moving devices.

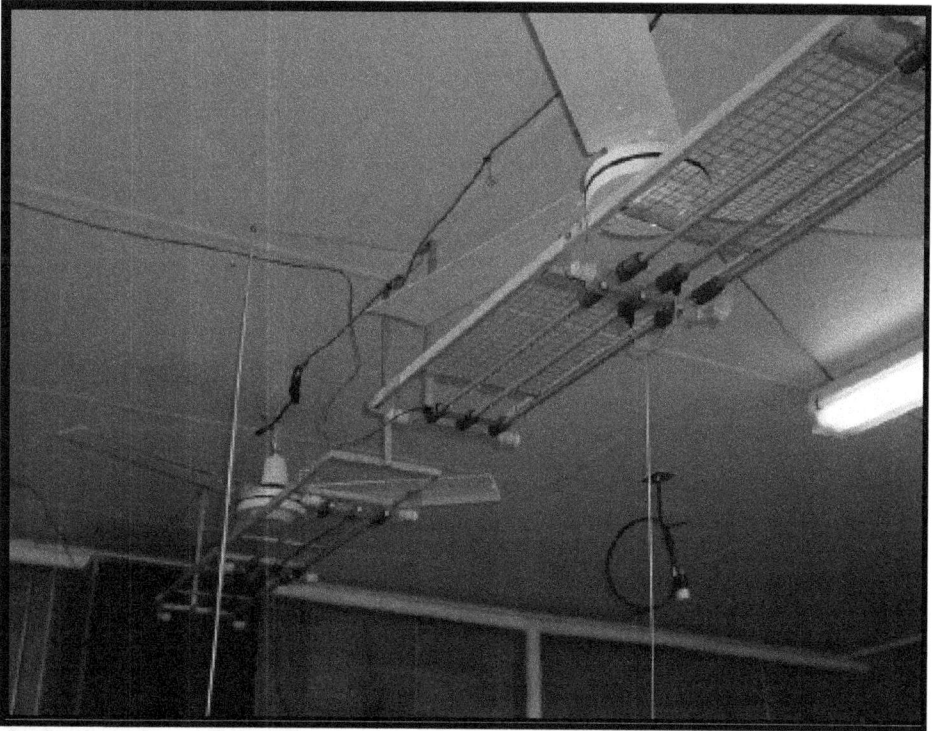

Figure 10. ESCS units suspended below the ceiling fans in the treatment room (Mitchell et al., 2004)

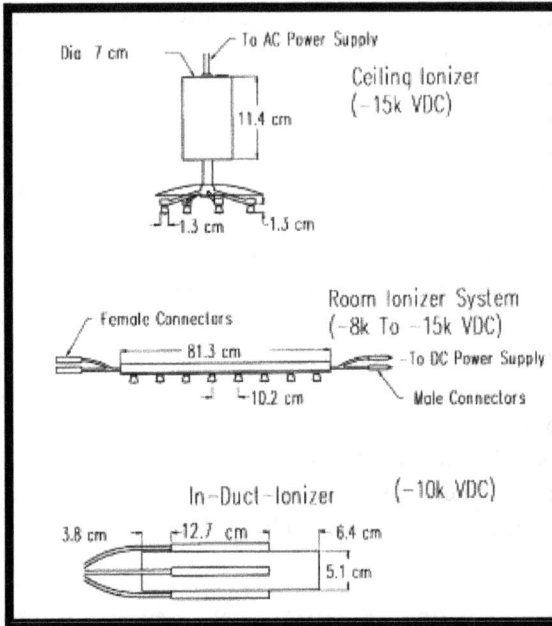

Figure 11. Ionizer configurations (Mitchell, 1997)

8.4. Oxidants

An oxidant can be defined as a chemical compound that readily transfers oxygen atoms or a substance that gains electrons in a chemical redox (short for reduction/oxidation reaction) reaction. Cleaning the air by oxidation has been used for decades using oxidizing agents such as ozone, potassium permanganate, chlorine and chlorine peroxide. Evaluation of an indoor ozone system for dust control effectiveness proved that the total dust concentrations decreased by 60 % at the fan exhaust under maximum tunnel ventilation compared with a nearby building without any ozone treatment (Ullman et al. 2004).

8.5. Windbreaks

A windbreak or shelterbelt is a plantation usually made up of one or more rows of trees planted in such a manner as to provide shelter from the wind. Two hundred operations in Taiwan have constructed walls downwind of tunnel-ventilated poultry buildings and had seen reduced dust emissions off-site. The effectiveness of module walls constructed of 3 x 3 m pipe frames covered securely with tarpaulins was determined by collecting aerial dust particles and demonstrating airflow from exhaust fans using smoke. An increase in the vertical height of the smoke plume subsequent to reaching the windbreak demonstrated the potential for reduced dust concentrations downwind of animal facilities. Elbows placed on exhaust fans designed to redirect fan airflow upward produced some plume rise. However, dispersion models indicated that tall stacks may offer further effectiveness (Ullman et al.

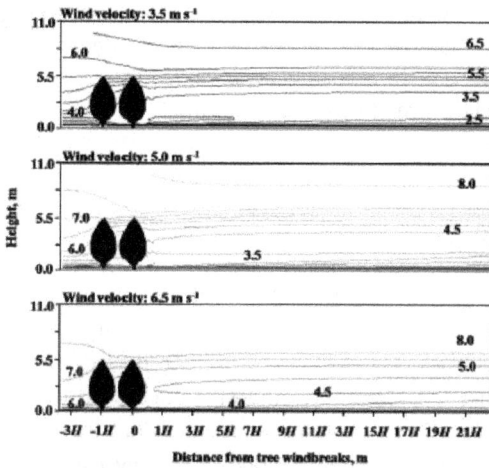

Figure 12. The velocity contour profile (ms⁻¹) at varied wind velocity of 3.5, 5.0 and 6.5 ms⁻¹ measured at 1.0 m height for two-rows of trees arranged alternately with 0.5m gap distance between trees, where H is the tree height (Bitog et al. 2012).

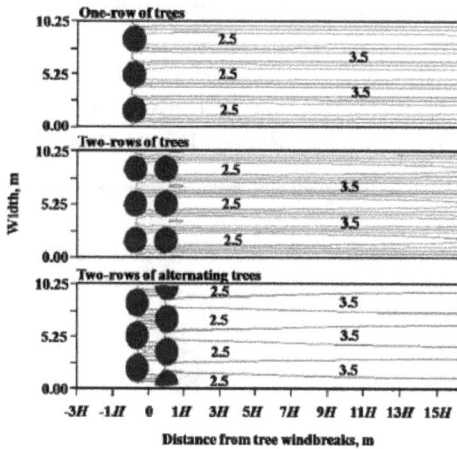

Figure 13. Visualisation of the velocity contour (ms⁻¹) of the tree windbreak at varied rows of trees and their arrangement (Top view at 1.0 m height; maximumvelocity is 8 ms⁻¹) Bitog et al. 2010.

2004). Windbreak walls placed at 3 and 6 m, respectively, from the building deflected the airflow from the exhaust fans in the upward direction similar to other wind barriers, thus providing surfaces for dust deposition. The vertical height at which the plume would flow over a downwind lagoon under low wind conditions was increased by building a windbreak wall. As a result, the dust levels in the area downwind from the windbreaks were lowered (Pedersen et al. 2000). Natural windbreaks such as trees are very efficient barriers to high velocity winds. The windbreaks exert drag force causing a net loss of momentum and thus disturb the characteristics of flow. The main factors which can affect the efficiency of the windbreaks are tree height, width, tree arrangement, porosity, etc. (Bitog et al. 2011). The simulation provides analysis of the effect of gaps between trees, rows of trees, and tree arrangements in reducing wind velocity. The simulations revealed that 0.5 m gap between trees were more effective in reducing wind velocity, figure (12). Two-rows of alternating trees were also found to be more effective than one-row and two-rows of trees as shown in figure 13 (Bitog et al. 2012).

8.6. Scrubbers and filters

The scrubbers consist of towers packed with a contact media, gas or liquid-driven venturi systems. These venturis and spray towers offer a more instantaneous removal of dust particles (Ullman et al. 2004). The wet pad scrubber placed in the animal house 1.2 m upwind from the exhaust fans achieved modest reduction in dust emission from animal buildings in warm weather. The results demonstrated that these control methods did not substantially challenge the existing ventilation systems by causing excessive resistance to airflow and they would therefore be practical and useful emission control methods (Pedersen et al. 2000). The bioscrubber system could be used to perform at a higher level of efficiency to reduce emissions in consideration of the huge amounts of airflow in poultry production. The working principle of the exhaust scrubber as shown in figure 14 is the continuous spraying of the partition grill with a high specific area. The spraying is done with three pumps (1.5 kWh) which discharge 75 m³/h of water at a height of 5 m. The cleaning water wets the synthetic partition grill evenly and causes the removal of dust particles. The efficiency of reduction for suspended dust is 45 % (Kosch et al. 2005).

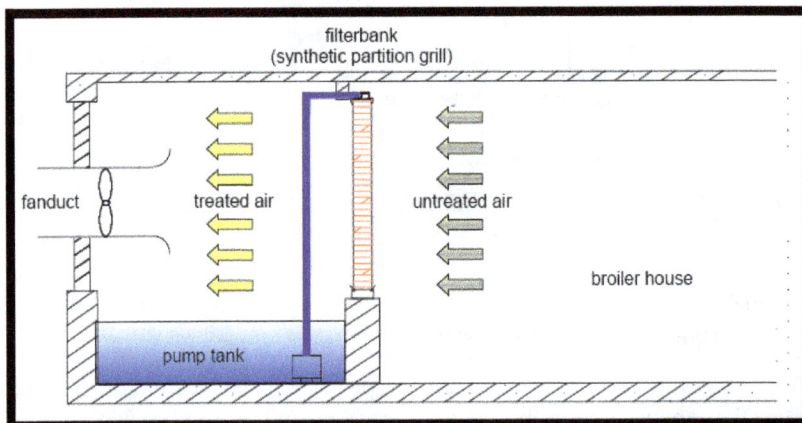

Figure 14. Schematic of the exhaust air cleaning system for a poultry house (Kosch et al., 2005)

The exhaust air cleaning system as shown in figure 15, based on a bioscrubber where the exhaust air flows horizontally to the house gable and passes through the fans to enter the filter which is located outside the stable. In the beginning the air is humidified and then flows into the first filter bank which consists of so-called pads. In this stage the dust is washed out of the air and transported downwards by the water and the air flows through the second filter bank. In this filter the ph value of the water is regulated by acid to eliminate NH₃, fine dust and odourous substances which cannot be washed out in the first filter bank. The water from both filters is collected and smoothed so the solid matter deposits on the ground of the basin and the water is then pumped up to flow over the pads again. The result of the dust concentration measurements shows that more than 80 % of the airborne dust was removed by the filter (Snell and Schwarz 2003).

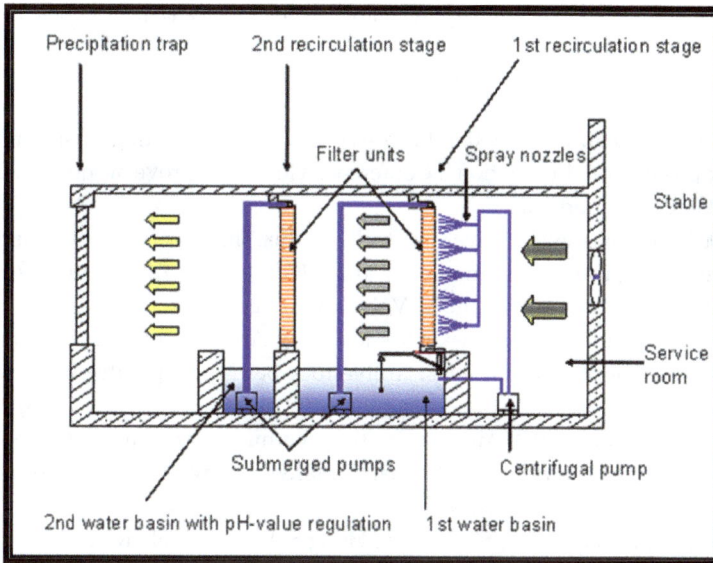

Figure 15. Schematic of the exhaust cleaning system (Snell and Schwarz, 2003)

The reduction of dust concentration in animal buildings using a filter provides an alternative method to air scrubbers for broiler operations. Dust became entrapped in fibers through a number of physical mechanisms. The traditional filter systems used in broiler operations reduced the dust content by up to 50 %. Clogging of traditional filter systems by dust and feathers in broiler facilities became problematic to a point when poultry operators found to forego filters over airconditioning units rather than deal with the required maintenance. To overcome such problems filters should be placed in a series with the first (i.e., upstream) filter consisting of a fairly coarse strainer primarily intended to remove feathers. The authors also used the Biofilter which operates by forcing air through a moist packing material to provide an alternative to traditional filter systems for broiler facility dust emission reduction. It is recommended that biofilters used at poultry facilities should be installed with dust removal equipment as dust accumulates on fans (Ullman et al. 2004). The reduction of dust emission rate from a pig barn with 515 pigs using two air scrubbers by recycling the air inside the building has been studied by Hölscher (2006). The measurements over a three month period indicated that emissions can be reduced on average by 54 % for total dust and by 51 % for PM10, compared with a reference pig barn. Indoor concentrations have been reduced on average by 63 % for total dust and by 60 % for PM10. Dry filter system has been presented by Mostafa and Buescher 2011 is showed high indoor dust reduction efficiency in comparison with wet filter system.

Author details

Ehab Mostafa
Agricultural Engineering Dept.,Faculty of Agriculture – Cairo University, Giza, Egypt

9. References

Aarnink A.J.A. and H.H. Ellen (2007): Processes and factors affecting dust emissions from livestock production. International Conference How to improve air quality. 23-24 April, Maastricht, the Netherlands.

Atsuo I. (2002): Ultra sonic sprayer controlling dust in experimental poultry houses. Agricultural Engineering International: the CIGR Journal of Scientific Research and Development. Manuscript BC 01 002. Vol. IV. October.

Bakutis B., E. Monstviliene, G. Januskeviciene (2004): Analyses of airborne contamination with Bacteria, Endotoxins and Dust in livestock barns and poultry houses. Acta. Vet. Brno 73: 283-289.

Banhazi T. and J. Seedorf (2007): Airborne particles within Australian and German piggeries – What are the differences?. International Conference How to improve airquality. 23 - 24 April, Maastricht, the Netherlands.

Bitog J.P., Lee I.-B., Hwang H.-S., Shin M.-H., Hong S.-W., Seo I.-H., Kwon K.-S., Mostafa E., Pang Z. (2012): Numerical simulation study of a tree windbreak. Biosystem Engineering Journal. 111: 40-48.

Bitog J. P., Lee I-B., Hwang H-S., Shin M.-H., Hong S-W., Seo I-H., Mostafa E., Pang Z. (2011): A wind tunnel study on aerodynamic porosity and windbreak drag. Forest Science and technology. Vol (7):1, 8-16.

Cox, C. S. und Wathes, C. M. (1995): Bioaerosols Handbook, Lewis Publishers, ISBN 087371-615-9.

DFG (Deutsche Forschungsgemeinschaft) (2006): MAK- BAT-Werte-Liste 2006. Senatskommission zur Prüfung gesundheitsschädlicher Arbeitsstoffe, Mitteilung 42. Wiley-VCH Verlagsgesellschaft mbh, Weinheim, Deutschland.

Ellen H. H., R.W. Bottcher, E. von Wachenfelt and H. Takai (2000): Dust levels and control methods in poultry houses. Journal of Agricultural Safety and Health 6(4).

Eugenija Z., J. Mustajbegovic, E. N. Schachter, J. Kern, N. Rienzi, S. Goswami, Z. Marom and S. Maayani (1995): Respiratory function in poultry works and pharmacologic characterization of poultry dust extract. Environmental research 70: 11-19.

GefStoffV (1999): Gefahrstoffverordnung vom 15.11.1999, BGBl. I S. 2233.

Guarino M., A. Caroli and P. Navarotto (1999): Dust concentration and mortality distribution in an enclose laying house. ASAE. Vol.42(4):1127-1133.

Gustafsson G. (1997): Investigations of factors affecting air pollutants in animal houses. Ann Agric Environ Med 4:203–215.

Haeussermann A., E. Hartung, A. Costa, M. Guarino, E. Vranken, and D. Berckmans (2007): Estimate Particulate Emissions by Intermittent Measurements: A Feasibility Study. International Conference How to improve air quality. 23 - 24 April, Maastricht, the Netherlands.

Hartung J. and M. Saleh (2007): Composition of dust and effects on animals. International – interdisciplinary – conference. Particulate matter in and from agriculture, 3-4 September, Braunschweig, Germany.

Hessel E. F. and H. F. A. Van den Weghe (2007): Airborne Dust (PM10) concentration in broiler houses as a function of fattening day, time of the day, and indoor light. International Conference How to improve air quality. 23 - 24 April, Maastricht, the Netherlands.

Hinz T., S. Linke, P. Bittner, J. Karlowski and T. Kolodziejczyk (2007): Measuring particle emissions in and from a polish cattle house. International – interdisciplinary – conference. Particulate matter in and from agriculture, 3-4 September,Braunschweig, Germany.

Hölscher R. (2007): Nachrüstlösungen zur Emissionsminderung dezentral entlüfteter Stallungen zur Schweinemast. Dissertation. VDI-MEG-Schrift 446.

ISO (1995): Air Quality - Particle Size Fraction Definitions for Health-related Sampling. ISO Standard 7708. International Organization for Standardization (ISO), Geneva.

IUPAC (1990): Glossary of atmospheric chemistry terms. International Union of Pure and Applied Chemistry, Applied Chemistry Division, Commission on Atmospheric Chemistry. Pure and Applied Chemistry 62 (11):2167-2219.

Iversen M., S. Kirychuk, H. Drost, L. Jacobson (2000): Human health effects of dust exposure in Animal confinement buildings. Journal of Agricultural Safety and Health 6(4): 283-288.

Jay H.D., R. Zhang and H. Xin. (1994): Human health concerns livestock and poultry housing. National Poultry Waste Management Symposium, Athens, Georgia, October 31. July 1994, AEN-159.

Kosch R., V. Siemers, H. Van den Weghe (2005): Efficiency of a bioscrubber system for the reduction of ammonia and dust emissions in a broiler house. ASAE Annual International Meeting, ASAE Tampa Convention Center Tampa, Florida.

Lim T.T., A. J. Heber, J.-Q. Ni, J. X. Gallien, and H. Xin (2003): Air quality measurements at a laying hen house: particulate matter concentrations andemissions. Proceedings of the 12-15 October 2003 Conference (Research Triangle Park, North Carolina USA), Air Pollution from Agricultural Operations III, ASAE Publication Number 701P1403, ed. H. Keener. pp. 249-256.

Mitchell B.W. (1997): Effect of airflow on ion distribution for potential dust reduction applications. Journal of Agricultural Safety and Health 3(2): 81-89.

Mitchell B. W. and J. W. Baumgartner (2007): Electrostatic Space Charge System for Reducing Dust in Poultry Production Houses and the Hatchery. International Conference How to improve air quality. 23 - 24 April, Maastricht, the Netherlands.

Mitchell B. W., L. J. Richardson, J. L. Wilson and C. L. Hofacre (2004): Application of an Electrostatic Space Charge System for Dust, Ammonia, and Pathogen Reduction in a Broiler Breeder House.Vol. 20(1): 87-93.

Mölter L. and M. Schmidt (2007): Advantages and limits of aerosol spectrometers for the particle size and particle quantity determination stables and air exhaust ducts. International – interdisciplinary – conference. Particulate matter in and from agriculture, 3-4 September, Braunschweig, Germany.

Mostafa E., and Buescher W. (2011): Indoor air quality improvement from particle matters for laying hen poultry houses. Biosystem Engineering journal. 109 (1):22-36.

Nannen C. and W. Büscher (2007) : Particle emissions from German livestock buildings - influences and fluctuation factors. International – interdisciplinary conference. Particulate matter in and from agriculture, 3-4 September, Braunschweig, Germany.

Pedersen S., M. Nonnenmann, R. Rautiainen, T. G. M. Demmers, T. Banhazi and M. Lyngbye (2000): Dust in pig buildings. Journal of Agricultural Safety and Health 6(4): 261-274.

Predicala B. Z. and R. G. Maghirang (2004): Measurement of particulate matter emission rates from mechanically ventilated. ASAE Vol. 47(2): 557-565.

Redwine J. S., R. E. Lacey, S. Mukhtar, J. B. Carey (2002): Concentration and emissions of ammonia and particulate matter in tunnel–ventilated broiler houses under summer conditions in Texas. Vol. 45(4): 1101–1109.

Reynolds S. J., D. Y. Chao, P. S. Thorne, P. Subramanian, P. F.Waldron, M. Selim, P. S. Whitten and W. J. Popendorf (1998): Field Comparison of methods for Evaluation of Vapor/Particle Phase Distribution of Ammonia in Livestock Buildings. Journal of Agricultural Safety and Health 4(2): 81-93.

Robert W.B. (2001): An environmental nuisance: odour concentration and transported by dust. Chem. Senses 26: 327-331.

Rosenthal E., W. Büscher and B. Diekmann (2007): Physical aspects of aerosol particle dispersion. International – interdisciplinary – conference. Particulate matter in and from agriculture, 3-4 September, Braunschweig, Germany.

Rosentrater K. (2004): Laboratory analysis of an electrostatic dust collection system. Agricultural Engineering International: the CIGR Journal of Scientific Researchand Development. Manuscript BC 03 008.

Scheuermann, H. (2004): Persönliche Schutzmaßnahmen. In: Luftgetragene biologische Belastungen und Infektionen am Arbeitsplatz Stall. KTBL-Schrift 436, S. 194-199

Schmitt-Pauksztat, G. (2006): Verfahren zur Bestimmung der Sedimentations geschwindigkeit von Stäuben und Festlegung partikelspezifischer Parameter für deren Ausbreitungssimulation. Dissertation. VDI-MEG-Schrift 440.

Seedorf J. and J. Hartung (2000): Emission of airborne particulates from animal production. Workshop 4 on sustainable animal production, Hannover.

Snell H.G.J. and A. Schwarz (2003): Development of an efficient bioscrubber system for the reduction of emissions. ASAE annual international meeting, ASAE, Las Vegas, Vevada, USA. 27-30 July.

Takai H. (2007): Factors influencing dust reduction efficiency of spraying of oil-water mixtures in pig buildings. International Conference How to improve air quality. 23 - 24 April, Maastricht, the Netherlands.

Takai H., K. Nekomoto, P. Dahl, E. Okamoto, S. Morita and S. Hoshiba (2002): Ammonia contents and desorption from dusts dollected inlivestock buildings.Agricultural Engineering International: the CIGR Journal of Scientific Research and Development. Manuscript BC 01 005. Vol. IV.

Takai, H., S. Pedersen, J.O. Johnsen, J.H.M. Metz, P.W.G.G. Koerkamp, G.H. Uenk, V.R. Phillips, M.R. Holden, R.W. Sneath, J.L. Short, R.P. White, J. Hartung, J. Seedorf, M. Schroder, K.H. Linkert, and C.M. Wathes (1998): Concentrations and emissions of airborne dust in livestock buildings in Northern Europe. J. Agric. Eng. Res. 70(1): 59-77.

TA-Luft (Technische Anleitung zur Reinhaltung der Luft) (2002): Erste Allgemeine Verwaltungsvorschrift zum Bundes–Immissionsschutzgesetz, Bundesministerium für Umwelt, Naturschutz und Reaktorsicherheit, 24. Juli 2002, GMBl. 2002, Heft 25 – 29, S. 511-605

Ullman J. L., S. Mukhtar, R. E. Lacey and J. B. Carey (2004): A Review of Literature Concerning Odours, Ammonia, and Dust from Broiler Production Facilities: 4. Remedial Management Practices. Journal of Appl. Poult. Res. 13:521–531

VDI 2066, Blatt 1 (1975): Messen von Partikeln; Staubmessungen in strömenden Gasen; Gravimetrische Bestimmung der Staubbeladung; Übersicht, Beuth-Verlag, Berlin.

VDI 2463, Blatt 1 (1999): Messen von Partikeln - Gravimetrische Bestimmung der Massenkonzentration von Partikeln in der Außenluft – Grundlagen, BeuthVerlag, Berlin.

Wang X., Y. Zhang and G. L. Riskowski (1999a): Dust spatial distribution in a typical swine building. In Proc. Int. Symposium on "Dust Control for Animal Production Facilities", Jatland, Denmark, 30 May–2 June 1999. pp: 48-55.

Wang X., Y. Zhang, L. Y. Zhao, G. L. Riskowski (2000): Effect of ventilation rate on dust spatial distribution in mechanically ventilated airspace. ASAE, vol. 43(6): 1877-1884.

Zhang Y., J. A. Polakow, Xinlei Wang, G.L. Riskowski, Y. Sun and S.E. Ford (2001): An aerodynamic deduster to reduce dust and gas emissions from ventilated livestock facilities. Proceedings of the sixth international symposium, 21-23 May (Louisville, Kentucky, USA) ASAE Publication Number 701P0201, eds. Richard R. Stowell, Ray Bucklin and Robert W. Bottcher. pp. 596-603.

Zhang Y., Z. Tan and X. Wang (2005): Aerodynamic deduster technologies for removing dust and ammonia in air streams. Proceedings of the seventh international symposium, 18-20 May (Beijing, China) ASAE Publication Number 701P0205, ed.T.   Brown-Brandl. pp. 224-229.

Zhu Z., H. Dong, X. Tao and H. Xin (2005): Evaluation of airborne dust concentration and effectiveness of cooling fan with spraying misting systems in swine gestation houses. Proceedings of the seventh international symposium, 18-20 May (Beijing, China) ASAE Publication Number 701P0205, ed. T. Brown-Brandl. pp. 224-229.

Chapter 14

Analysis of Halogenated Polycyclic Aromatic Hydrocarbons in the Air

Takeshi Ohura, Yuta Kamiya, Fumikazu Ikemori,
Tsutoshi Imanaka and Masanori Ando

Additional information is available at the end of the chapter

1. Introduction

ClPAHs are a class of compounds with one or more chlorines attached to the aromatic rings of a PAH. In studies of ClPAHs in the environment, naphthalenes with four to eight chlorines attached—polychlorinated naphthalenes (PCNs)—have been comparatively well investigated [1-4]. Technical PCN mixtures known as Halowax and Nibren wax have physical and chemical properties similar to those of polychlorinated biphenyls (PCBs), so that they have been widely used flame retardant as alternative material of PCBs. In addition to their emission as byproducts of product manufacture, PCNs are released to the environment from waste incineration processes, in slag residues from copper ore smelters, and from some chloro-alkali processes [5-8]. PCNs are also present as impurities in PCB mixtures [9]. In contrast, because there have been only limited environmental studies of the larger (>3-ring) ClPAHs, the environmental occurrence of these compounds has not yet been fully investigated. One of the reasons is that reference substances for almost all the larger ClPAHs are not generally available commercially. Therefore, researchers who want to investigate the environmental occurrence and properties of the ClPAHs have to synthesize their own reference substances. Nevertheless, some energetic researchers have performed environmental studies of the ClPAHs. Recently, Ohura et al. succeeded to synthesize a variety of ClPAHs with 3 to 5 ring as analytical standards (Figure 1) [10-12]. It has been advanced the studies on not only environmental behaviors but also biological effects of ClPAHs. Indeed, the AhR-mediated activities of ClPAHs have been determined by using yeast and CULUX assay systems [13, 14]. Although the activities were observed for all the 3- to 5-ring ClPAHs, which were at levels considerably lower (>80-times) than that of TCDD. Also, the intensities of activities depended on the molecular size: relatively high molecular weight ClPAHs (> 4-ring) were lower than those of the parent compounds, for the relatively low molecular weight ClPAHs, the activity tended to increase with increasing chlorine substitution [13]. For mutagenicity of some ClPAHs, there were some reports in which the toxicities were carried out with *Salmonella typhimurium* TA98 and TA100 in the presence or absence of S9 activation enzyme system [15-17]. In addition, certain ClPAHs are reported to have other toxic effects, such as tumorigenicity and oncogene activation, as reviewed by Fu et al. [18]. Here we describe the analytical and environmental studies on 3- to 5-ring Cl-/Br-PAHs.

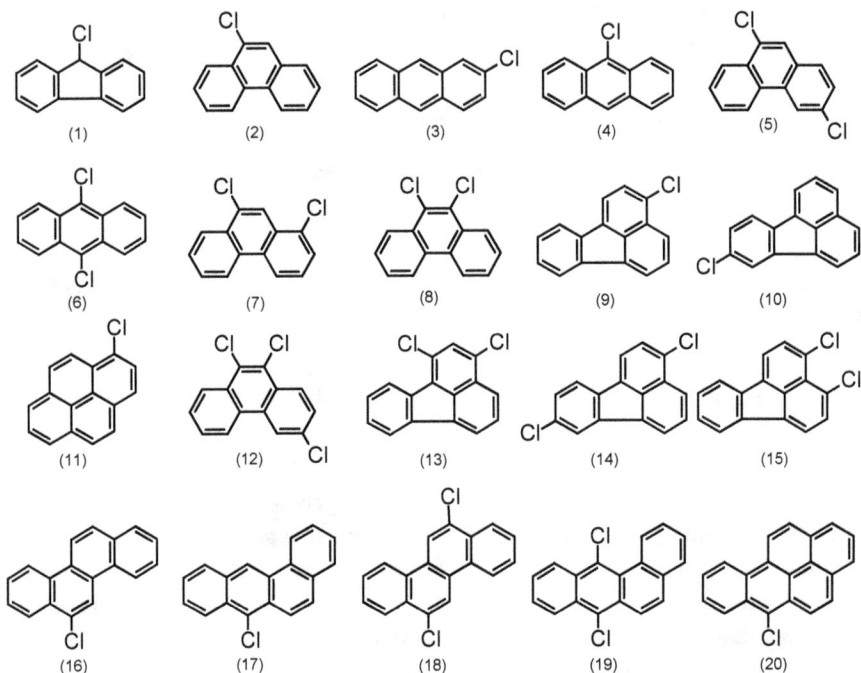

Figure 1. Structures of ClPAHs. The number noted is corresponded to Table 1.

2. Analytical methods of HPAHs

Comparing halogenated aromatics such as PCB and dioxins, there are fewer reports in regard to HPAHs in the environment. In the late 1970s to 1980s, Oyler et al. identified various chlorinated PAHs with 2~4-ring as the products of aqueous chlorination reactions of PAHs using reversed phase HPLC with UV detector, GC equipped with photoionization detector, and GC/MS [19, 20]. The study would be first report that ClPAHs were detected from environmental sample, whereas that would be artificial environment of water-treatment including chlorination reaction of PAHs.

Nilsson and Colmsjo investigated in detail the analytical methods of not only chloro-substituted PAHs, here defined as ClPAHs, but also chloro-added PAHs using normal phase HPLC [21, 22]. In the study, three stationary phases, i.e. aminopropylsilica, cyanopropyldimethylsilica and silica were used, resulting that ClPAHs among congeners showed similar retention behaviors among these columns. On the other hand, the retention behaviors of chloro-added PAHs showed stronger than ClPAHs. It was dismissed as due to the differences of steric hindrance caused by the large chlorine atoms.

For analysis of ambient halogenated PAHs, authors developed the methods for ClPAHs and BrPAHs using GC/qMS and HRGC/HRMS. We investigated the analytical method of 20 and 11 species of ClPAHs and BrPAHs, respectively, and applied to the surveys in the air. As GC methods, those conditions relatively adapt to those of PAH methods such as EPA method 8275A. The usage of non-polar GC column such as DB-5 could achieve good separation of HPAHs. Figure 2 and 3 shows the GC/qMS chromatographs of 20 species of ClPAHs and 11 species of BrPAHs standard mixtures, respectively. The common conditions are as follows: the HPAHs were analyzed on 7890A gas chromatograph (Agilent) coupled to a JMS-Q1000GC mass selective detector (JEOL) operated in selective ion monitoring mode. The column used was a 30 m x 0.25 mm i.d. Inertcap-5MS/NP capillary column with a film thickness of 0.25 μm (GL Sciences). Helium was used as the carrier gas and the flow rate of 1.0 ml/min at constant. The injection volume was 2.0 μl and was a pulsed splitless injection at 300 °C. The temperature program began at 100 °C, held for 1min, increased 200 °C at 25 °C/min, finally increased to 300 °C at 5 °C /min, and was held for 5 min (total 30 min). The MSD system was run in the electron impact mode. The ion source temperature was 300 °C, and the ionization energy and current was 70 eV and 200 μA, respectively. Table 1 and 2 show the retention time (RT) and monitored ions of individual ClPAH and BrPAH targeted, respectively.

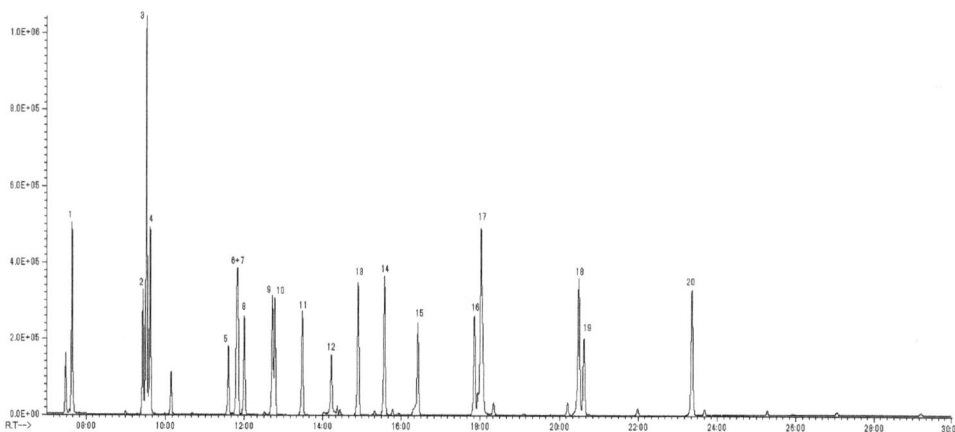

Figure 2. GC/qMS chromatogram of 20 ClPAHs. The number noted in chromatogram is corresponded to Table 1.

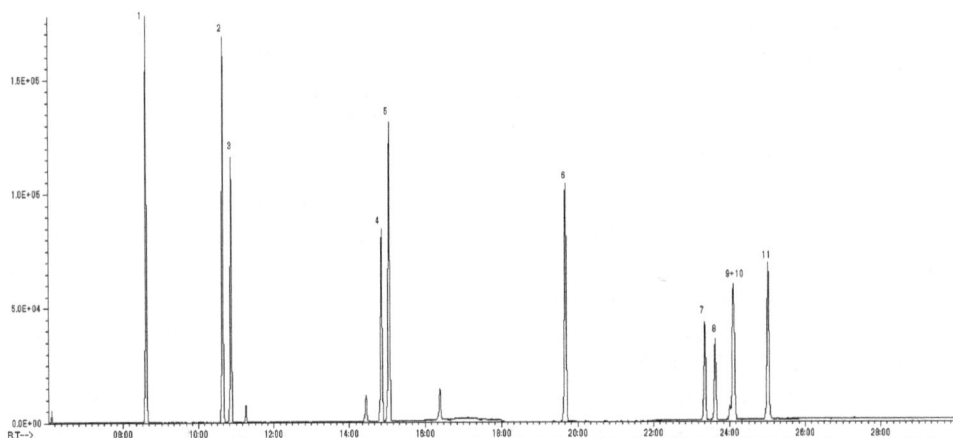

Figure 3. GC/qMS chromatogram of 11 BrPAHs. The number noted in chromatogram is corresponded to Table 2.

No	ClPAH	Abbrev.	RT[a] (h:m:s)	Ions monitored	
1	9-chlorofluorene	9-ClFlu	0:07:45	200	202
2	9-chlorophenanthrene	9-ClPhe	0:09:35	212	214
3	2-chloroanthracene	2-ClAnt	0:09:40	212	214
4	9-chloroanthracene	9-ClAnt	0:09:46	212	214
5	3,9-dichlorophenanthrene	3,9-Cl2Phe	0:11:46	246	248
6	9,10-dichloroanthracene	9,10-Cl2Ant	0:12:01	246	248
7	1,9-dichlorophenanthrene	1,9-Cl2Phe	0:12:01	246	248
8	9,10-dichlorophenanthrene	9,10-Cl2Phe	0:12:11	246	248
9	3-chlorofluoranthene	3-ClFluor	0:12:54	236	238
10	8-chlorofluoranthene	8-ClFluor	0:12:58	236	238
11	1-chloropyrene	1-ClPy	0:13:41	236	238
12	3,9,10-trichlorophenanthren	3,9,10-Cl3Phe	0:14:26	280	282
13	1,3-dichlorofluoranthene	1,3-Cl2Fluor	0:15:07	270	272
14	3,8-dichlorofluoranthene	3,8- Cl2Fluor	0:15:47	270	272
15	3,4-dichlorofluoranthene	3,4- Cl2Fluor	0:16:39	270	272
16	6-chlorochrysene	6-ClChry	0:18:05	262	264
17	7-chlorabenz[a]anthracene	7-ClBaA	0:18:15	262	264
18	6,12-dichlorochrysene	6,12-Cl2Chry	0:20:44	296	298
19	7,12dichlorobenz[a]anthracene	7,12-Cl2BaA	0:20:52	296	298
20	6-chlorobenzo[a]pyrene	6-ClBaP	0:23:36	286	288

[a] RT: retention time.

Table 1. Abbreviations of ClPAH standard and analytical performance of GC/qMS

No	BrPAH	Abbrev.	RT[a] (h:m:s)	Ions monitored	
1	2-bromofluorene	2-BrFlu	0:08:36	244	246
2	9-bromophenanthrene	9-BrPhe	0:10:40	256	258
3	9-bromoanthracene	9-BrAnt	0:10:52	256	258
4	9,10-bromoanthracene	9,10-Br2Ant	0:14:51	336	338
5	1-bromopyrene	1-BrPy	0:15:04	280	282
6	7-bromobenz[a]anthracene	7-BrBaA	0:19:41	306	308
7	7,11-dibromobenz[a]anthracene	7,11-Br2BaA	0:23:21	386	388
8	7,12-dibromobenz[a]anthracene	7,12--Br2BaA	0:23:37	386	388
9	5,7-dibromobenz[a]anthracene	5,7-Br2BaA	0:24:06	386	388
10	4,7-dibromobenz[a]anthracene	4,7-Br2BaA	0:24:06	386	388
11	6-bromobenzo[a]pyrene	6-BrBaP	0:25:01	330	332

[a] RT: retention time.

Table 2. Abbreviations of BrPAH standard and analytical performance of GC/qMS

Air sample collection for HPAHs might be generally performed using high volume air sampler, which is fundamentally similar to the case of PAHs and dioxins in the air. The differences of sampling efficiency between quartz fiber filters and glass fiber filters for HPAHs analysis have been not confirmed. The extraction process is also fundamentally same to the case of PAHs. Dichloromethane will be useful solvent in the Soxhlet extraction because of the solubility and handling ability. In the case of toluene as the solvent, it requires attention such as protection from light because toluene has highly reactivity compared to dichloromethane. Also, authors have adopted a ultrasonic extraction (20 min) in the case of small size of sampling because the method was also obtained suitable recovery rates.

Concerning pretreatment of ClPAH analysis from suspended particle matters, Nilsson and Ostman investigated the cleanup method and revealed that the behaviors of chloro-added PAHs showed the different from chloro-substituted PAHs in detail [23]. It represents that after silica cleanup, treatment using HPLC back-flush elution with cyanopropyldimethylsilica was effective method to separate chloro-added PAHs from chloro-substituted PAHs, parent PAHs, and aliphatics. For the extraction process, artifact formation of ClPAHs was also investigated from comparison of extraction solvents, dichloromethane and benzene. It shows that no artifact formation of ClPAHs takes place by using chlorine-involving solvent, dichloromethane. Note that the cleanup process after extraction should use the column packing material appropriated for PAHs but not dioxins because the usage of such sulfate treatment column for elimination of organic compounds could lead to disappearance of HPAHs. We believe that the cleanup could successfully satisfy by using only silica gel column. Furthermore, the possibility of artifact formation to the difference of sampling filters, glass fiber filters and Teflon filters, were also investigated, showing that there was no difference between the two types of filter. In short, required information from sampling to analysis of ClPAHs in the air could be based on those of PAHs.

Recently, Ieda et al. reported environmental analysis of HPAHs using comprehensive two-dimensional GC coupled to high-resolution time-of-flight mass spectrometry (GC x GC-HRTOF-MS) [24]. The GC x GC-HRTOF-MS method allowed highly selective group type analysis in the two-dimensional (2D) mass chromatograms with a very narrow mass window, accurate mass measurements for the full mass range (m/z 35-600) in GC x GC mode, and the calculation of the elemental composition for the detected HPAHs congeners in the real-world sample. This tool combines high sensitivity and selectivity for organic compounds including HPAHs detection by which higher chlorinated PAHs (penta, hexa and hepta substitution) will be possible to be detected in the environmental samples.

3. HPAHs in the air

First report concerning ClPAHs in the air was achieved by Nilsson and Östman, which were performed as long ago as 1990s [23]. Firstly, they synthesized some standard compounds of ClPAH. At that time, the environmental study on ClPAHs had been behind compared to the cases of PAHs and dioxins. That is, one of the top priorities for study on HPAHs could be difficult to get the standard compounds. Despite of such problems, they detected nine species of ClPAHs from urban air and road tunnel (Table 3). The concentrations in urban air ranged from 0.4 pg/m³ for 7-chlorobenz[a]anthracene (7-ClBaA) to 10.4 pg/m³ for 1,6- and 1,8-dichloropyrene (1,6-/1,8-Cl2Py) in particle phase, and from <0.3 pg/m³ for 7-ClBaA and 6-chlorobenzo[a]pyrene (6-ClBaP) to 17.5 pg/m³ for 1,6-/1,8-Cl2Py in gas phase. In addition, the concentrations in road tunnel ranged from 2.0 pg/m³ for 1,3-dichloropyrene to 19.1 pg/m³ for 4-chloropyrene (4-ClPy) in particle phase, and from <0.3 pg/m³ for 7-ClBaA and 6-chlorobenzo[a]pyrene to 30.7 pg/m³ for 4-ClPy in gas phase.

From the 2000's, Ohura and co-workers have been investigated the occurrences and behaviors of ClPAHs and BrPAHs in the air [10-12, 25]. Of the detected ClPAHs, the concentrations are summarized in Table 3. The magnitudes of ClPAHs concentrations were extent of ~100 pg/m³, which were middle range between dioxins and PAHs (Figure 4). The concentrations of almost ClPAHs in the air showed typical seasonal variation that were elevated in winter and decreased in summer, observed in the case of PAHs. Also, the characteristics of gas/particle partitioning of ClPAHs in the air showed the similar trends of PAHs: highly gaseous contents of relatively low-molecular weights ClPAHs and highly particulate contents of relatively high-molecular weights ClPAHs. It indicates that the partitioning could be significantly affected by the characteristics of corresponding parent PAHs even if being high molecular weights by substitution of chlorine. On the other hand, the concentration of 6-ClBaP was relatively high (<3.2 ~ 137 pg/m³) in compared to other ClPAHs and showed a characteristic trend: elevated suddenly in summer. The level was significantly different from the early work that was performed by Nilsson and Östman [23]. Currently, there is no data that account nicely for the mysterious phenomenon. We have suggested that the trend, elevated in summer, of 6-ClBaP might be due to secondary formation of BaP and activated chlorine in the particles. Kitazawa et al. have investigated temporal trends of particulate ClPAHs in an urban air, and relationships to the parent PAHs are also investigated [11]. Over the study period, the concentrations of the ClPAHs, except

6-ClBaP, remained almost constant, whereas the parent PAH concentrations declined moderately. In addition there was significant correlation between the concentrations of the ClPAHs, except 3,9- and 9,10-Cl2Phe, and the concentrations of the corresponding parent PAHs. This finding could indicate that the formation of ClPAHs associated with particles proceeds via that of PAHs, which provide clues as to the underlying emission sources and environmental behaviors. To ravel the occurrences, behaviors, and sources of ClPAHs in the air, it will need to investigate in various sites to clarify the concentrations.

Site/source	Air					Product			
	urban street	road tunnel	urban air	urban air	flue gas	fly ash	bottom ash	PVC	PVDC
9-ClFlu	na	na	na	0.62	na	nd	nd	na	na
2-ClAnt	na	na	na	1	1.2	nd~15	nd~0.68	140	58
9-ClAnt	na	na	na	9.9	13	nd~230	nd~1.0	700	29
9-ClPhe	na	na	3.5	19	67	nd~240	nd~16	9,600	31
3,9-Cl2Phe	na	na	2.7	1.4	2.8	nd~46	nd~3.1	na	nd
9,10-Cl2Phe/Ant	5.6	na	5.1	2.9	7.2	nd~290	nd~17	1,200	7.9
3-ClFluor	4.6	19.1	4.5	11	26	nd~1,000	nd~4.0	960	2.3
8-ClFluor	1.3	8.6	nd	3	6.4	nd~110	nd~1.1	na	nd
1-ClPy	11.8	39.6	7.5	13	310	nd~1,100	nd~2.6	1,100	nd
4-ClPy	8.1	49.8	na	na	na	na	na	4,500	na
1,3-Cl2Py	5.6	4	na	na	na	na	na	na	na
1,6/1,8-Cl2Py	27.9	42.7	na	na	na	na	na	na	na
6-ClChry	na	na	na	1.9	6.8	nd~1,200	nd~7.9	na	nd
6,12-Cl2Chry	na	na	na	2.1	2.7	nd~870	nd~2.5	na	nd
7-ClBaA	0.4	2.3	2.4	5.6	16	nd~870	nd~0.92	na	nd
7,12-Cl2BaA	na	na	na	14	2.8	nd~1,100	nd~2.9	na	nd
6-ClBaP	1.9	6.1	5.6	7.1	120	nd~5,300	nd~0.68	na	nd
unit	pg/m^3	pg/m^3	pg/m^3	pg/m^3	ng/Nm^3	ng/g	ng/g	ng/g	ng/g
Material	G, P	G, P	P	G, P	G	S	S	G, P	G, P
Reference	[23]	[23]	[10]	[12]	[28]	[26]	[26]	[30]	[29]

[a] na: not analyzed, nd: not detected. [b] air samples released from combustion (800°C) of polyvinylchloride (PVC). [c] G: gas-phase, P: particle matters, R: residues

Table 3. ClPAH concentrations in the air and emission sources[a]

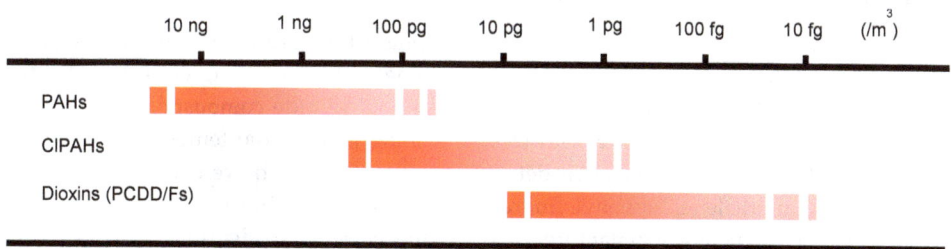

Figure 4. Concentration levels of PAHs, ClPAHs, and dioxins in the air.

Based on the findings of ClPAHs, brominated PAHs (BrPAHs) were expected to be present in the environment. There was, however, no the study at all. Ohura et al. tried to synthesis BrPAHs with 3- to 5-ring as standard compounds, which were newly detected from the urban air samples in Japan [25]. Of the BrPAHs detected, 5,7-Br2BaA was most abundant (mean concentration, 8.7 pg/m^3), followed by 7,12-Br2BaA (6.3 pg/m^3) and 6-BrBaP (3.3 pg/m^3). The mean concentrations of total BrPAHs, ClPAHs, and PAHs detected were 8.6 pg/m^3, 15.2 pg/m^3, and 1.2 ng/m^3, respectively, which showed that concentrations of such HPAHs tended to be approximately 100-fold lower than PAHs. For the moment, this study is only report in which BrPAHs were investigated in the air. Besides the environmental analysis, the occurrence and profiles of BrPAHs in waste incinerators were investigated by Horii et al. [26]. 1-Brominated pyrene was predominant in the fly ash sample among BrPAHs targeted. The reports regarding environmental surveys of BrPAHs have been limited rather than those of ClPAHs, so that the environmental behaviors and sources remain unclear.

4. Emission sources of ambient HPAHs

In light of the production mechanisms established by studies of PAHs and dioxins, it is no wonder that the principal sources of ClPAHs in the air are traffic and incineration facilities. In a previous study by Nilsson and Östman, it was suggested that dichloroethane as a scavenger contained in leaded gasoline may be the source of chlorine in ClPAHs [23]. Indeed, some ClPAHs have been detected from not only urban air but also in snow and vehicle exhaust gas, although they have not been quantified [27].

The air and residues samples emitted from incinerators were also investigated to evaluate the occurrences of HPAHs [26, 28]. For flue gas samples collected in actual operating incinerators, a number of ClPAHs were detected at extremely high levels compared to the air samples (Table 3). Dioxins were also detected from the flue gas samples, levels of which showed significant correlations to the certain ClPAHs levels. These findings show that ClPAHs in air are possible to emit from incinerators, and could be indicator of dioxins production. Horii et al. have, as described in above, investigated the occurrences of ClPAHs rather than BrPAHs in fly ash and bottom ash from waste incinerators [26]. The total concentrations of ClPAHs in ash samples ranged from <0.06 to 7000 ng/g, in which 6-ClBaP and 1-ClPy were dominant compounds.

Studies of the production of ClPAHs in other combustion processes have investigated emissions from the combustion of chlorine-containing materials. Wang et al. investigated the mechanism of formation of ClPAHs in the polyvinylchloride combustion process by a laboratory-scale tube-type furnace with electric heating [29]. At the temperature range of 600-900 °C, about 18 Cl-PAHs were determined, most of which were monochlorinated derivatives of naphthalene, biphenyl, fluorene, phenanthrene, anthracene, fluoranthene and pyrene. Only two dichlorophenanthrenes or anthracenes were identified. As the other experiment, ClPAHs emissions were also investigated by polyvinylidene chloride plastic wrap over a flame of the gas burner [30]. Only seven of 27 target ClPAHs were detected, which were all relatively low molecular weight ClPAHs with 2~4-ring. As the other interesting emission sources,

Considering the concentrations of individual ClPAHs, there were significant correlations among them, suggesting that these compounds might be emitted from common sources [10-12]. Also, factor analysis based on the experimental data of ClPAHs and PAHs in air suggest that the emission sources for ClPAHs except for 6-ClBaP may be similar to the sources of the parent PAHs [11]. As is suggested by the above discussions, the production of ClPAHs and dioxins through the combustion of organic materials requires chlorine. The occurrence of ClPAHs and dioxins simultaneously should be investigated further to give us more information on their production processes in the environment and their possible sources.

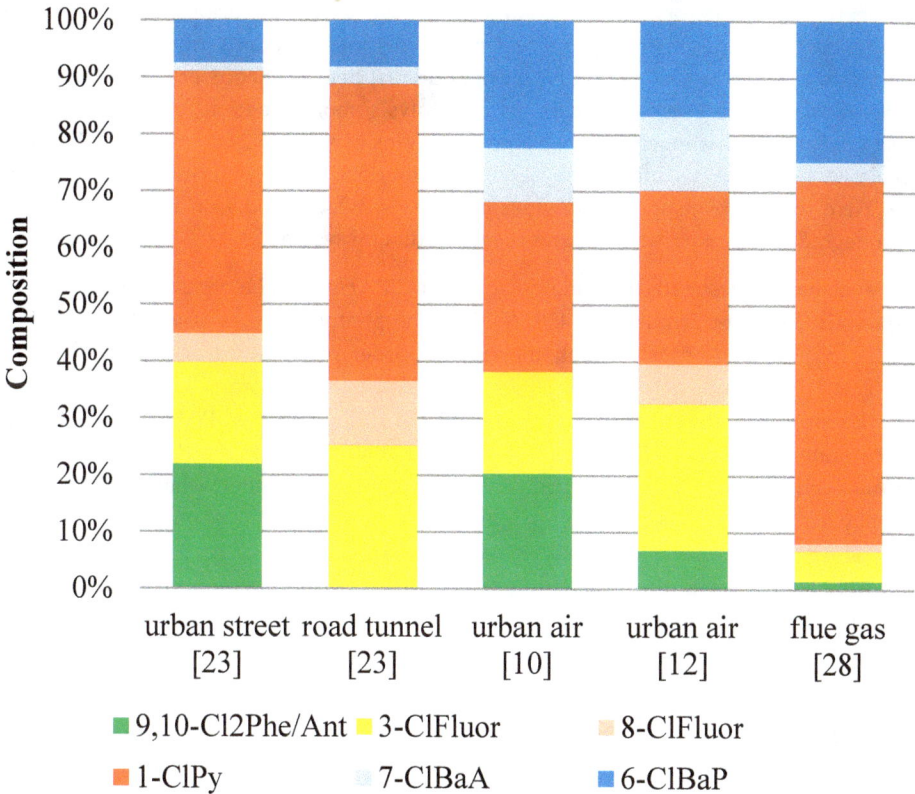

Figure 5. Profiles of typical ClPAH concentrations in the air and flue gas samples.

Comparing the compositions of targeted compounds in among the different samples could be useful technique to understand the origins and behaviors in the environment. Here we compared the profiles of typical ClPAH (9,10-Cl2Phe/Ant, 3-ClFluor, 8-ClFluor, 1-ClPy, 7-ClBaA, and 6-ClBaP) concentrations in among the samples from air and flue gas from incinerators (Figure 5). 1-ClPy was most abundant among the group of ClPAHs. The profiles were somewhat same between samples between urban street and road tunnel, suggested that the ClPAHs detected in urban air could be significantly contributed to traffic sources. The profiles obtained from air samples collected in Japan were also similar ones, whereas those sampling date were different. This suggests remaining unchanged of the emission sources for long periods of time in the area. On the other hand, the characteristic profile of flue gas sample was observed that the contribution of 1-ClPy was much larger than those of air samples. These findings suggest the diversity of emission sources of ClPAHs in the air.

5. Photoreaction of HPAHs

Atmospheric behaviors of organic pollutants are affected by various factors: climate conditions and chemical/physical properties of own organic pollutants. In the case of PAHs, the significant factors fall into two broad categories: (i) heterogeneous processes involving particle-associated compounds such as photolysis/photooxidation and gas–particle interactions; and (ii) homogeneous gas-phase reactions of volatile 2- and 3-ring PAHs and semivolatile 4-ring PAHs, initiated by OH (daytime) and NO_3 (night-time) radicals and ozone. Here we focus on photoreactivities of HPAHs, which model on the photoreaction (phootodegradation) on particle surfaces using irradiation system.

Ohura et al. have investigated the photoreaction of ClPAHs in some organic solvents under high-pressure Hg lamp irradiation (450 W) [10]. The photolysis rates of all of the ClPAHs fitted the pseudo-first-order reaction model. The order of decay rates of mono chloro-substituted PAHs such as 1-ClPy, 7-ClBaA, 6-ClBaP, etc. were somewhat consistent with that of the corresponding parent PAHs. In the case of polychlorinated PAHs (Cl_nPAH, n>2), the decay rates decreased with increasing extent of chlorination of the PAHs whereas the trend has been observed in only chlorinated phenanthrenes. It suggests that chlorination of PAH may be more photostable compared to parent PAHs. The photoproducts of the photolytic reactions were also tentatively identified [10]. The photolyses of ClPhe and 7-ClBaA were confirmed to proceed by initial abstraction of chlorine, followed by oxidative degradation. Although it is not clear whether the oxidized byproducts of ClPAHs are ubiquitous in the environment or body tissues, the toxicities of the byproducts should also be determined.

Concerning BrPAHs, the photolysis rates were relatively faster than the corresponding ClPAHs [25]. This is could be due to weak bond energy of C-Br (276 kJ/mol) than that of C-Cl (330 kJ/mol). Comparing the ambient profiles among the PAH congeners suggested that ambient BrPAHs that came from the specific local emission sources differed from ClPAHs and PAHs, and/or could be driven by various seasonal factors, including photodecay processes.

6. Conclusion

The study on halogenated PAHs in the environment could be at the beginning stage. Here we note mainly about the analytical methods and environmental data of HPAHs that are still insufficiency in comparison of PAHs and dioxins. Therefore, the hazardous contributions of HPAHs in the environment have yet to be revealed. The concentrations of HPAHs in the air were middle range between dioxins and PAHs. Because the structures of HPAHs are combination of those of dioxins and PAHs, the production mechanism of HPAHs may be also follow to the characteristics of them. The ambient levels of HPAHs, therefore, could be middle range of them. This finding may be useful characteristic for environmental and risk monitoring of such hazardous aromatic compounds. That is, monitoring HPAHs in the environment may provide the levels of both dioxins and PAHs. The attempt may be also capable of adapting to the risk assessment.

Author details

Takeshi Ohura* and Yuta Kamiya
Meijo University, Shiogamaguchi, Nagoya, Japan

Fumikazu Ikemori
Nagoya City Institute For Environmental Science, Toyoda, Nagoya, Japan

Tsutoshi Imanaka
GL Sciences Inc. Sayamagaoka, Iruma, Japan

Masanori Ando
Musashino University, Shinmachi, Nishitokyo, Japan

Acknowledgement

We thank the Japan Society for the Promotion of Science (Grant-in-Aid for Scientific Research (B); No. 23310011) and the Ministry of the Environment, Japan (the Environment Research and Technology Development Fund; No. RFb-1103) for their supports of this work.

7. References

[1] Dörr G, Hippelein M, Hutzinger O (1996) Baseline contamination assessment for a new resource recovery facility in Germany. Part V: Analysis and seasonaliregional variability of ambient air concentrations of polychlorinated naphthalenes (PCN). Chemosphere 33: 1563-1568

---

* Corresponding Author

[2] Falandysz J (1998) Polychlorinated naphthalenes: an environmental update. Environ. Pollut. 101: 77-90

[3] Harner T, Lee R G M, Jones K C (2000) Polychlorinated Naphthalenes in the Atmosphere of the United Kingdom. Environ. Sci. Technnol. 34: 3137-3142

[4] Egebäck A-L, Wideqvist U, Järnberg U, Asplund L (2004) Polychlorinated Naphthalenes in Swedish Background Air. Environ. Sci. Technnol. 38: 4913-4920

[5] Weber R, Iino F, Imagawa T, Takeuchi M, Sakurai T, Sadakata M (2001) Formation of PCDF, PCDD, PCB, and PCN in de novo synthesis from PAH: Mechanistic aspects and correlation to fluidized bed incinerators. Chemosphere 44: 1429-1438

[6] Helm P A, Bidleman T F (2003) Current Combustion-Related Sources Contribute to Polychlorinated Naphthalene and Dioxin-Like Polychlorinated Biphenyl Levels and Profiles in Air in Toronto, Canada. Environ. Sci. Technnol. 37: 1075-1082

[7] Stevens J L, Northcott G L, Stern G A, Tomy G T, Jones K C (2003) PAHs, PCBs, PCNs, Organochlorine Pesticides, Synthetic Musks, and Polychlorinated n-Alkanes in U.K. Sewage Sludge: Survey Results and Implications. Environ. Sci. Technnol. 37: 462-467

[8] Helm P A, Bidleman T F, Li H H, Fellin P (2004) Seasonal and Spatial Variation of Polychlorinated Naphthalenes and Non-/Mono-Ortho-Substituted Polychlorinated Biphenyls in Arctic Air. Environ. Sci. Technnol. 38: 5514-5521

[9] Yamashita N, Kannan K, Imagawa T, Miyazaki A, Giesy J P (2000) Concentrations and Profiles of Polychlorinated Naphthalene Congeners in Eighteen Technical Polychlorinated Biphenyl Preparations. Environ. Sci. Technnol. 34: 4236-4241

[10] Ohura T, Kitazawa A, Amagai T, Makino M (2005) Occurrence, profiles, and photostabilities of chlorinated polycyclic aromatic hydrocarbons associated with particulates in urban air. Environ. Sci. Technol. 39: 85-91

[11] Kitazawa A, Amagai T, Ohura T (2006) Temporal trends and relationships of particulate chlorinated polycyclic aromatic hydrocarbons and their parent compounds in urban air. Environ. Sci. Technol. 40: 4592-4528

[12] Ohura T, Fujima S, Amagai T, Shinomiya M (2008) Chlorinated polycyclic aromatic hydrocarbons in the atmosphere: seasonal levels, gas-particle partitioning, and origin. Environ. Sci. Technol. 42: 3296-3302

[13] Ohura T, Morita M, Makino M, Amagai T, Shimoi K (2007) Aryl hydrocarbon receptor-mediated effects of chlorinated polycyclic aromatic hydrocarbons. Chem. Res. Toxicol. 20: 1237-1241

[14] Horii Y, Khim J S, Higley E B, Giesy J P, Ohura T, Kannan K (2009) Relative potencies of individual chlorinated and brominated polycyclic aromatic hydrocarbons for induction of aryl hydrocarbon receptor-mediated responses. Environ. Sci. Technol. 43: 2159-2165

[15] Bhatia A L, Tausch H, Stehlik G (1987) Mutagenicity of chlorinated polycyclic aromatic compounds. Ecotoxicol. Environ. Saf. 14: 48-55

[16] Colmsjö A, Rannug A, Rannug U (1984) Some chloro derivatives of polynuclear aromatic hydrocarbons are potent mutagens in Salmonella typhimurium. Mutat. Res. 135: 21-29

[17] Löfroth G, Nilsson L, Agurell E, Sugiyama T (1985) Salmonella/microsome mutagenicity of monochloro derivatives of some di-, tri- and tetracyclic aromatic hydrocarbons. Mutat. Res. 155: 91-94

[18] Fu P P, Von Tungeln L S, Chiu L-H, Own Z Y (1999) Halogenated-polycyclic aromatic hydrocarbons: a class of genotoxic environmental pollutants. Environ. Carcinogen. Ecotoxicol. Rev. C17: 71-109

[19] Oyler A R, Liukkonen R J, Lukasewycz M K, Cox D A, Peake D A, Carlson R M (1982) Implications of treating water containing polynuclear aromatic hydrocarbons with chlorine: a gas chromatographic-mass spectrometric study. Environ. Health Perspect. 46: 73-86

[20] Oyler A R, Liukkonen R J, Lukasewycz M T, Helkklla K E, Cox D A, Carlson R M (1983) Chlorine disinfection chemistry of aromatic compounds. Polynuclear aromatic hydrocarbons: rates, products, and mechanisms. Environ. Sci. Technol. 17: 334-342

[21] Nilsson U, Colmsjö A (1991) Retention Characteristics of Chlorinated Polycyclic Aromatic Hydrocarbons in Normal Phase HPLC I. Chloro Added PAH. Chromatographia 32: 334-340

[22] Nilsson U, Colmsjö A (1992) Retention Characteristics of Chlorinated Polycyclic Aromatic Hydrocarbons in Normal Phase HPLC. II - Chloro Substituted PAHs. Chromatographia 34: 115-120

[23] Nilsson U L, Oestman C E (1993) Chlorinated polycyclic aromatic hydrocarbons: method of analysis and their occurrence in urban air. Environ. Sci. Technol. 27: 1826-1831

[24] Ieda T, Ochiai N, Miyawaki T, Ohura T, Horii Y (2011) Environmental analysis of chlorinated and brominated polycyclic aromatic hydrocarbons by comprehensive two-dimensional gas chromatography coupled to high-resolution time-of-flight mass spectrometry. J. Chromatogr. A. 1218: 3224-3232

[25] Ohura T, Sawada K, Amagai T, Shinomiya M (2009) Discovery of novel halogenated polycyclic aromatic hydrocarbons in urban particulate matters: occurrence, photostability, and AhR activity. Environ. Sci. Technol. 43: 2269-2275

[26] Horii Y, Ok G, Ohura T, Kannanct K (2008) Occurrence and profiles of chlorinated and brominated polycyclic aromatic hydrocarbons in waste incinerators. Environ. Sci. Technol. 42: 1904-1909

[27] Haglund P, Alsberg T, Bergman A, Jansson B (1987) Analysis of halogenated polycyclic aromatic hydrocarbons in urban air, snow and automobile exhaust. Chemosphere 16: 2441-2450

[28] Ohura T, Kitazawa A, Amagai T, Shinomiya M (2007) Relationships between chlorinatedpolycyclic aromatic hydrocarbons and dioxins in urban air and incinerators. Organohalogen Compd. 69: 2902-2905

[29] Wang D, Xu X, Chu S, Zhang D (2003) Analysis and structure prediction of chlorinated polycyclic aromatic hydrocarbons released from combustion of polyvinylchloride. Chemosphere 53: 495-503

[30] Fujima S, Ohura T, Amagai T. (2006) Simultaneous determination of gaseous and particulate chlorinated polycyclic aromatic hydrocarbons in emissions from the scorching of polyvinylidene chloride film. Chemosphere 65: 1983-1989

Chapter 15

H₂S Pollution and Its Effect
on Corrosion of Electronic Components

Benjamin Valdez Salas,
Michael Schorr Wiener, Gustavo Lopez Badilla,
Monica Carrillo Beltran, Roumen Zlatev,
Margarita Stoycheva, Juan de Dios Ocampo Diaz,
Lidia Vargas Osuna and Juan Terrazas Gaynor

Additional information is available at the end of the chapter

1. Introduction

The microelectronic industry applies materials with good electrical and corrosion resistance properties for the manufacturing of the microelectronic devices. Silver and copper are materials with this characteristics used for that purpose. Some of their main applications are as high thermal conductive die attach paste that had silver flakes as conductive filler material, silver plated over copper frames, Sn-Ag and Sn-Cu alloys for solder paste used in Surface Mount Technology (SMT) process, conductive internal copper layers in printed circuit boards, copper wire bonding, and more. Other metals and alloys, widely used for support frames, heat diffusers and case wares in the electronics industry are tin, nickel, aluminum, carbon steel and galvanized steel.

Corrosion in microelectronics depends on several variants such as the package type, materials involved, assembly processes, moisture, inorganic and organic contaminants, atmospheric pollutants, temperature, thermal stress and electrical bias.(P. Roberge, 2000, J. Payer, 1990, M.Reid et al., 2007, M. Tullmin and P. Roberge, 1995, M. McNeil and B. Little, 1992, X. Lin and J. Zhang, 2004)

The plastic packages are the more widely used because of its size, cost and manufacturability advantages but, compared to other hermetic package systems like ceramics, are non-hermetic due to its polymeric materials that permit the permeation of moisture and corrosive gases, allowing in this manner the appearance of other problems associated with corrosion of the inside components. Therefore, the advances in the microelectronics technology that have promoted the developing of devices with smaller and thin components, and the exposition of these devices to environments where temperature, humidity and atmospheric pollutants like chlorides, NO_x, SO_x, COS and hydrogen sulfide (H_2S) which are not completely controlled, can favor atmospheric corrosion on their metallic components. Small quantities of corrosion products are enough to induce reliability issues and even catastrophic failures in the microelectronic devices due to the formation of insulating layers by the corrosion film.

In order to minimize the risk of corrosion failure it is important to be aware of corrosion damage during the design stage, reliability evaluations and qualifications, assembly processes, storage, shipping, and in the final use of the microelectronic devices.

Atmospheric contaminants such as H_2S and carbonyl sulfide (COS), that promote the corrosion of silver and copper, are dissolved in the thin layer of electrolyte over metals as a consequence of even low relative humidity (RH) and produce the HS^- ion which is their main reduced sulphur constituent at neutral pH. When silver is exposed in an environment containing a minimum concentration of these contaminants (<1ppm), the corrosion product formed is silver sulfide (Ag_2S). A similar process of sulphidation occurs when the exposed metal is copper that reacts with H_2S producing copper sulfide (Cu_2S). Several studies have been done on the indoor corrosion of silver and copper and revealed that silver sulfide is the main corrosion product on silver that had been exposed indoors. These corrosion films can form an insulating layer on the contact surfaces causing electrical failures on the microelectronic devices.

When silver alloys are exposed in a sulphur-rich environment, corrosion products of the most reactive metal are produced. That is the case of Ag-Cu, where the principal corrosion product is Cu_2S or in alloys of silver with palladium where the corrosion product is Ag_2S. (T. Graedel, 1992, D. Rice et al., 1981, C. Yang et al., 2007, H. Kim, 2003, M. Watanabe et al., 2006, C. Kleber et al., 2008, J. Franey et al., 1985, M. Watanabe et al., 2005, S,. Sharma, 1978, P. Vassiliou and C. Dervos, 1999, g. Russ, 1970, L. Veleva et al., 2008)

Corrosion kinetics depends on the type of metal and also is related to the nature of the electrolyte, atmospheric contaminants and corrosion products. Two kinetic corrosion laws are known for silver kinetics in indoor environments. Silver sulphide, which is a film with low corrosion resistance, obeys a linear corrosion law while AgCl a more protective corrosion layer presents a parabolic behavior. (L. Veleva et al., 2008)

Morphology of corrosion film on silver does not tend to be uniform because of the presence of dendrites or whiskers. Dendrites are fern-shaped and grow across the surface of the metal as a consequence of moisture capable of dissolving the metal and then the ions are redistributed by electro migration in presence of an electromagnetic field. When there is a thick layer of corrosion products, thin filaments projected at a right angle to the surface, called whiskers, begin to grow spontaneously even at room temperature and without an applied electric field. Several cases have been reported where tin whiskers caused failures. (B. Chudnovsky, 2002)

This chapter describes the more relevant results got from the study of silver corrosion at indoor conditions in companies dedicated to assembly and functional test of microelectronics devices. To achieve this, silver coupons and silver plated copper leadframes were exposed in two sites of the assembly process. In addition these materials were exposed in a test chamber that simulates indoor conditions of a plant with no controls for outdoor atmospherics contaminants.

On the other hand, in the electronics industry of Mexicali city located in the State of Baja California, Mexico, there are a variety of devices and electronic equipment inside the plants which are exposed to environments with no climate control and air pollution.

The electronics equipment suffers from corrosion, as humidity levels, sources emitting pollutants such as CO, NOx and sulphide penetrate thought cracks or air conditioning systems. Corrosion phenomena affect connections of electronic equipment (Frankel, 1995) and other electronics components protected with plastic or metallic materials. Atmospheric corrosion is an electrochemical phenomenon that occurs in the wet film formed on metal surfaces by climatic factors (Table I). The corrosion products form dendrites or whiskers in the metallic joints and connectors (Nishikata, et al 1995, Nishimura et al, 2000).

There are obvious differences in outdoor and indoor environments and consequent differences between outdoor and indoor corrosion behavior (Lyon et al, 1996). The corrosion of metals as copper in indoor environments may be viewed as a variation of outdoor atmospheric corrosion. In contrast to outdoor exposure, in an indoor environment the wet film on the metal surface is thinner and it is often governed by relatively constant controlled humidity conditions. Sometimes the indoor environment temperature and RH are controlled and as a consequence, the amount of adsorbed water on surfaces is minimal and is constrained within reasonably tight limits. Since atmospheric corrosion occurs when moisture is formed on the metal surface and depends on its duration and corrosion intensity increases (Veleva et al, 2008).

Factors	Measuring instrument	Unit
Humidity	Hygrometer	%
Temperature	Thermometer	°C
Atmospheric pressure	Barometer	mmHg
Solar radiation	Pyranometer	W / m²
Pluvial precipitation	Rain gauge	mm
Wind direction	Wind vane	°Grade
Wind speed	Anemometer	m/seg

Table 1. Climatic factors and their measurement

2. Chemistry of sulphidic corrosion

H₂S is a weak, reducing acid, soluble in water with a ionic dissociation:

$$H_2S \leftrightarrow H^+ + HS^- \leftrightarrow 2H^+ + S^{2-}$$

H₂S attacks steel:

$$Fe+H_2S \rightarrow FeS+H_2$$

The FeS layer of steel is not stable, it is removed from the steel surface in an acidic environment, forming again H₂S, enhancing corrosion:

$$FeS+2H \rightarrow Fe^{+2}+H_2S$$

H₂S reduces the Fe^{+3} present in rust: $Fe_2O_3 \cdot n\ H_2O$:

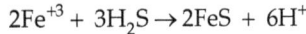

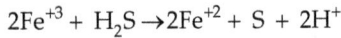

$$2Fe^{+3} + 3H_2S \rightarrow 2FeS + 6H^+$$

$$2Fe^{+3} + H_2S \rightarrow 2Fe^{+2} + S + 2H^+$$

Other metals, applied an electronic device e.g. Ag, Cu, Sn, undergo similar reactions forming unstable metallic sulfides.

Under oxidizing conditions H₂S is converted into sulphuric acid, a strong corrosive agent:

$$H_2S + 2O_2 \rightarrow H_2SO_4$$

For silver and copper at atmospheric conditions, the general reaction are explained as follows:

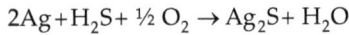

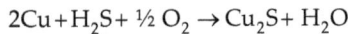

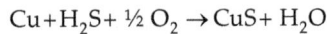

$$2Ag+H_2S+ \tfrac{1}{2}O_2 \rightarrow Ag_2S+ H_2O$$

$$2Cu+H_2S+ \tfrac{1}{2}O_2 \rightarrow Cu_2S+ H_2O$$

$$Cu+H_2S+ \tfrac{1}{2}O_2 \rightarrow CuS+ H_2O$$

3. H₂S toxicity

It is appropriate to report in the context of the present paper the toxicity of H₂S since this also affects quality of the environment and human health, central issues of modern society. H₂S gas emitted into the atmosphere from municipal sewage, industrial plants, animal farms, geothermal wells and polluted sewers and ports (M. Schorr and B. Valdez, 2005) causes inflammation of the eyes, skin burns and respiratory diseases such as rhinitis, bronchitis and pneumonia. When inhaled in small amounts, the gas produces headaches and nausea; a large amount produces paralysis. H₂S is very toxic, rapid death ensues from exposure to air containing > 1000 ppm H₂S owing to asphyxiation since it paralyses the Fe-phorphirin molecule in the human respiratory system. Lower doses cause dizziness and excitement because of damage to the central nervous system. Causes of the death of workers following the release of H₂S from sewage installations and from plant for the removal of sulphur from natural gas have been reported. (M. Schorr et al., 2006; S.E. Manahan, 1993)

4. Case studies

4.1. Corrosion behavior of silver and silver plated copper leadframes in H₂S polluted outdoor and indoor environments

In order to evaluate their corrosion behavior, metallic silver coupons and silver plated copper leadframes were displayed per triplicate during a period of 60 days between the summer months of July, August and September in three different sites. Two of the sites were inside and outside the clean room along the assembly process of a microelectronic company and the third one was on a sheltered test chamber (105 x 45 x 65cm in size of aluminum sheet 0.6mm thick material) to simulate indoor environment at uncontrolled flow of atmospheric contaminants O. Vargas et al, 2009).

The test chamber was located in a ventilated place at 10 m over the ground level. In parallel an exposition of metallic silver coupons was followed during a period of 12 and 24 months in the test chamber exposure site to study silver corrosion over a long period of time.

The exposure assembly was developed in Mexicali, Baja California, which is an urban semi-arid zone near to the Cerro Prieto geothermal power plant, the largest and oldest geothermal field in México.Prior to their exposure, the rectangular metallic silver coupons (99.95% pure and 10 x 5 x 1mm in size) were polished with silicon carbide (SiC) abrasive papers to 1200 grit, then were fine polished on a nylon cloth using a 3μm diamond suspension and finally were rinsed with deionized water and dried with nitrogen gas flow (Figure 1).

The leadframes materials, used for a wide range of microelectronics devices, consist in a copper frame with a silver electroplated die paddle and leads to provide mechanical support to the die during the assembly process and for external electrical connection. Dimension of silver plated area is 6x6 mm with a thickness of 6.5μm (Figure 2).

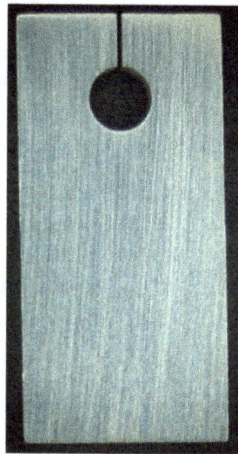

Figure 1. Metallic silver coupon

Figure 2. Silver plated on copper frame

The temperature, relative humidity (%RH) and main atmospherics contaminants were monitored and recorded during all the testing time. The morphology of samples was obtained with a JEOL JSM-6360 Scanning Electron Microscope (SEM) and the elemental microanalysis of corrosion products was performed using an EDAX brand Energy Dispersive X-ray Spectroscopy (EDS) detector attached to the SEM.

H_2S is a critical air pollutant in Mexicali. The formed thin AgS layer on the Ag surface in presence of H_2S prevents its adhesion toward the solder material thus making problematic the low resistance contact formation after the soldering process causing an increase of the device failure percentage. To determine the rate of corrosion provoked by H_2S of the silver frames used in microelectronics devices production, the Quartz Crystal Microbalance (QCM) technique was applied at controlled laboratory conditions simulating the indoor conditions. This technique allows the real time determination of the corrosion rate and the formed AgS layer average thickness. The mass gain is calculated applying the Sauerbrey equation:

$$\Delta F = -Cf \cdot \Delta m$$

with: ΔF = frequency change in Hz; Cf = sensitivity factor of the crystal in Hz/ng/cm2 ; Δm = mass change per unit area in g/cm2 . For 1 inch diameter 5 MHz quartz crystal Cf = 0.0566 Hz/ng/cm2 according to the data provided by QCM producer Maxtek (USA).

The Sauerbrey equation allows the real time determination of the mass gain and hence the corrosion rate of Ag provoked by H_2S as well as the thickness of the formed AgS layer. Family of curves in coordinates: QCM frequency - time was registered for different H_2S concentrations while the temperature was held constant and the relative humidity was monitored.

5. Corrosion process

Average values of temperature and RH during exposure of samples are shown in Table 2. These values were similar in the two sites inside the assembly process of the plant and were more aggressive in the test chamber used to simulate indoor conditions.

Atmospheric corrosion frequently occurs in the presence of a thin moisture layer that forms on the metal under certain environmental conditions. The layer may vary from monomolecular thickness to clearly visible water films. Above the critical value of 50% RH at room temperature, the metal is covered by physical adsorption with more than 3 molecular water monolayers (Table 3). This aqueous layer acts as an electrolyte and allows the incorporation of atmospheric contaminants. The critical RH for different metals in sulfur rich environments has been reported to be between 50 and 90%. (P. Marcus, 2002)

The time of wetness (TOW) is the period of time during which, due to the atmospheric conditions, is formed the moisture layer on the surface of metal and the corrosion process can occurs. When relative humidity raises the 90% and temperature is $0° \leq t \leq 25°C$ the dew point is reached and the humidity surface layer on metal becomes thicker. (ISO 9223, 1992)

According the RH registered data and temperature during the exposure, the three exposure sites presented the conditions for atmospheric corrosion of metallic silver coupons and silver plated copper frames.

Exposure Site	Temperature (°C)		Relative humidity (%)	
	Min	Max	Min	Max
Inside clean room	19.44	27.83	32.7	64.2
Outside clean room	19.16	27.33	36.5	65.3
Test chamber	24.36	46.46	10.83	89.9

Table 2. Temperature and relative humidity values during the exposition time

RH%	Water monolayers
90	8 on Ag
80	5-10
60	2-5
40	1.5-2
20	1

Table 3. Approximate number of water monolayers on different metals versus relative humidity

During the exposure time hydrogen sulfide (H₂S) was present, because the activities of vapor exploitation at Cerro Prieto geothermal power plant. When the geothermal fluid is processed to produce electricity, emissions of non-condensable gases are released into the atmosphere. The main involved gases normally are carbon dioxide (CO₂) at around 90%, followed by H₂S with only 2-3% by weight of total gases and, in lower proportion methane, ammonia, nitrogen, hydrogen, mercury and radon. H₂S is a pollutant with a characteristic odor of rotten eggs even at low concentrations, which affects the air quality, induces health damages and it is very corrosive. (H. Puente and L. Hernandez, 2005)

Anthropogenic activities in the region contribute to the increase of H₂S. An automatic air pollutant monitoring station that belongs to the California Environmental Protection Agency located in the nearest has shown an average concentration of 0.9 ppm during the year. Other important atmospheric contaminants monitored by this station are listed in Table 3. (L. Veleva et al., 2008)

Indoor corrosivity indexes (IC2-IC3) for silver and copper has been reported for the Mexicali urban semi-arid environment in evaluations performed in a sheltered test chamber. (L. Veleva et al., 2008, ISO 11844-1, 2000)

Appearance of tarnish film on silver coupons and silver plated copper frames at the end of 60 days of exposure in the three sites can be appreciated in Figure 3 and are described in Table 5.

Contaminant	Average of Min. and Max. Concentration (ppm)
Carbon monoxide (CO)	8.39 to 12.04
Nitrogen oxide (NO)	0.029 to 0.061
Ozone (O$_3$)	0.03 to 0.10
Sulfur dioxide (SO$_2$)	0.029 to 0.086
Hydrogen sulphide (H$_2$S)	0.1 to 0.5

Table 4. Average of annual concentrations of atmospheric contaminants

Exposure Site	Silver Coupons	Silver plated over leadframe
Inside clean room	Royal blue coloration in the center and slightly surrounded by purple in the edges.	Light gold coloration surrounded by purple and blue in the edges.
Outside clean room	Gray coloration in the center and slightly surrounded by purple in the edges.	Purple and blue coloration in the center surrounded by squared shaped lines of purple and royal blue. Small stains were observed in the purple centered area.
In test chamber	Uniform gray coloration.	Gray coloration surrounded by royal blue in the edge. Several purple and dark stains were observed along the silver plated surface.

Table 5. Appearance description of samples after 60 days of exposure

The coloration of the tarnish film depends on its thickness. As the silver sulfide film becomes thicker it turns also darker. C. Yang et al, 2007)

After each exposure time, the samples were analyzed by SEM and EDS in order to observe the corrosion products morphology and chemical composition. In the case of metallic silver coupons, it was observed an uniform morphology with absence of dendrites or whiskers, and the presence of silver sulfide (Figure 4). For the case of silver plated on copper frames, they presented dendrites along the corrosion film (Figure 5). Punctual microanalysis was performed over representative areas with and without dendrites and was noted that, the areas with dendrites presented just a little higher silver sulfide composition but in general composition in both areas was similar in the case of samples coming from inside and outside the clean room.

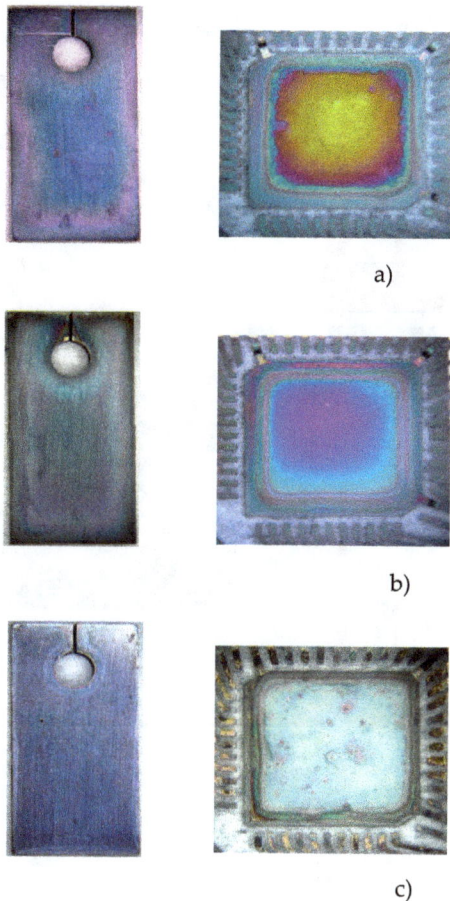

a)

b)

c)

Figure 3. Tarnishing appearance of silver coupons and silver plated copper frames, after 60 days of exposure: a) inside the clean room, b) outside the clean room and c) in test chamber.

Dendrites can cause failures in electrical equipment by short circuits when they bridge across components or between pads. Dendrite growth depends on the applied voltage, the quantity of contamination and surface moisture. This growth involves an anodic dissolution of the metal, electro migration of the ions and a subsequent cathodic deposition.

In the test chamber samples microanalysis it was detected in the corrosion film, the presence of copper sulfide in addition to silver sulfide. Figure 6 shows corrosion products of copper that comes from underneath the silver through the porosity of the plated layer. EDS microanalysis confirmed the composition of rich zones with corrosion products of copper. This could be

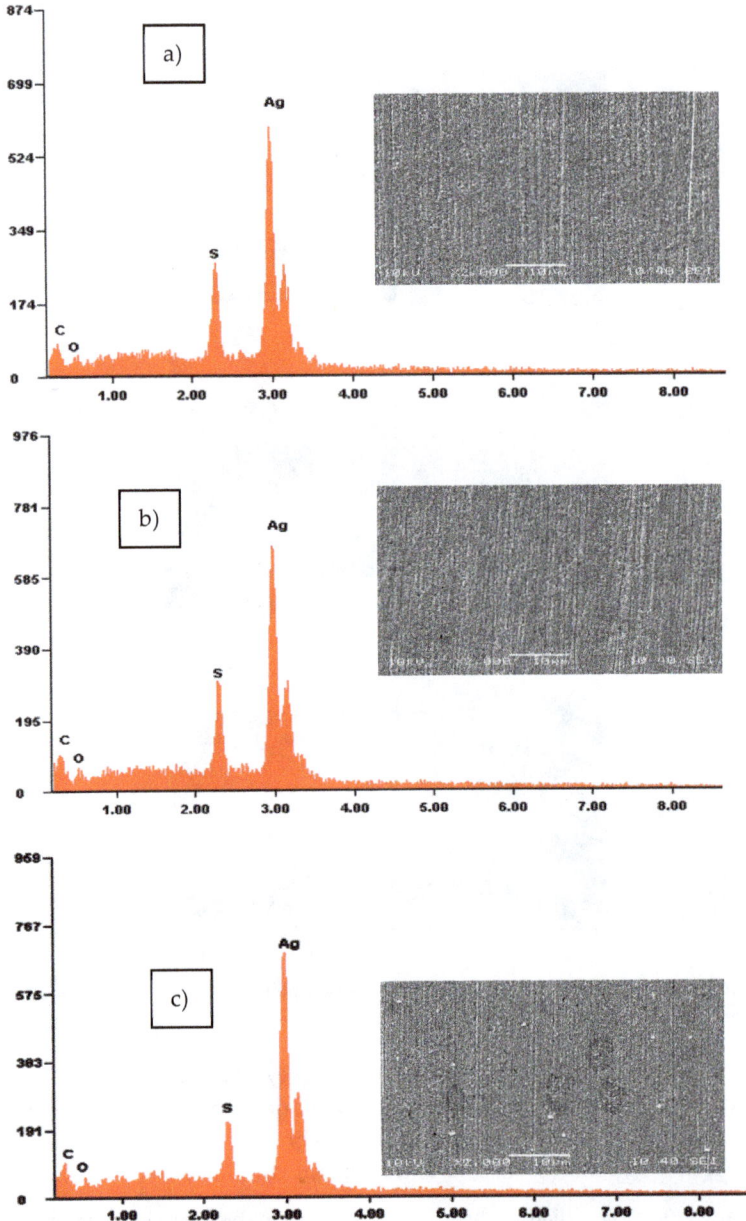

Figure 4. Results of SEM and EDS analysis of metallic silver coupons after 60 days of exposure: a) Inside clean room b) Outside clean room and c) Test chamber.

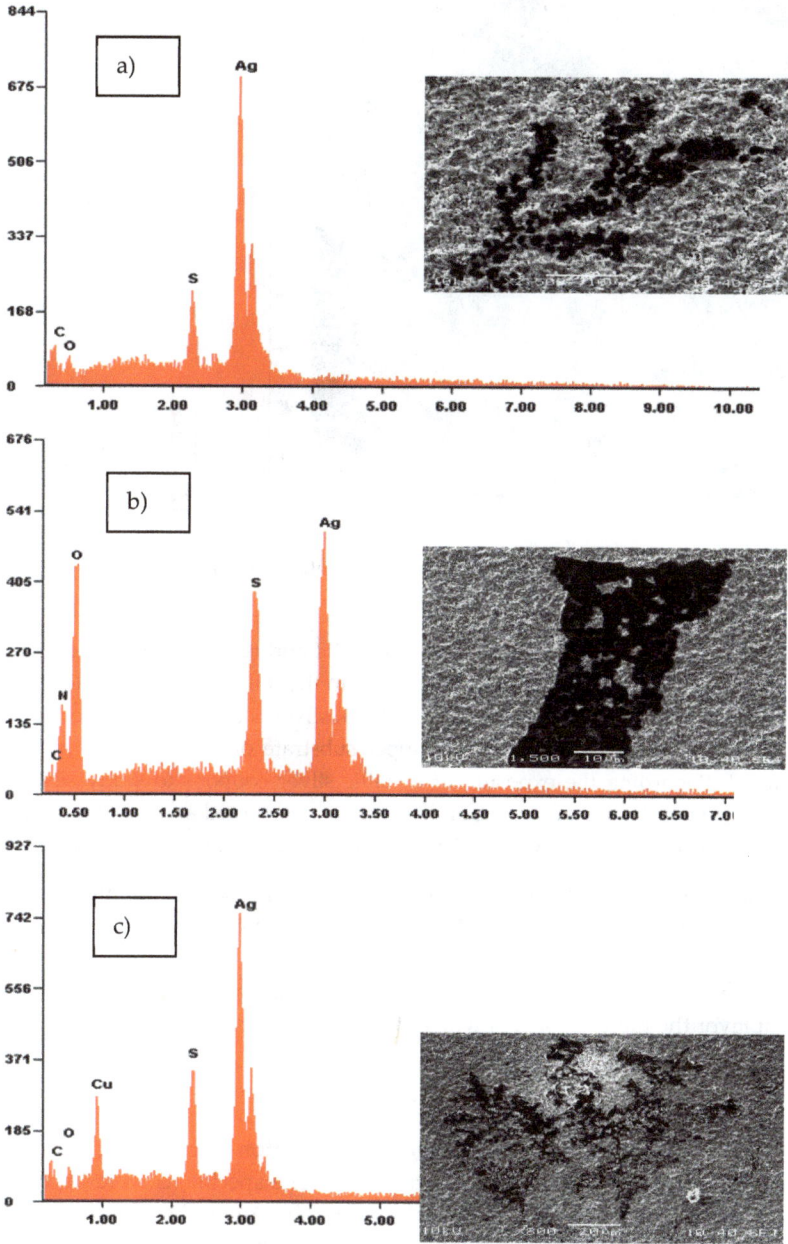

Figure 5. Results of SEM and EDS analysis of silver plated on copper frames after 60 days of exposure: a) Inside clean room b) Outside clean room and c) Test chamber.

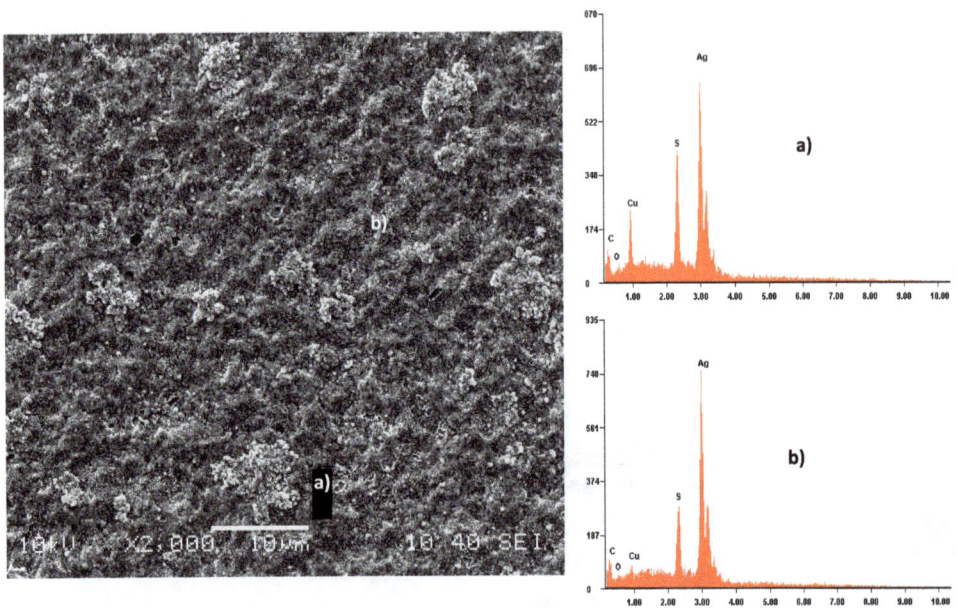

Figure 6. Corrosion products of copper in two different points (a and b) that comes from underneath the silver through the porosity of the plated layer.

happen in this site due to the higher conditions of RH and temperature. The thickness of the silver plated over copper is a very important factor in the reliability of microelectronics, when they are exposed in corrosive environments. A thick silver coating without porosity is highly recommended to avoid influences of the copper substrate on the corrosion mechanism for these materials.Regarding the exposure of metallic silver coupons to study silver corrosion over a long period of time, the presence of dendrites was observed after 12 months and after 24 months, thin branched whiskers were formed on the surface of silver (Figure 7 and 8).

Since whiskers are elongated single crystals of pure metal, they are highly conductive and can cause short circuits and arcing in the electronic and microelectronic devices. There are different shapes of whiskers, they can be straight, kinked, hooked, or forked, and some are reported to be hollow. High temperature and certain thickness of silver sulfide (Ag_2S) are factors that favor the rapid growth of whiskers. (B. Chudnovsky, 2002)

6. QCM technique application for mass gain

This QCM technique was applied for real time determination of the mass gain and hence the corrosion rate and the AgS layer average thickness. The specimens were Maxtek, 5 MHz, one inch in diameter polished quartz crystals covered by Ag on the active surface having a diameter of ½ inch. The temperature was held constant (25 °C) and the RH were monitored applying a Hygro-Thermometer Data logger during the experiment. The average registered RH value was 38.7 % with fluctuations during the experiment of +/-5.6 %.

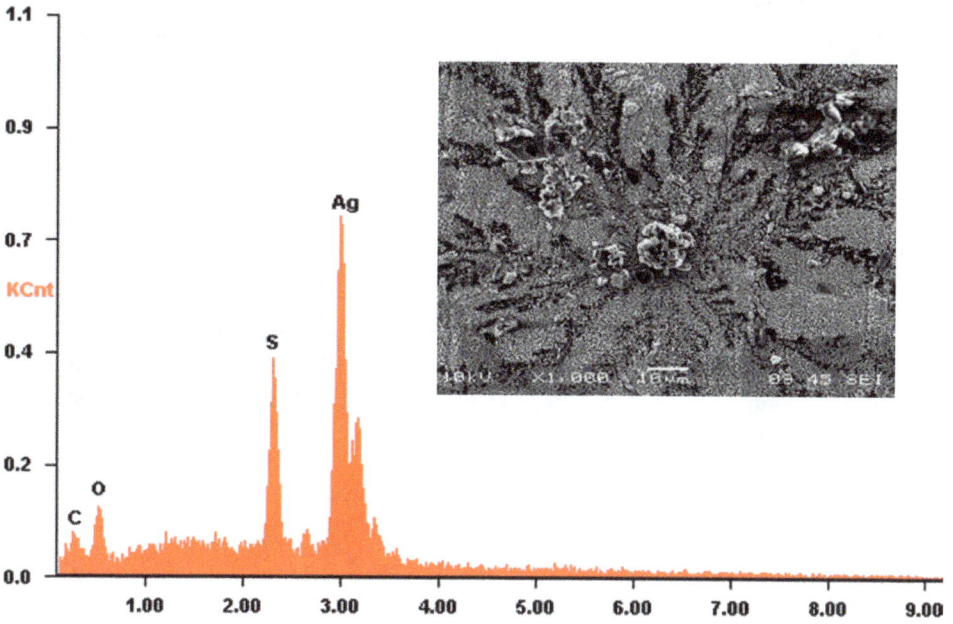

Figure 7. SEM and EDS analysis results for silver after 12 months of exposure in the test chamber.

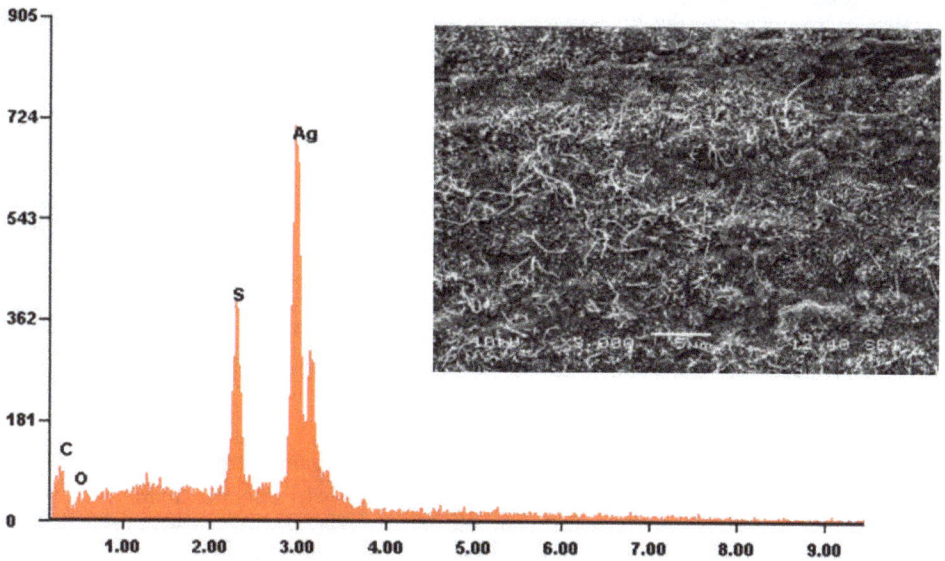

Figure 8. SEM and EDS analysis of metallic silver coupon after 24 months of exposure in the test chamber.

The controlled H2S concentrations were achieved by additions of known volumes of H2S gas to the transparent plastic chamber containing, the covered by Ag one inch in diameter QCM quartz crystal serving as a specimen. A set of Frequency – Time curves were registered for H2S concentrations in the range 0.05 to 1 ppm which values represent the minimal and the maximal concentrations of this pollutant, while the average annual value for Mexicali was found to be 0.1 ppm (B.G. Lopez et al, 2007). The frequency – time curve for 0.5 ppm H2S concentrations is shown on Figure 9. The change of the curve slope about the 1000th minute can be explained with the thickness increasing of the formed AgS layer, thus making difficult the further H2S gas diffusion through it in order to reach the fresh Ag surface. Using the curve shown on Figure 9 it was calculated that the rate of the AgS formation for the first 700 s represent 6.14 ng/cm.s while for the next part of the curve it falls almost twice to 3.27 ng/cm.s. The calculated rate of AgS thickness increasing is 6 nm/s, for the first part of the curve and 3.19 nm/s for the second part. These values showed that the AgS layer formation is very fast even at ambient temperature defining the need of measures to be taken for Ag surface protection.

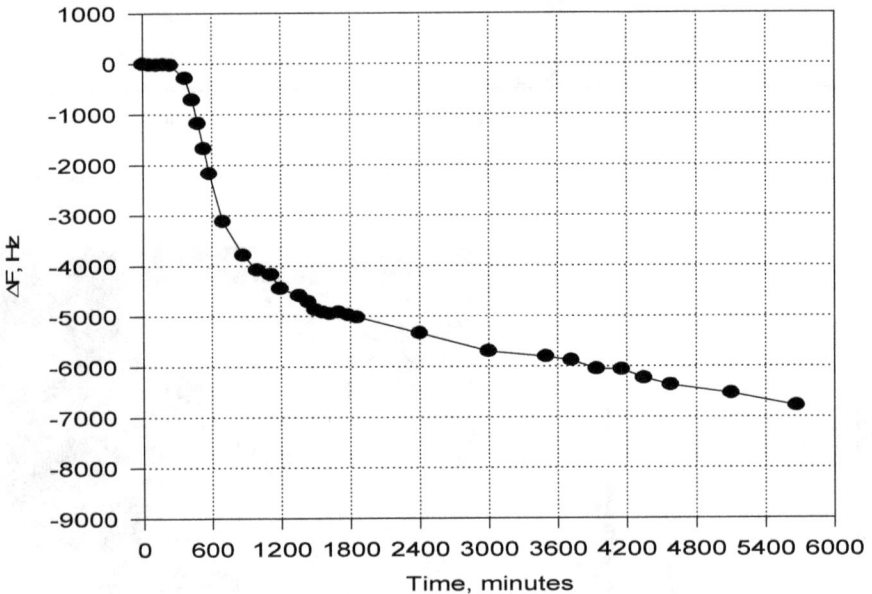

Figure 9. QCM Frequency – Time curve of active Ag surface at 25 °C, 38.7% RH and 0.5 ppm H2S

Results of exposed silver coupons and silver plated copper frames during 60 days revealed that the indoor environment in the plant can induce silver corrosion, due to the presence of hydrogen sulfide high RH and temperature that favor the corrosion process. There was no difference between the corrosion behavior of samples inside and outside the clean room environment. In the two groups of samples a silver sulfide film was detected. Silver plated copper frame samples presented in addition, growth of dendrites in the corrosion film.

On the samples exposed in the sheltered test chamber corrosion products of copper were noted due to the porosity of the silver plated copper frame.

Regarding the silver coupons exposed in the test chamber used to study the corrosion behavior in a long period of time, growth of dendrites after 12 months and growth of whiskers after 24 months was observed.

The results obtained at the plant, where the study was carried out, indicate that it has not adequate controls to avoid the penetration of outdoor atmospheric contaminants into the assembly process, including the clean room.

7. Corrosion of copper, carbon steel, tin, nickel and silver in H₂S polluted outdoor and indoor environments

7.1. Exposure of metal specimens

Rectangular specimens with an approximate area of 6.45 cm² were prepared by polishing with silicon carbide paper to SiC 600 grit, cleaned in deionised water, degreased in ethanol, dried and stored in plastic bags and placed in a dessicator.

The specimens for corrosion tests were carbon steel, copper, tin, nickel and silver, with exposure periods of 1, 3, 6, 12 and 24 months in an upright position (ASTM G 1, G 4). An aluminum metal cabin, which allows free circulation of air, was built to simulate the conditions of indoor environments (J. Flores et al, 2006) (Figure 10), and installed 10 meters above the ground on the roof of a building. Subsequently the specimens were weighed on an analytical balance.

Figure 10. Simulation chamber for corrosion: left. Front view with installed specimens, Right. Side view with the ventilation device.

The classification of the corrosivity categories according to TOW was established following the standards ISO 9223, ISO 9224, ISO 9225, ISO 9226, ISO 1184-1 and ISO 1184-2 (Table 6). The deterioration of the metal specimens was evaluated by the gravimetric method; it was correlated with the minimum, average and maximum RH and outdoor temperature in different seasonal periods, which affect the plants indoor climate.

The electrical properties of a material are at least partially a function of the amount of humidity and pollutants present in the indoor environment, because the corrosive effect increases after moisture and ionic compounds are mixed G. Lopez, et al, 2007).

Categories [a]	TOW [b] h/year
τ_1	≤ 10
τ_2	10 to 250
τ_3	250 to 2500
τ_4	2500 to 5500
τ_5	> 5500

Source. Environmental Deterioration of Materials, A. Moncmanova, WIT Press, 2007. [a] According to ISO 9223. [b]TOW. time of wetness; RH 80%, Temp > 0° C.

Table 6. Level categories of time of wetness in metals

7.2. Corrosion measurement

To determine the rates of indoor corrosion in the electronics industry and their relationship with outdoor conditions in a desertic region, a comparative measurement was performed exposing samples at indoor conditions and other metallic coupons in a test chamber. ASTM G50 was used to evaluate the corrosion rate in seasonal periods and in different environments (B.G. Lopez et al, 2007). The surfaces of the corroded metals were analyzed by scanning electron microscopy coupled to an electron disperse X-ray analyzer in order to characterize their morphology and chemical composition.

7.3. Gravimetric analysis

The corrosion test specimens were installed for periods of 1, 3, 6, 12 and 24 months and exposed to air pollutants from outside sources. After each exposure period the specimens were removed and weighed to obtain the weight gain due to the corrosion process.

The corrosion products morphology was observed in a optical microscope before being cleaned and reweighed to obtain the mass loss on an analytical balance to the nearest 0.0001 g. The simulation chamber was fabricated with pre-coated aluminum and had a volume of $0.1m^3$ with two air inlets blinds coupled to metallic filters in order to permit the penetration of gases with the flow of air and to prevent the penetration of dust to avoid mistakes in the weightings. The Scanning Electron Microscope (SEM) was used and the chemical composition was determined by Electron Dispersive X-ray (EDS).

7.4. Numerical analysis

MATLAB a numerical computing environment and programming language, which allows easy matrix manipulation, creation of graphs of functions and data, implementation of algorithms, creation of user interfaces between operations and other program in different languages was utilized for the numerical analysis.

The analysis gives the correlation of climatic parameters, environmental and corrosion of metal specimens evaluated in the test chamber and in the companies. The graphics are of three-dimensions, which indicate areas of greater relationship between the variables of weather, air pollutants and corrosion rate, in order to determine the causes of the generation of corrosion in metallic materials used in the electronics industry and suggest methods of protection required to increase their lifetime.

The data were evaluated to determine the relationship between environmental parameters and corrosion rates. Linear regression analyses were performed to get the best fit models for experimental data and the trend of corrosion rate.

7.5. Corrosion of the metals at electronics plants conditions

Several techniques were used to obtain and organize information from the indoor environment, where mainly sulfates and chlorides were monitored and evaluated for TOW with RH levels and temperatures greater than 80% and 0 º C.

In the analysis data of RH higher than 70%, temperatures higher than 35 ºC and air pollutants such as CO, NOx, O₃ and SO₂, in the periods that overpass the air quality standards in Mexicali, were considered (G. Lopez, 2008). In summer season the corrosion rate increased in the temperature range of 30 ° C to 42 ° C and RH levels of 35% to 65. Moreover, in the winter, with a range of temperature from 2 ° C to 12 º C and RH levels ranging between 35% and 70%, there was a higher incidence of water condensation on the metal surface.

The Figure 11 shows the evaluation of corrosion of copper at 2 ° C to 13 ° C and RH of 34% to 70% indicating that the corrosion rate diminishes. At temperatures of 23 ° C to 30 ° C, with RH levels of 30% and 75%, the corrosion rates increases. This methodology was applied for the corrosion behavior of the other metals tested and the results showed a very good fit in all the cases. Nevertheless, it is necessary to analyze data from several environmental monitoring stations in order to establish a map of the main pollutants behavior by season, zone, climate, economic activities, temperature and RH for at least 5 years. This map must be the most possible closer to the real conditions.

7.6. Corrosivity levels

The main sources of corrosive emissions in Mexicali are cars, fine dust from agricultural fields of an arid zone and the thermal and geothermal power plants (Valdez et al, 2006, J. Flores et al., 2003). Figures 12 and 13 show the corrosion rate values for both simulation chamber and industrial plants tests, it is clear that carbon steel is suffering an accelerated deterioration regarding to silver, copper, tin and nickel, in this order.

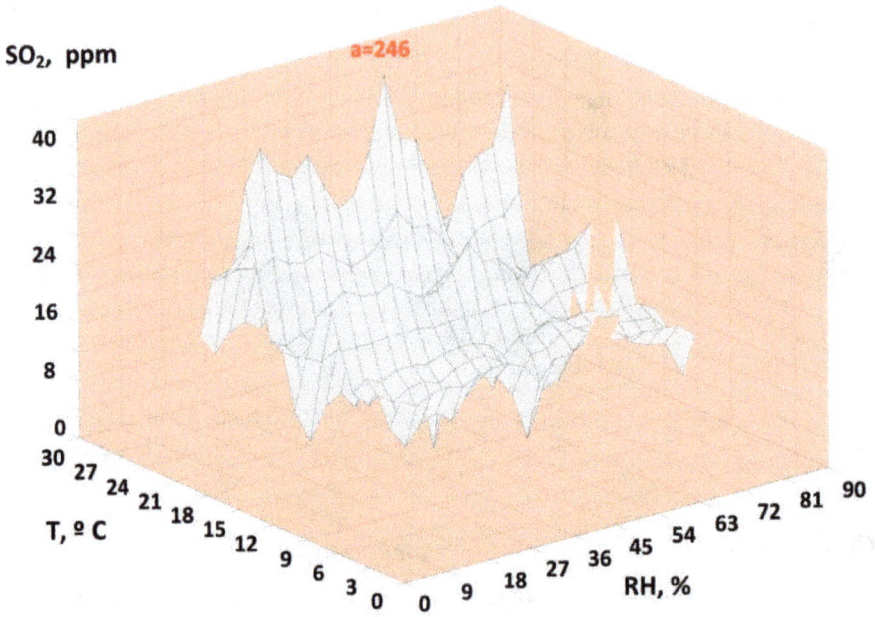

Figure 11. Correlation of temperature, RH and SO_2, a) represents maximum corrosion rate $(mg/m^2.year)$ of copper

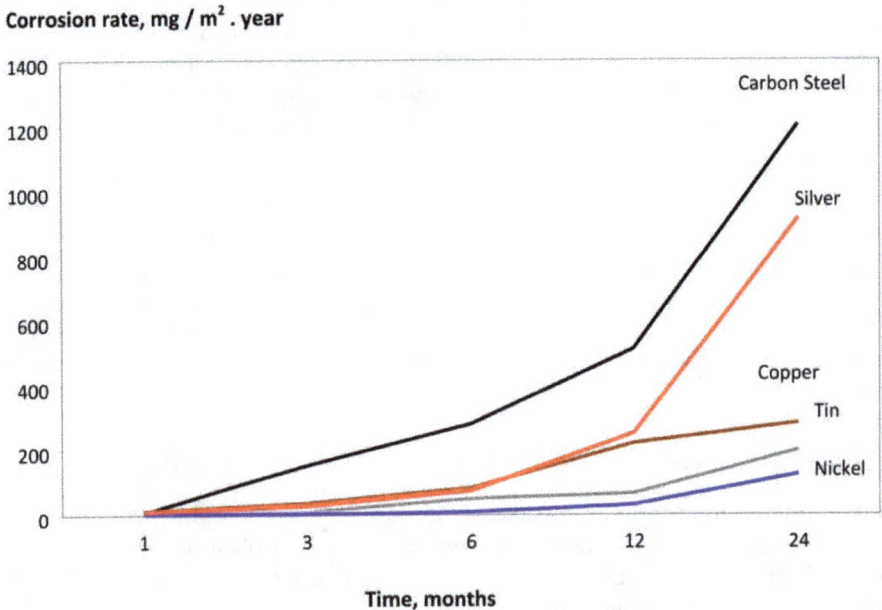

Figure 12. Corrosion rate of metals in the simulation chamber (2003 to 2005)

Corrosion rate, mg / m² . year

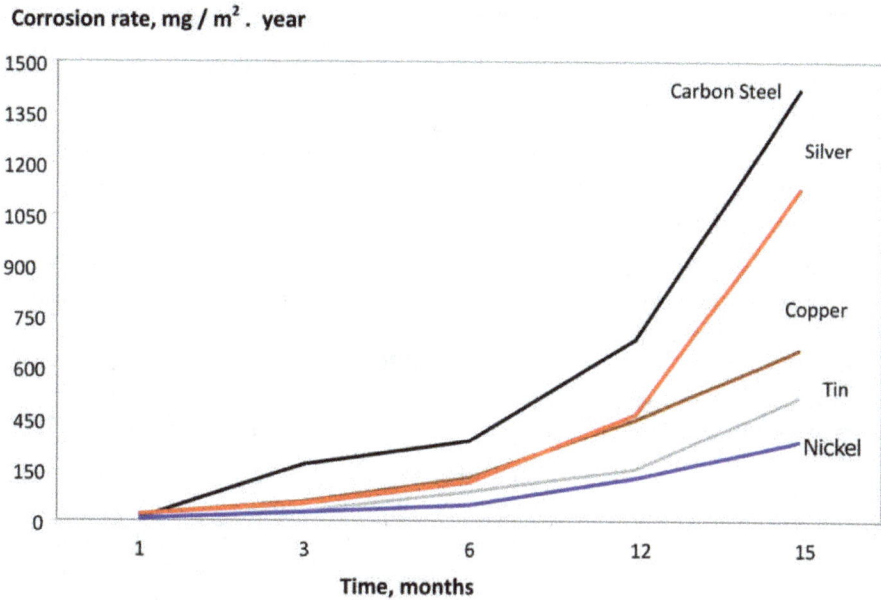

Figure 13. Corrosion rate of metals in industrial plants (2003 to 2005).

The exposure time in the chamber was 24 months and 16 months in the companies. The concentration of several atmospheric pollutants: CO, NOx, O₃ and SO₂ were monthly recorded by an automatic air pollutants monitoring station that belongs to the California Environmental Protection Agency-CALEPA network.

The electronic instruments used to monitoring air pollutants were equipped with filters to trap and detect gases and particles of air pollutants. To determine the concentration of sulfate in indoor of industrial plants the sulfatation plate technique was used inside industrial plants (ASTM G 91). Sulfates penetrate the plant environment by air currents, cracks, access doors and air conditioning systems that do not have special filters.

The exposure periods at indoor industrial plants conditions were determined according to the aggressiveness of the indoor environment, by day, week and month, seasonal and annual period. The wet candle method is applied to measure the speed of atmospheric deposition of chloride salts in a given area per unit time (ASTM G 140).

The information obtained from the deposition of chloride is used to classify the level of corrosivity of a specific area, such as conditions in the electronics industry. The standard is based in the ISO 9223 with reference to ASTM standards G 91 and G 140, measured in mg/m².year. This test is done because the levels of chlorides increase indoor atmospheric corrosion and is an important factor of corrosion phenomena. Sometimes the addition of chloride ions to metallic surfaces generate and increase the corrosion intensity, as in the case of copper and zinc, where these salts do not dissolve easily during corrosion in some areas of the metal causing pitting (Moncmanova et al, 2007).

In the beginning the corrosion rates were low for all the metals, with a variation of the RH from 40% to 60% and temperatures ranging from 25 ºC to 35 ºC (Zlatev et al, 2009). The surface spots presents on the surfaces of carbon steel, copper and silver were indicatives of an incipient corrosion process on these metals.

After 100 days exposure the weather turns warm and the temperature increases in the range from 30 ºC to 40 ºC and RH lower than 70%, the corrosion of copper, carbon steel and silver was pronounced, while nickel remains unaffected (G. Lopez et al, 2009).

The corrosion rate of all the metals was more severe after six month exposure. Ni and Sn show the best behavior compared with carbon steel, copper and silver which are covered with uniform layers of corrosion products. In this period the RH varied between 50% and 75 % and the average range of temperature was 30 ºC to 40 ºC and all the pollutants measured were higher than the permissible concentrations by the air quality standards. After one year the corrosion rate increases rapidly at approximately twice the value of to the previous period.

For specimens exposed during two years, the corrosion rate of copper coupons increases exponentially in a clear activated process. In general, all the metals showed corrosion damage and deterioration of their surfaces by the corrosion process effect. It was confirmed that SO_2 and NO_x were the predominant gaseous air pollutants. Fig. 14 shows the microstructures of corrosion products formed on the copper surface observing the presence of aggregates and pitting; the EDX analysis indicates that the main corrosion product is copper sulphide. Portable hydrogen sulphide monitors were used to measure the concentration of this gas at indoor conditions in different electronics industrial plants and in the interior of the simulation chamber. The H_2S was detected at indoor conditions with concentrations close to 1 ppm and an average value of 0.1 ppm (B.G. Lopez et al, 2007).

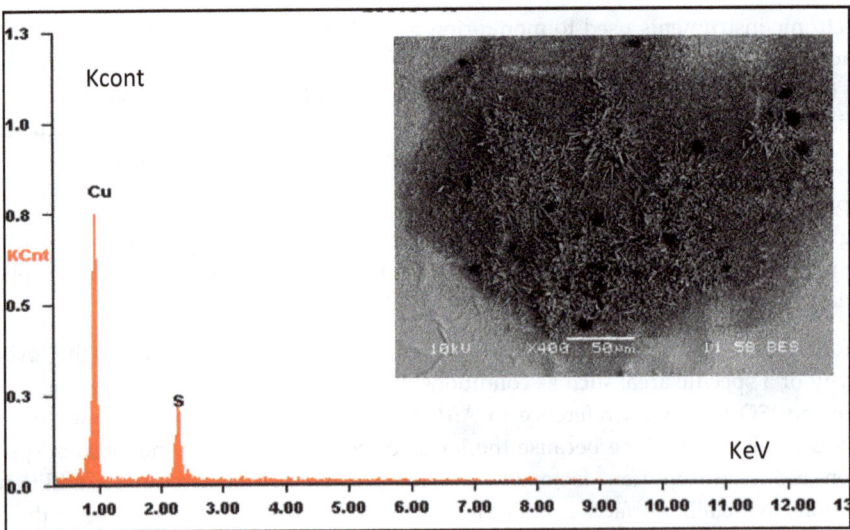

Figure 14. SEM microphotograph of copper after 6 months of exposure at indoors and EDX analysis.

8. Conclusions

All the metals tested used in the manufacturing of electronic devices, were deteriorated by atmospheric corrosion after two years exposure. Sulphide was the main component in the corrosion product layer, due the presence of SO_2 and H_2S produce by human activities and the geothermal field of Cerro Prieto, which promotes the contamination of indoor places.

Carbon steel, copper, nickel, silver and tin exposed to air pollutants reveal that an increase on their concentrations at outdoor conditions has a critical impact on the indoor corrosion process. Long exposure periods with RH values higher than 75 %, and concentrations of pollutants that exceeds those levels established in the regulations, promotes the corrosion of the metals evaluated.

In a descendent order, the most susceptible metals were carbon steel, copper, silver, tin and nickel. The use of 3D plots for multivariable systems generated by the MathLab software represents a useful tool for the monitoring of corrosion at indoor plants conditions. The 3D plots indicate the presence of pollutants in different levels of RH and temperature, and corrosion rates of the metallic probes. These plots and the Mathlab software constitute a very useful tool to study these phenomena.

Author details

Benjamin Valdez Salas, Michael Schorr Wiener,
Monica Carrillo Beltran, Roumen Zlatev and Margarita Stoycheva
Instituto de Ingenieria, Departamento de Materiales, Minerales y Corrosion,
Universidad Autonoma de Baja California, Mexico, Mexicali, Baja California

Gustavo Lopez Badilla, Lidia Vargas Osuna and Juan Terrazas Gaynor
Universidad Politecnica de Baja California, Calle de la claridad SN,Col. Plutarco Elias Calles,
Mexicali, California

Juan de Dios Ocampo Diaz
Facultad de Ingenieria, Universidad Autonoma de Baja California, Mexico, Mexicali, Baja California

9. References

A. Moncmanova; Environmental Deterioration of Materials, Ed. WIT Press, 2007, pp 108-112.

Annual Book of ASTM Standards G 1, Practice for preparing, cleaning and evaluation corrosion test specimens.

Annual Book of ASTM Standards G 4, Guide for conducting corrosion coupon tests in field applications.

Annual Book of ASTM Standards G140, Test method for determination chloride deposition rate by wet candel method.

Annual Book of ASTM Standards G50, Practice for conducting atmospheric corrosion test on metals.

Annual Book of ASTM Standards G91, Practice for monitoring atmospheric SO2 using the sulfatation plate technique.

B. H. Chudnovsky, Degradation of Power Contacts in Industrial Atmosphere: Silver Corrosion and Whiskers, Proceedings of the 48th IEEE Holm Conference on Electrical Contacts, p.140-150, 2002.

B.G. Lopez, S.B. Valdez, K.R. Zlatev, P.J. Flores, B.M. Carrillo and W.M. Schorr, Corrosion of metals at indoor conditions in the electronics manufacturing industry, Anti-Corrosion Methods and Materials, Vol. 54/6 p. 354–359, 2007.

C. J. Yang, C. H. Liang, X. Liu, Tarnishing of silver in environments with sulphur contamination, Anti-Corrosion Methods and Materials Vol. 54 No. 1, p. 21–26, 2007.

Ch. Kleber, R. Wiesinger, J. Schnöller, U. Hilfrich, H. Hutter, M. Schereiner, Initial oxidation of silver surfaces by S2- and S4+ species, Corrosion Science, Vol 50, p. 1112-1121, 2008.

Corrosion of metals and alloys – Classification of atmospheres with low corrosivity – Determination and estimation of corrosivity, ISO 11844-1, International Organization for Standardization, Geneva, Switzerland, 2000.

Corrosion of metals and alloys – Classification of corrosivity of atmospheres, ISO 9223, International Organization for Standardization, Geneva, Switzerland, 1992.

D. W. Rice, P. Peterson, E. B. Rigby, P. B. P. Phipps, R. J. Cappell, R. Tremoureux, Atmospheric corrosion of copper and silver, J. Electrochem. Soc., Vol 128, No.2, p. 275-284, 1981.

Flores P.J.F., Valdez S. B. and Schorr W. M.; "Cabina de investigacion de corrosion para la industria electronica en interiores"; Ingenierias, Vol. VI, No. 21, 2006. (Spanish).

Frankel R.P. in 'Uhlig's Corrosion Handbook', Second Edit., (ed. R. Winston), John Wiley & Sons, Inc., 1995, N. Y., 941-947.

G. Lopez B., B. Valdez S., R. Zlatev K., J. Flores P., M. Carrillo B. and M. Schorr W. Corrosion of metals at indoor conditions in the electronics manufacturing industry. Anti-Corrosion Methods and Materials, United Kingdom, Vol. 54, N0. 6, 354-359, 2007

G. Russ, Electrical characteristics of contacts contaminated with silver sulfide film, IEEE Transactions on Parts, Materials and Packaging, Vol. 6, No. 4, p.129-137, 1970.

H. G. Puente, L. Hernandez, H2S Monitoring and emission control at the cerro prieto geothermal field, Mexico, Proceedings World Geothermal Congress, 2005.

H. Kim, Corrosion process of silver in environments containing 0.1 ppm H2S and 1.2 ppm NO2, Materials and Corrosion 54, p. 243-250, 2003.

ISO 11844 PART 1. Corrosion of metals and alloys- Classification of low corrosivity of indoor atmospheres- Determination and estimation of indoor corrosivity.

ISO 11844 PART 2. Corrosion of metals and alloys- Classification of low corrosivity of indoor atmospheres- Determination and estimation attack in indoor atmospheres.

ISO 9223:1992, Corrosion of metals and alloys, Corrosivity of Atmospheres, Classification.

ISO 9224:1992, Corrosion of metals and alloys, Corrosivity of Atmospheres, Standard values for corrosivity categories.

ISO 9225: 1992, Corrosion of metals and alloys, Corrosivity of Atmospheres, Measurement of pollution.

ISO 9226: 1992, Corrosion of metals and alloys, Corrosivity of Atmospheres, Determination of corrosion rate of standard specimens for the purpose of evaluation corrosivity.

J. F. Flores, B. Valdez, M. Schorr, Cabinet to study corrosion behavior of metals used in the electronic industry, Ingenierias, Vol VI, No. 21, p.33-37, 2003.

J. P. Franey, G. W. Kammlott, T. E. Graedel, The corrosion of silver by atmospheric sulfurous gases, Corrosion Science, Voo. 25, No. 2, p. 133-143, 1985.

J.H. Payer, Corrosion processes in the development of thin tarnish films, Proceedings of the Thirty-Sixth IEEE Holm Conference on Electrical Contacts and the Fifteenth International Conference on Electrical Contacts, p. 203-211, 1990.

L. Veleva, B. Valdez, G. Lopez, L. Vargas, J. Flores, Atmospheric corrosion of electro-electronics metals in urban desert simulated indoor environment, Corrosion Engineering, Science and Technology, Vol. 43, No. 2, p.149-155, 2008.

López Badilla Gustavo; Ph.D. Thesis; Caracterización de la corrosión en materiales metálicos de la industria electrónica en Mexicali, B.C., 2008 (Spanish).

Lyon S.B., Wong C. W. and Ajiboye P.; Analysis of atmospheric corrosion in indoor conditions; ASTM STP 1239; Philadelphia, PA, American Society for Testing and Materials; 1996.

M. B. McNeil, B. J. Little, Corrosion mechanisms for copper and silver objects in near surface environments, JAIC, Vol. 31, No.3, p.355-366, 1992.

M. Reid, J. Punch, C. Ryan, J. Franey, G. E. Derkits, Jr., W. D. Reents, Jr., L. F. Garfias, The corrosion of electronic resistor, IEEE Transactions on Components and Packaging Technologies, Vol. 30, No.4, p. 666-672, 2007.

M. Schorr, B.V. Salas, M. Quintero and R. Zlatev, Effect of H₂S on corrosion in polluted waters: A review, Corros. Sci. Eng. Sci. and Technol., 41 (3), 2006, pp. 221-227.

M. Schorr and B.Valdez, Corrosion of the marine infrastructure in polluted ports, Corros. Sci. Eng. Sci. and Technol., 40 (2), 2005, pp. 137-142.

M. Tullmin, P. R. Roberge, Corrosion of metallic materials, IEEE Transactions on Reliability, Volume 44, No. 2, p. 271-278, 1995.

M. Watanabe, A. Hokazono, T. Handa, T. Ichino, N. Kuwaki, Corrosion of copper and silver plates by volcanic gases, Corrosion Science, 2005.

M. Watanabe, S. Shinozaki, E. Toyoda, K. Asakura, T. Ichino, N. Kuwaki, Y. Higashi, T. Tanaka, Corrosion products formed on silver after a one-month exposure to urban atmospheres, Corrosion, Vol. 62, No. 3, p.243-250, 2006.

Nishikata A. and Ichihara Y; The effect of time of wetness (TOW) in metallic components; Corrosion Science; 1995, No. 37.

Nishimura T., Katayama H., Noda K. and Kodama T.; Effect of Co and Ni on the corrosion behaviour of low alloys steels in wet-dry environments; Corrosion Science, 2000.

O. L. Vargas, S. B. Valdez, M. L. Veleva, K. R. Zlatev, W. M. Schorr and G. J. Terrazas, Corrosion of silver at indoor conditions of assembly processes in the microelectronics industry, Anti-Corrosion Methods and Materials, United Kingdom, Vol. 56, N0. 4, 218 - 225, 2009.

P. Marcus: Corrosion mechanisms in theory and practice, Marcel Dekker, New York, 2002, p. 534.

P. R. Roberge, Handbook of corrosion engineering, New York, NY, Mc Graw Hill, 2000.

P. Vassiliou, C.T. Dervos, Corrosion effects on the electrical performance of silver metal contacts, Anti-Corrosion Methods and Materials, Vol. 46, No. 2, p.85-94, 1999.

S.E. Manahan, Fundamentals of Environmental Chemistry, Lewish Publishers, Boca Raton, PP. 380-390; 415-422, 1993.

S. P. Sharma, Atmospheric corrosion of silver, copper and nickel, J. Electrochem. Soc., Vol. 125, No. 12, 2005-201, 1978.

T.E. Graedel, Corrosion Mechanisms for silver exposed to the atmosphere, J. Electrochem. Soc., Vol 139, No. 7, 1992.

T.M.H. Saber, A. A. El Warraky, AES and XPS study on the tarnishing of silver in alkaline sulphide solutions, Materials Science, Vol. 23, p. 1496-1501, 1988.

Valdez B. y Schorr M.; El control de la corrosión en la industria electrónica; Revista Ciencia; 2006 (Spanish).

X. Lin, Ji. Zhang, Dust corrosion, Proceedings of the 50th IEEE Holm Conference on Electrical Contacts and the 22nd International Conference on Electrical Contacts, p. 255-262, 2004.

Zlatev R., Valdez B., Stoytcheva M., Vargas L., Lopez G., Schorr M.; Symposium 16: NACE "Corrosion and Metallurgy"; IMRC 2009, Cancun, Mexico.

Permissions

The contributors of this book come from diverse backgrounds, making this book a truly international effort. This book will bring forth new frontiers with its revolutionizing research information and detailed analysis of the nascent developments around the world.

We would like to thank Dr. Gustavo Lopez Badilla, Dr. Benjamin Valdez and Dr. Michael Schorr, for lending their expertise to make the book truly unique. They have played a crucial role in the development of this book. Without their invaluable contribution this book wouldn't have been possible. They have made vital efforts to compile up to date information on the varied aspects of this subject to make this book a valuable addition to the collection of many professionals and students.

This book was conceptualized with the vision of imparting up-to-date information and advanced data in this field. To ensure the same, a matchless editorial board was set up. Every individual on the board went through rigorous rounds of assessment to prove their worth. After which they invested a large part of their time researching and compiling the most relevant data for our readers. Conferences and sessions were held from time to time between the editorial board and the contributing authors to present the data in the most comprehensible form. The editorial team has worked tirelessly to provide valuable and valid information to help people across the globe.

Every chapter published in this book has been scrutinized by our experts. Their significance has been extensively debated. The topics covered herein carry significant findings which will fuel the growth of the discipline. They may even be implemented as practical applications or may be referred to as a beginning point for another development. Chapters in this book were first published by InTech; hereby published with permission under the Creative Commons Attribution License or equivalent.

The editorial board has been involved in producing this book since its inception. They have spent rigorous hours researching and exploring the diverse topics which have resulted in the successful publishing of this book. They have passed on their knowledge of decades through this book. To expedite this challenging task, the publisher supported the team at every step. A small team of assistant editors was also appointed to further simplify the editing procedure and attain best results for the readers.

Our editorial team has been hand-picked from every corner of the world. Their multi-ethnicity adds dynamic inputs to the discussions which result in innovative outcomes. These outcomes are then further discussed with the researchers and contributors who give their valuable feedback and opinion regarding the same. The feedback is then collaborated with the researches and they are edited in a comprehensive manner to aid the understanding of the subject.

Apart from the editorial board, the designing team has also invested a significant amount of their time in understanding the subject and creating the most relevant covers. They scrutinized every image to scout for the most suitable representation of the subject and create an appropriate cover for the book.

The publishing team has been involved in this book since its early stages. They were actively engaged in every process, be it collecting the data, connecting with the contributors or procuring relevant information. The team has been an ardent support to the editorial, designing and production team. Their endless efforts to recruit the best for this project, has resulted in the accomplishment of this book. They are a veteran in the field of academics and their pool of knowledge is as vast as their experience in printing. Their expertise and guidance has proved useful at every step. Their uncompromising quality standards have made this book an exceptional effort. Their encouragement from time to time has been an inspiration for everyone.

The publisher and the editorial board hope that this book will prove to be a valuable piece of knowledge for researchers, students, practitioners and scholars across the globe.

List of Contributors

Dhananjai S. Borwankar, William A. Anderson, and Michael Fowler
Department of Chemical Engineering, University of Waterloo, Canada

Davide Astiaso Garcia and Franco Gugliermetti
DIAEE (Dipartimento di Ingegneria. Astronautica, Elettrica ed Energetica – Department of Astronautical, Electrical and Energy Engineering) of the Sapienza University of Rome, Italy

Fabrizio Cumo
DATA (Dipartimento Design, Tecnologia dell'Architettura, Territorio e Ambiente –Design, Architectural Technolog, Territory and Environment Department) of the Sapienza, University of Rome, Italy

Motoya Hayashi
Miyagigakuin Women's University, Japan

Yishinori Honma
Iwate Prefectural University, Japan

Haruki Osawa
National Institute of Public Health, Japan

Silvije Davila, Ivan Bešlić and Krešimir Šega
Institute for Medical Research and Occupational Health,
Environmental Hygiene Unit, Zagreb, Croatia

Yasuko Yamada Maruo and Akira Sugiyama
NTT Energy and Environment Systems Laboratories, Japan

D.G. Gajghate, P.Pipalatkar and V.V. Kharparde
Air Pollution Control Division, National Environmental Engineering Research Institute (NEERI), Nehru Marg, Nagpur, India

Manju Mohan, Shweta Bhati and Preeti Gunwani
Centre for Atmospheric Sciences, Indian Institute of Technology, Delhi, India

Pallavi Marappu
Center for Global and Regional Environmental Research, The University of Iowa, Iowa City, USA

Gustavo Lopez Badilla
Universidad Politecnica de Baja California, Calle de la Claridad SN,Col. Plutarco Elias Calles, Mexicali, B.C., Mexico

Benjamin Valdez Salas, Michael Schorr Wiener and Carlos Raúl Navarro González
Instituto de Ingenieria, Departamento de Materiales, Minerales y Corrosion, Universidad Autonoma de Baja California, Mexicali, Baja California, Mexico

Małgorzata Kowalska
Medical University of Silesia, Poland

Edmilson de Souza
Mato Grosso do Sul State University, UEMS, Brazil

Josmar Davilson Pagliuso
University of Sao Paulo, USP, Brazil

Zoran Mijić, Andreja Stojić, Mirjana Perišić, Slavica Rajšić and Mirjana Tasić
Institute of Physics, University of Belgrade, Serbia

Giuseppe Petrone and Giuliano Cammarata
Department of Industrial Engineering, University of Catania, Catania, Italy

Carla Balocco
Energy Engineering Department, University of Firenze, Firenze, Italy

Ehab Mostafa
Agricultural Engineering Dept.,Faculty of Agriculture – Cairo University, Giza, Egypt

Takeshi Ohura and Yuta Kamiya
Meijo University, Shiogamaguchi, Nagoya, Japan

Fumikazu Ikemori
Nagoya City Institute For Environmental Science, Toyoda, Nagoya, Japan

Tsutoshi Imanaka
GL Sciences Inc. Sayamagaoka, Iruma, Japan

Masanori Ando
Musashino University, Shinmachi, Nishitokyo, Japan

Benjamin Valdez Salas, Michael Schorr Wiener, Monica Carrillo Beltran, Roumen Zlatev and Margarita Stoycheva
Instituto de Ingenieria, Departamento de Materiales, Minerales y Corrosion, Universidad Autonoma de Baja California, Mexico, Mexicali, Baja California

Gustavo Lopez Badilla, Lidia Vargas Osuna and Juan Terrazas Gaynor
Universidad Politecnica de Baja California, Calle de la claridad SN,Col. Plutarco Elias Calles, Mexicali, California

Juan de Dios Ocampo Diaz
Facultad de Ingenieria, Universidad Autonoma de Baja California, Mexico, Mexicali, Baja California

www.ingramcontent.com/pod-product-compliance
Lightning Source LLC
Chambersburg PA
CBHW070727190326
41458CB00004B/1066